Springer
Proceedings in Physics 83

Springer
Berlin
Heidelberg
New York
Barcelona
Budapest
Hong Kong
London
Milan
Paris
Santa Clara
Singapore
Tokyo

Springer Proceedings in Physics

Managing Editor: H. K. V. Lotsch

Volumes 1–82 are listed at the end of the book

D. P. Landau   K. K. Mon   H.-B. Schüttler (Eds.)

Computer Simulation Studies in Condensed-Matter Physics X

Proceedings of the Tenth Workshop
Athens, GA, USA, February 24–28, 1997

With 109 Figures

Springer

Professor David P. Landau, Ph. D.
Professor K. K. Mon, Ph. D.
Professor Heinz-Bernd Schüttler, Ph. D.
Center for Simulational Physics
The University of Georgia
Athens, GA 30602, USA

Library of Congress Cataloging-in-Publication Data

Computer simulation studies in condensed-matter physics X: proceedings of the tenth workshop Athens, Ga, USA, February 24–28, 1997 / D. P. Landau, K. K. Mon, H.-B. Schüttler, eds. p. cm. – (Springer proceedings in physics, ISSN 0930-8989; 83) includes bibliographical references.
ISBN 3-540-63510-6 (hc: alk. paper)
1. Condensed matter–Computer simulation–Congresses.   2. Condensed matter–Mathematical models–Congresses. 3. Monte Carlo method–Congresses. I. Landau, David, P. II. Mon, K. K. (Kin K.), 1950– . III. Schüttler, Heinz-Bernd, 1956– IV. Series: Springer proceedings in physics; v. 83.
IN PROCESS 530.4'1'0113–dc21 97-39888 CIP

ISSN 0930-8989
ISBN 3-540-63510-6 Springer-Verlag Berlin Heidelberg New York

Typesetting: Camera ready copy from the authors/editors
SPIN: 10645200   54/3144 - 5 4 3 2 1 0 - Printed on acid-free paper

Preface

Eleven years ago, because of the phenomenal growth in the power of computer simulations, the University of Georgia formed the first institutional unit devoted to the use of simulations in research and teaching: the Center for Simulational Physics. As the simulations community expanded further, we sensed a need for a meeting place for both experienced simulators and neophytes to discuss new techniques and recent results in an environment that promoted extended discussion. As a consequence, the Center for Simulational Physics established an annual workshop on *Recent Developments in Computer Simulation Studies in Condensed Matter Physics*. This year's workshop was the tenth in this series, and the interest shown by the scientific community demonstrates quite clearly the useful purpose the series has served. The latest workshop was held at the University of Georgia, February 24–28, 1997, and these proceedings form a record published with the goal of timely dissemination of the material to a wider audience.

This year's workshop was held in honor of Prof. Masuo Suzuki, who has just retired from the University of Tokyo. Although not personally involved in simulations, Prof. Suzuki has developed many of the ideas that have made novel, new simulations possible. In addition, many of his students have entered the world of computation and gone on to join the international elite of simulational physicists.

We also wish to offer a special thanks to IBM Corporation for their generous support of this year's workshop.

This volume is composed of three parts. The first part contains invited papers that deal with simulational studies of classical systems. The second part of the proceedings is devoted to invited papers on quantum systems, including new results for strongly correlated electron and quantum spin models. The final part contains a large number of contributed presentations.

We hope that each reader will benefit from specialized results and profit from exposure to new algorithms, methods of analysis, and conceptual developments. We have already learned that fruitful collaborations and new research projects have resulted from these gatherings at the University of Georgia, and we hope that these proceedings will further expand this tendency.

Athens, GA, U.S.A.   *D.P. Landau*
May 1997   *K. K. Mon*
                                                                   H.-B. Schüttler

Contents

Part III Contributed Papers

Computer Simulation Studies in Condensed Matter Physics: An Introduction

D. P. Landau, K. K. Mon, and H.-B. Schüttler
Center for Simulational Physics, The University of Georgia, Athens, GA 30602-2451, USA

The 10th Annual Workshop Proceedings are divided into three parts. The first two parts contain the invited presentations and include some pedagogical material. The final part contains short, contributed papers that cover a wide range of simulational studies.

To open the first part of this volume, Hu et al. describe histogram Monte Carlo simulations of several models, which are then used to test for the existence of universal finite-size scaling functions. Examples include the bulk critical behavior of various continuum percolation models and the dynamic (relaxational) critical behavior for Ising models on different lattices. Grest et al. then present results from molecular dynamics simulations of polymers. Recent progress is reviewed for off-lattice models with particular emphasis on polymer surfaces and interfaces. Bead-spring model results for end-grafted polymers under shear are found to agree well with recent experiments. The authors also show how the measured pressure tensor can be used to obtain the surface tension for the interface between immiscible polymer blends. In the next paper, Schmid et al. study a coarse-grained continuum model for Langmuir monolayers. This is a model of monolayers of amphiphilic molecules of stiff chains of beads with one end grafted to a polar planar substrate. Constant-pressure Monte Carlo simulations are used and the resulting phase diagrams are analyzed. The authors conclude that their Monte Carlo studies of very simplified models can address, and partly answer, fundamental questions, but do not make quantitative predictions. Next, Blöte and Heringa use cluster Monte Carlo algorithms to simulate hard-core lattice gases. This method is efficient and suppresses critical slowing down. Both simple-cubic and body-centered-cubic lattices are considered. Finite-size-scaling analyses of the data are consistent with the Ising universality class and disagree with published simulation data analyses; the disagreement is explained in terms of large corrections to scaling. To conclude this part, Kaski and Heino discuss computer simulations of fracture in disordered visco-elastic systems by using two-dimensional models. Two models have been used. The first model uses a network of dissipative Born springs and a molecular dynamics method with the Verlet algorithm. The second model considers a finite-element method with a similar dissipative force relaxation mechanism and the Adams–Moulton predictor–corrector scheme. The disorder appears either as topological disorder or as nonuniform mass distribution. The

Springer Proceedings in Physics, Volume 83
Computer Simulation Studies in Condensed-Matter Physics X
Eds.: D. P. Landau, K.K. Mon, H. -B. Schüttler

results of the computer simulations are similar to those from recent experiments on crack branching.

Several interesting new developments in computational studies of quantum systems have been presented in this workshop. The paper by Bhatt summarizes recent Monte Carlo studies of quantum phase transitions in random magnets. The behavior of the nearest-neighbor Ising spin glass as well as the random Ising ferromagnet is treated in detail. Bhatt finds that rare fluctuations in the bond distribution are essential to the understanding of the response. Next, Imada reports on projector quantum Monte Carlo studies of the Mott-metal–insulator transition in the two-dimensional Hubbard model near half-filling. He compares the results to the predictions of scaling theory and presents evidence for the existence of a new universality class which is characterized, on the metallic side, by an unusual suppression of quasi-particle coherence, resulting from a large dynamical critical exponent of value $z=4$. José then discusses recent quantum Monte Carlo studies on quantum zero-point fluctuations in models for ultrasmall Josephson junctions arrays. Results near the semi-classical critical temperature are successfully compared to experimental data. Quantum Monte Carlo and analytical results for the existence of a quantum-induced low-temperature phase transition in a capacitance-dominated model at zero magnetic field are presented. Similarities in the phase structure of capacitively coupled Josephson junction arrays and in the fractional quantum Hall effect are pointed out. In the fourth paper in this section, Huscroft et al. describe determinant fermion quantum Monte Carlo investigations of the effects of randomness on the long-range order in two-dimensional Hubbard models at half-filling. For the repulsive Hubbard model, they find that both random on-site energies and random nearest-neighbor hybridization tend to destroy long-range antiferromag-netic correlations. They also suggest that a bimodal random distribution of repulsive on-site Hubbard repulsions might allow separation of the magnetic and the Mott metal-insulator transition. In the attractive version of the model, they find that random on-site energies tend to favor the superconducting over the charge density wave phase. To close this part, Troyer and Imada report on a high-precision quantum Monte Carlo study of the quantum critical point in a 1/5th depleted planar antiferromagnet. Employing the recently developed continuous, imaginary time loop algorithm, they obtain precise estimates for the critical exponents governing the transition between the zero-temperature ordered and quantum disordered phases. Their results are fully consistent with the conjecture that the quantum Heisenberg antiferromagnet and the O(3) nonlinear sigma model share the same universality class and suggest that Berry phases are not relevant for the critical behavior in the quantum antiferromagnet.

A quite rich variation of very interesting contributed papers follows. In the first of these, Cheon and Chang use a large-cell Monte Carlo real-space renormalization group method to investigate scaling behavior for binary fragmentation of percolation clusters. They provide evidence for a scaling relation in both two and three dimensions. Avgin et al. use a combination of analytic and numerical techniques to examine magnons in random bond Heisenberg chains. They calculate the low-temperature specific heat and compare the results with those of real-space renormalization group theory. Ledue et al. examine the $q=8$ Potts model on the quasiperiodic octagonal lattice. They find important effects from the free boundary and conclude that the transition is 1st order. Hansmann and Okamoto turn to the problem of simulations of peptides and proteins that have complicated energy landscapes. They show that use of generalized ensemble simulations is a promising approach to tackling this class of problems. Schweika and Landau describe a Monte Carlo study of order–disorder transitions in A_3B binary alloys. They find possible new ordering cases that could lead to surface-induced ordering. Somer et al. present preliminary results of a computational study of the "inherent structures" associated with equilibrium two-dimensional Lennard–Jones systems. Their results distinguish between the crystal and isotropic liquid phases and support the inherent-structures concept of Stillinger and Weber. Takayama et al. discuss Monte Carlo simulation on aging phenomena in the SK spin-glass model. The system is quenched rapidly from an infinite temperature state to the spin-glass phase. Various results are reported and support the growth of a quasi-equilibrium domains (GQED) scenario of aging process in the model. Sandler et al. develop a continuum quasi-equilibrium growth model to study the unusual "seahorse"-like growth patterns, revealed in recent experiments using ionized-cluster-beam deposition. The model results are consistent with the experiments, and the authors conclude that a chiral instability in the model is responsible for this agreement. Next, Rapaport considers an interactive modeling of granular flow. He presents several examples in which visualization makes it possible to examine features which are not readily characterized by other means. The problems presented are granular flow from a silo, inclined-chute flow and surface waves in a thin, vertically vibrated layer. Kosterlitz and Simkin have studied the random gauge XY system as a model for a superconducting glass. They find that the weak disorder is marginal in two and probably irrelevant in three dimensions. For strong disorder they find flow towards a nonsuperconducting glass in two, and towards a superconducting glass in three dimensions. Their results agree with recent analytical work, but contradict earlier predictions of a very low temperature re-entrant transition. Enjalran et al. present Monte Carlo results for coupled two-dimensional classical square lattice Heisenberg antiferromagnet layers, a model system relevant to the recently discovered manganese pnictide-oxide materials. They show that, in the presence of frustrating interplanar couplings, arising from layer offset, the magnetic order in the planes can become orthogonal even without explicit symmetry breaking terms. Schöppe and Heermann present a molecular dynamics study of a coarse-grained polymer chain model made up of ellipsoidal "building blocks". They examine both static and dynamic properties and conclude that with this model it is possible to see effects at moderate chain length that can be seen for only very long chain models of other types. Creswick and Kim have used the microcanonical transfer matrix method to study the distribution of Yang–Lee zeros in $Q=2$ (Ising) and $Q=3$ Potts models. Their method provides a new approach for studying finite two-dimensional systems, but in three dimensions it is limited to relatively small system sizes although it may be extended by combining it with recently developed Monte Carlo methods. Adler et al. evaluate the roughening temperatures for three facets of the HCP lattices and also directly estimate the surface tension for many temperatures. The model uses both nearest- and next-nearest-neighbor interactions. The results are in good agreement with experiment. Scholten and King then describe a Monte Carlo study of the three-dimensional 4-state chiral clock model. As the chiral parameter Δ is varied, they find evidence for a new chiral phase that was not previously predicted. Baker reports on a Monte Carlo study of finite-size effects in the critical region of the two-dimensional Ising model. He examines finite-size scaling properties using a variable that is the ratio of the lattice size to the finite lattice correlation length and finds great differences in the behavior of different quantities. Kolesik et al. study magnetization switching dynamics in the kinetic Ising model. They map the Monte Carlo dynamics onto a one-dimensional absorbing Markov chain and extract estimates for metastable lifetimes. A size extrapolation scheme is used to predict large system behavior. Krech studies the short-time evolution of a growing interface using both analytical and numerical techniques. Dynamic exponent estimates are extracted for both the RSOS model and for ballistic deposition. Janke then presents high-resolution data for the two-dimensional XY model in the Villain representation. He concludes that when logarithmic corrections are included in the analysis, the data are consistent with the Kosterlitz–Thouless predictions, but the correction exponent is ill defined. In the final presentation, Bunker et al. report Monte Carlo and analytical studies for a classical pair potential model of dense fluid hydrogen. They obtained results for the pair distribution function, the degree of dissociation, and the effects of dissociation on the proton–proton pair distribution. The equation of state for their model is found to be in good agreement with experimental data.

Part I
Classical Systems

Monte Carlo Approaches to Universal Finite-Size Scaling Functions

Chin-Kun Hu[1,+], Jau-Ann Chen[1], Chai-Yu Lin[2], and Fu-Gao Wang[3,]*

[1]Institute of Physics, Academia Sinica, Nankang, Taipei 11529, Taiwan
[2]Department of Physics, National Tsing Hua University, Hsinchu 300, Taiwan
[3]Center for Simulational Physics, University of Georgia, Athens, GA 30602, USA

Abstract. The universality of critical exponents in critical phenomena has been well known for long time and it is generally believed that systems within a given universality class have different finite-size scaling functions. In 1984, Privman and Fisher proposed the idea of universal finite-size scaling functions (UFSSF) and nonuniversal metric factors for static critical phenomena. From 1984 to 1994, the progress of research in this direction was very slow. In this paper, we review recent developments relating to universal finite-size scaling functions in static and dynamic critical phenomena. The topics under discussion include: 1. UFSSF of the existence probability E_p and the percolation probability P in lattice percolation models, 2. UFSSF of the probability for the appearance of n percolating clusters W_n in lattice percolation models, 3. UFSSF of E_p and W_n in continuum percolation models, and 4. UFSSF in dynamic critical phenomena of the Ising model.

1 Introduction

Universality and scaling are two important concepts in the theory of critical phenomena [1, 2]. The former dates from the work on Yang [3] and Chang [4]. In 1952, Yang [3] derived the exact spontaneous magnetization M of the Ising model on a square lattice with isotropic interactions and found that the critical exponent β of M is 1/8. In the same year, Chang [4] derived the exact spontaneous magnetization M of the Ising model on a square lattice with anisotropic interactions, i.e. the coupling constants in the horizontal direction J_1 and in the vertical direction J_2 are different. Chang [4] found that, for $0 < J_1/J_2 < \infty$, β is always equal to 1/8. Chang [4] conjectured that for other planar lattices β is also equal to 1/8, which was confirmed by later calculations: this marked the beginning of theory of the universality of critical exponents. Now it is generally believed that for the Ising model on all planar lattices, including the square (sq), the plane triangular (pt), the honeycomb (hc) lattices, etc, the specific heat exponent α, the spontaneous magnetization exponent β, the magnetic susceptibility exponent γ, and the correlation length exponent ν are 0 (logarithmic divergence), 1/8, 7/4, and 1, respectively [1]. It is also believed that for site and bond percolation on all planar lattices, the correlation length exponent ν, the percolation probability exponent β, and the mean cluster size exponent γ are 4/3, 5/36, and 43/18, respectively [5].

Springer Proceedings in Physics, Volume 83
Computer Simulation Studies in Condensed-Matter Physics X
Eds.: D. P. Landau, K.K. Mon, H. -B. Schüttler
© Springer-Verlag Berlin Heidelberg 1998

Another important concept in the theory of critical phenomena is scaling [1, 2]. For example, in a ferromagnetic system, e.g. $CrBr_3$, for temperatures T near the critical temperature T_c (also called the Curie temperature in ferromagnetic systems), if we plot $\sigma/|\epsilon|^\beta$ as a function of $h/|\epsilon|^{\beta+\gamma}$, where σ is the magnetization, $\epsilon = (T - T_c)/T_c$, and h is the external magnetic field, then the experimental data for different temperatures collapse on a single curve, called the scaling function [1]. In this paper, we consider another kind of scaling, called finite-size scaling.

According to the theory of finite-size scaling [5, 6, 7, 8, 9], if the dependence of a physical quantity Q of a thermodynamic system on the parameter ϵ, which vanishes at the critical point $\epsilon = 0$, is of the form $Q(\epsilon) \sim |\epsilon|^a$ near the critical point, then for a finite system of linear dimension L, the corresponding quantity $Q(L, \epsilon)$ is of the form:

$$Q(L, \epsilon) \approx L^{-a y_t} F(\epsilon L^{y_t}), \tag{1}$$

where y_t $(=\nu^{-1})$ is the thermal scaling power and $F(x)$ is the finite-size scaling function. It follows from (1) that the scaled data $Q(L, \epsilon) L^{a y_t}$ for different values of L and ϵ can be described as a single function of the scaling variable $x = \epsilon L^{y_t}$. Thus it is important to know general features of the finite-size scaling function under various conditions.

In 1984, Privman and Fisher proposed the idea of universal finite-size scaling functions (UFSSF) and nonuniversal metric factors for static critical phenomena [7] for T near T_c and the external magnetic field h near 0. Specifically, they proposed that, near $\epsilon = 0$ and $h = 0$, the singular part of the free energy for a ferromagnetic system can be written as

$$f_s(\epsilon, h, L) \approx L^{-d} Y(C_1 \epsilon L^{1/\nu}, C_2 h L^{(\beta+\gamma)/\nu}), \tag{2}$$

where d is the spatial dimensionality of the lattice, Y is a universal finite-size scaling function, and C_1 and C_2 are adjustable nonuniversal metric factors [7] which depend on the specific lattice structure. From Eq.(2) and the scaling relations $\nu d = 2 - \alpha$ and $\alpha + 2\beta + \gamma = 2$ [1], one obtains the scaling expression for the finite-size magnetization [7]

$$m = -\frac{\partial}{\partial h} f_s(\epsilon, h, L) \approx C_2 L^{-\beta/\nu} Y^{(1)}(C_1 \epsilon L^{1/\nu}, C_2 h L^{(\beta+\gamma)/\nu}), \tag{3}$$

which is the order parameter of the system. From 1984 to 1994, the progress in research on UFSSF was very slow. In this paper, we briefly review recent developments from Monte Carlo approaches to UFSSF in static and dynamic critical phenomena.

This paper is organized as follows. In Sec. 2 we review histogram Monte Carlo approaches to UFSSF for the existence probability E_p and the percolation probability P in lattice percolation models. In Sec. 3 we review histogram Monte Carlo approaches to UFSSF of the probability for the appearance of n percolating clusters, W_n, in lattice percolation problems. In

Sec. 4 we present our Monte Carlo results for UFSSF in continuum percolation of soft disks and hard disks. In Sec. 5 we present our Monte Carlo results
for UFSSF in dynamic critical phenomena of the Ising model on lattices. In
Sec. 6 we summarize our results and give final comments.

2   UFSSF for E_p and P of Lattice Percolation Models

In 1992, Hu proposed a histogram Monte Carlo simulation method (HMCSM)
[10, 11], which was then used to calculate the finite-size scaling functions for
the existence probability E_p and the percolation probability P of the percolation model and the q-state bond-correlated percolation model corresponding
to the q-state Potts model [12, 13, 14, 15, 16, 17, 18]. Here $E_p(G, p)$ is the
probability that the system percolates and $P(G, p)$ is the probability that a
given lattice site belongs to a percolating cluster. Now we briefly review the
HMCSM for the bond percolation [10, 11, 18] on an $L_1 \times L_2$ lattice G and define related quantities. The extension to other lattices and to site percolation
[15, 16] is straightforward. In bond percolation on a lattice G with N sites,
$N = L_1 \times L_2$, and E bonds, each bond of G is occupied with a probability
p, where $0 \leq p \leq 1$. There are several different rules to define percolating
clusters, called spanning rules, first discussed by Reynolds, Stanley, and Klein
[19]. In R_1, a cluster percolates if it extends from the top row of G to the
bottom row of G; in R_2, a cluster percolates if it extends from the top row
to the bottom row and from left boundary to right boundary of G [19]. In a
given spanning rule, a subgraph which contains a percolating cluster is a percolating subgraph and denoted by G'_p. Then we have following definitions for
the existence probability $E_p(G, p)$ and the percolation probability $P(G, p)$.

$$E_p(G, p) = \sum_{G'_p \subseteq G} p^{b(G'_p)}(1 - p)^{E - b(G'_p)}, \tag{4}$$

$$P(G, p) = \sum_{G'_p \subseteq G} p^{b(G'_p)}(1 - p)^{E - b(G'_p)} N^*(G'_p)/N, \tag{5}$$

where $b(G'_p)$ is the number of occupied bonds in G'_p and $N^*(G'_p)$ is the total
number of sites in the percolating clusters of G'_p. The summations in (4) and
(5) are over all subgraphs G'_p of G.

To carry out histogram Monte Carlo simulations, we first choose w different values of p. For a given $p = p_j$, $1 \leq j \leq w$, we generate N_R different
subgraphs G'. The data obtained from the wN_R different G' are then used to
construct three arrays of numbers of length E with elements $N_p(b)$, $N_f(b)$,
and $N_{pp}(b)$, which are, respectively, the total numbers of percolating subgraphs with b occupied bonds, nonpercolating subgraphs with b occupied
bonds, and the sum of $N^*(G'_p)$ over subgraphs with b occupied bonds. After
a sufficient number of simulations, these arrays can be used to approximate
E_p and P for any value of the bond occupation probability p [10, 15]:

$$E_p(G,p) = \sum_{b=0}^{E} p^b (1-p)^{E-b} C_b^E \frac{N_p(b)}{N_p(b) + N_f(b)}, \qquad (6)$$

$$P(G,p) = \frac{1}{N} \sum_{b=0}^{E} p^b (1-p)^{E-b} C_b^E \frac{N_{pp}(b)}{N_p(b) + N_f(b)}, \qquad (7)$$

where $C_b^E = E!/(E-b)!b!$. Once we have histogram data, we can calculate E_p and P as continuous functions of p. This is different from traditional Monte Carlo methods [5].

As $L \to \infty$, E_p is 0 for $p < p_c$ and is 1 for $p > p_c$; if we write $E_p \sim (p-p_c)^a$ just above p_c, then the critical exponent a of E_p is 0 [5]. On the other hand, $P \sim (p - p_c)^\beta$ just above p_c. According to Eq.(1), we may write $E_p = F(x)$ and $PL^{\beta/\nu} = S(x)$ with $x = (p - p_c)L^{1/\nu}$, where $F(x)$ and $S(x)$ are scaling functions. We found that E_p and $PL^{\beta/\nu}$ have very good scaling behavior and the finite-size scaling functions depend sensitively on boundary conditions and aspect ratios of the lattice and spanning rules for percolating clusters [15, 16, 17, 18]. Equation (1) for E_p implies that E_p for all models in the same universality class must be equal at the critical point in order to have universal finite-size scaling functions. In 1992, Ziff found that E_p=0.5 for site and bond percolation on large square lattices with free boundary conditions [20] and Langlands, Pichet, Pouliot, and Saint-Aubin (LPPS) proposed that when aspect ratios for the square (sq), honeycomb (hc), and plane triangular (pt) lattices have the relative proportions $1{:}\sqrt{3}{:}\sqrt{3}/2$, then site and bond percolation on such lattices have the same value of E_p at the critical point [21]. In 1992, Cardy used a conformal theory to write down a formula for the critical E_p as a function of aspect ratio for percolation on lattices with free boundary conditions [22]. Cardy and LPPS did not discuss the values of E_p for $p \neq p_c$.

In 1995, Hu, Lin and Chen (HLC) [23] applied the HMCSM [10, 11] to calculate E_p and P of site and bond percolation on finite 512×512 sq, 433×250 hc, and 433×500 pt lattices, i.e. they used 512/512:433/250:433/500 to approximate the proportions $1{:}\sqrt{3}{:}\sqrt{3}/2$ of aspect ratios for sq, hc, and pt lattices considered by LPPS. Plotting E_p as a function of $x = D_1(p-p_c)L^{1/\nu}$ and $D_3 PL^{\beta/\nu}$ as a function of $x = D_2(p - p_c)L^{1/\nu}$, where D_1, D_2 and D_3 are nonuniversal metric factors, they found that the six percolation models have very nice universal finite-size scaling functions for E_p and P. Within numerical uncertainties $D_1 = D_2$ and the nonuniversal metric factors for periodic boundary conditions are consistent with those for free boundary conditions, although the scaling functions are quite different. We also found that the nonuniversal metric factors are independent of changes in aspect ratios holding the ratio between them constant [24]. These results indicate for each percolation model we need only two nonuniversal metric factors, i.e. D_1 and D_3.

After Ref. [23] by HLC was published, Okabe and Kikuchi obtained universal finite-size scaling functions for the two-dimensional Ising model [25] and Hovi and Aharony (HA) [26] calculated the scaling function $f(x)$ for bond and site percolation on the square lattice with both free (f) and periodic (p) boundary conditions (bc). HA found that their $f(x)$ for fbc is consistent with HLC's $f(x)$, but their $f(x)$ for pbc is quite different from HLC's $f(x)$, i.e. HA obtained $f(0) = 0.63665 \pm 0.0008$ and HLC obtained $f(0) = 0.93(4)$. Hu [27] conjectured that the difference was because HA considered pbc only in one direction, while HLC considered pbc in both horizontal and vertical directions. This conjecture was confirmed by numerical calculations [27]. This result provided another evidence that finite-size scaling functions sensitively depends on the boundary conditions [15].

3 UFSSF for W_n of Lattice Percolation Models

In low-temperature measurements of quantum Hall effects (QHE), when the external magnetic field is increased from small values to large values, the conductivity σ_{xy} moves from one plateau with $\sigma_{xy} = \sigma_1$ to another plateau with the value $\sigma_{xy} = \sigma_2$ and the conductivity σ_{xx} has a maximum σ_{xx}^{max} in the transition region. In a recent theory of QHE, Ruzin, Cooper, and Halperin (RCH) [28] proposed that the number of percolating clusters in the sample at the critical point is useful for understanding σ_{xx}^{max}. Therefore, the number of percolating clusters in percolation problems is an interesting quantity and we should know more about its behavior.

To mimic the Corbino disk often used in experimental studies of quantum Hall effects [28], Hu used the HMCSM to study bond percolation on $L_1 \times L_2$ square [29] lattices G with pbc in the horizontal L_1 direction and fbc in the vertical L_2 direction [30]. A cluster which extends from the top row of G to the bottom row of G is a percolating cluster. A subgraph which contains at least one percolating cluster is a percolating subgraph and denoted by G'_p. It should be noted that the definition of G'_p in [30, 31] and this section is different from that of [15, 16, 17, 23, 24] in which only the largest cluster is used to define G'_p. A percolating subgraph which contains exactly n percolating clusters is denoted by G'_n. Now we have the definitions

$$W_n(L_1, L_2, p) = \sum_{G'_n \subseteq G} p^{b(G'_n)}(1 - p)^{E - b(G'_n)}, \tag{8}$$

where $b(G'_n)$ is the number of occupied bonds in G'_n. The summation in (8) is over all subgraphs G'_n of G. To use the HMCSM to evaluate W_n, in addition to $N_p(b)$ and $N_f(b)$ considered in Sec. 2, we also evaluate $N_n(b)$, $0 \leq b \leq E$, which is the number of percolating subgraphs with b occupied bonds and n percolating clusters. After a large number of simulations, the probability $W_n(L_1, L_2, p)$ at any value of the bond occupation probability p can be calculated approximately from the following equation [10, 30]:

$$W_n(L_1, L_2, p) = \sum_{b=0}^{E} p^b (1-p)^{E-b} C_b^E \frac{N_n(b)}{N_p(b) + N_f(b)}. \qquad (9)$$

It is obvious that $W_0(L_1, L_2, p) = 1 - E_p$. Hu found that W_n as a function of $z = (p - p_c)L^{1/\nu}$ has very good scaling behavior. Hu also considered fbc in both horizontal and vertical directions and found that the scaling functions for W_n depend sensitively on boundary conditions [30].

Using the HMCSM [10, 30], Hu and Lin (HL) calculated W_n for bond and site percolations on sq, hc, and pt lattices with pbc in the horizontal direction and fbc in the vertical directions; the aspect ratios of sq, hc, and pt lattices are 4, $4\sqrt{3}$, $2\sqrt{3}$, respectively. Using nonuniversal metric factors of [23], HL found that these percolation models have UFSSF for W_n [31].

Hu and Halperin (HH) [32] considered bond percolation with bond probability p on an $L_1 \times L_2$ self-dual square lattice with pbc in the horizontal direction and fbc in the vertical direction. HH defined the number M of alternating percolating clusters as the minimum of n_{p} and n_{n}, where n_{p} is the number of independent percolating clusters connecting sites on the top and bottom edges, and n_{n} is the number of percolating clusters in the complementary configuration on the dual lattice, a bond being present in the complementary configuration if and only if it is absent in the original configuration. They used the HMCSM [10, 30] to evaluate the probability $W_M^{\mathrm{a}}(L_1, L_2, p)$ for finding a given value of M and found that, for a given aspect ratio L_1/L_2 all data of $W_M^{\mathrm{a}}(L_1, L_2, p)$ near the critical point p_c fall on the same scaling function F_M^{a}, which is symmetric with respect to the scaling variable for all M. The values of $W_M^{\mathrm{a}}(L_1, L_2, p)$ at the critical point are useful for understanding σ_{xx}^{max} in the quantum Hall effects [28, 32, 33].

4 UFSSF for Continuum Percolation of Disks

Many interesting quantities and problems in solid state physics, e.g. σ_{xx}^{max} in QHE, conductor-insulator transition, etc., are related to continuum percolation [28, 34]. However, to study continuum percolation is much more difficult than to study lattice percolation. People usually study lattice percolation rather than continuum percolation. The problem is to what extent the quantities, e.g. critical exponents and finite-size scaling functions, obtained from lattice percolation models (LPM) may be applied to continuum percolation models (CPM). Hu and Wang (HW) have tried to answer this interesting and important question [35].

HW considered both soft disks and hard disks. Typical configurations of soft disks and hard disks are shown in Fig. 1(a) and Fig. 1(b), respectively. In the general case, HW considered (hard and soft) disks on an $L_1 \times L_2$ continuum space C of rectangular domain with linear dimension L_1 in the horizontal direction and linear dimension L_2 in the vertical direction, where

L_1 and L_2 are integers. The space C is divided into $L_1 \times L_2$ covering meshes, which are (1×1) unit squares. The squares (meshes) are labeled by integers 1, 2, 3, ..., $L_1 \times L_2$. A disk belongs to a square if and only if the center of the disk is in that square. The disks have a radius $R = \sqrt{2}/2$, so that at most one hard disk is allowed in one unit square. Two hard disks are in the same cluster if and only if their separation is smaller than or equal to $2\sqrt{2}$. Such a definition of clusters was considered by Hu [36] and Kratky [37] before. Two soft disks are in the same cluster if and only if they overlap. More than one soft disk may be in a given unit square in which case they are always in the same cluster. HW extended the multiple-labeling technique of Hoshen and Kopleman [38] to label unit squares which have disks. The label for a unit square is also the label for the disks belong to that unit square. This multiple-labeling technique for CPM was used to study critical properties and scaling functions for soft disks and hard disks.

Fig. 1(a)   Fig. 1(b)

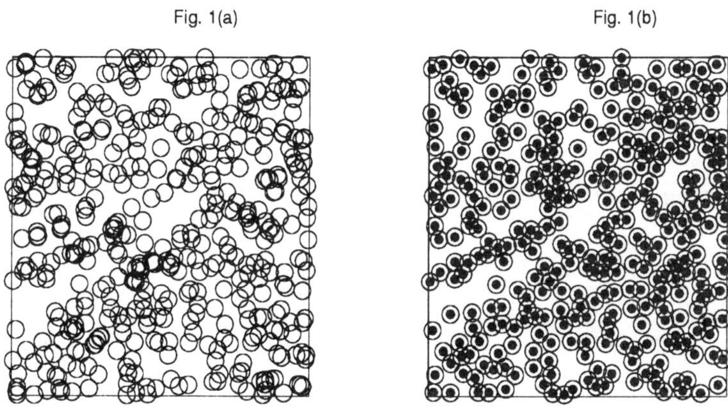

Fig. 1. (a). A configuration of soft disks for continuum percolation. (b). A configuration of hard disks for continuum percolation.

HW used a random deposition process to generate configurations of disks. In the deposition process, if the hard cores of two disks overlap, then the attempt to put the second hard disk is abandoned. The concentration of disks is defined by $\eta = \pi R^2 N/L^2$, where N is the number of the disks in the system and $L = \sqrt{L_1 L_2}$ is the linear dimension of the system. At a given η, the number of percolating configurations observed divided by the total number of generated configurations gives the existence probability E_p. The calculated E_p as a function of η for continuum percolation of soft disks on $L \times L$ space with free boundary conditions in both horizontal and vertical directions are shown in Fig. 2(a). The intersection of curves in Fig. 2(a) give the critical point η_c and the critical existence probability $E_p(\eta_c)$, which are 1.1302 ± 0.0008 and 0.50 ± 0.01, respectively. The former is consistent with the result of Gawlinski and Stanley [39] and the latter is consistent with

13

the result of LPM [20, 21, 23]. From the slopes of curves at η_c, HW used a percolation renormalization group method [10] to find $\nu = 1.39 \pm 0.07$, which is consistent with the exact $\nu = 4/3$ for LPM on planar lattices [5]. The data of Fig. 2(a) as a function of the scaling variable $x = (\eta - \eta_c)L^{1/\nu}$ with $\nu=4/3$ are shown in Fig. 2(b), which shows that E_p has very good scaling behavior. HW found similar results for systems of hard disks and systems with pbc in the horizontal direction and fbc in the vertical direction. Typical results for these boundary conditions are presented in Fig. 2(c), which shows that E_p of soft disks, hard disks, and lattice site percolation have a universal finite-size scaling function.

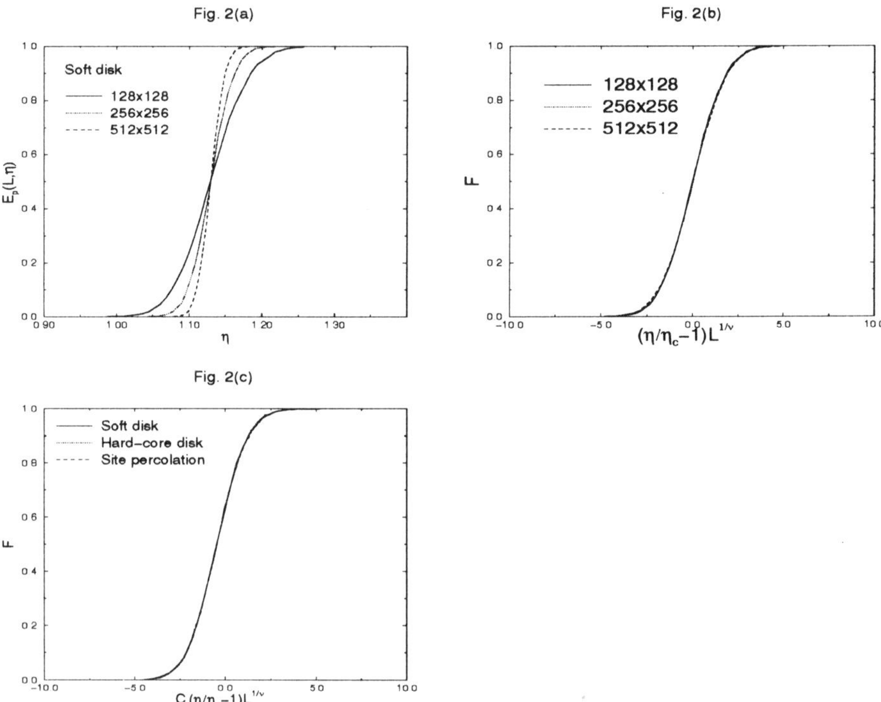

Fig. 2. (a) The calculated E_p as a function of η for the continuum percolation of soft disks on $L \times L$ space with free boundary conditions in the horizontal and vertical directions, where $L =128$, 256, and 512. The number of different η is between 50 and 100. The numbers of independent configurations for $L=128$, 256, and 512 are 40000, 10000, and 5000, respectively. (b) The data of (a) are plotted as a function of the scaling variable $x = (\eta - \eta_c)L^{1/\nu}$, where $\nu=4/3$. The scaling function is $F(x)$. (c) The universal finite-size scaling function of E_p for soft disks, hard disks and site percolation on a square lattice. The numbers of independent configurations for hard disks and site percolation are two and eight times of that for soft disks, respectively. The non-universal metric factors for soft disks, hard disks, and lattice site percolation are $C_1 = 1$, $C_2 = 0.897 \pm 0.029$, and $C_3 = 1.60 \pm 0.07$, respectively.

HW also calculated the probability W_n for the appearance of n percolating clusters for soft disks and hard disks on an $L_1 \times L_2$ space with pbc in the horizontal direction and fbc in the vertical direction. Typical calculated results are presented in Figs. 3(a), 3(b) and 3(c). Figure 3(b) shows that W_n has very good scaling behavior and Figure 3(c) shows that W_n of soft disks, hard disks, and LPM have universal finite-size scaling functions. It is of interest to note that the nonuniversal metric factors of Fig. 3(c) are the same as those of Fig. 2 (c), which is similar to the case of lattice percolation [31].

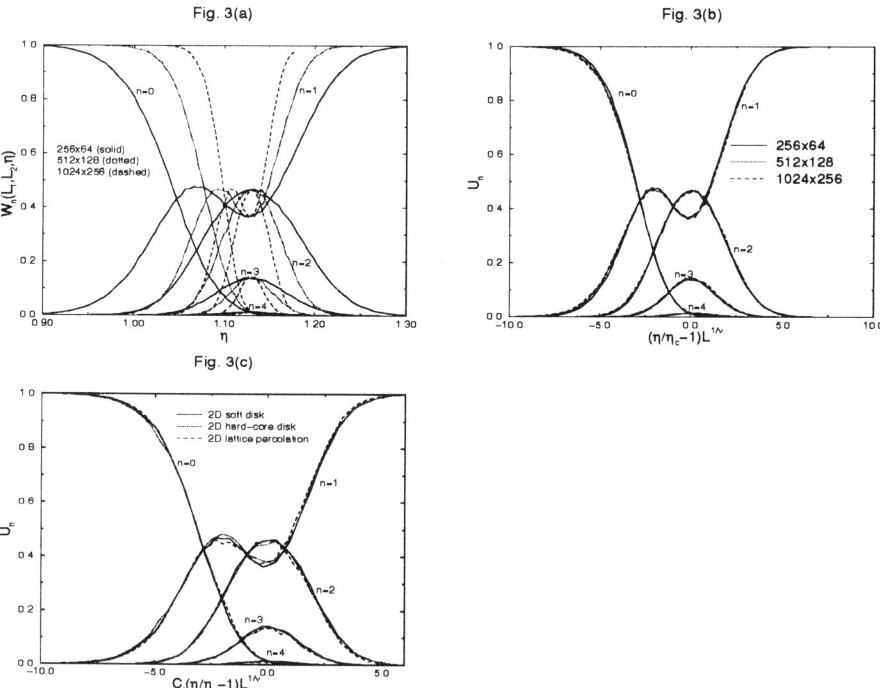

Fig. 3. (a) $W_n(L_1, L_2, \eta)$ as a function of η for continuum percolation of soft disks on 256×64, 512×128 and 1024×256 space, which are represented by solid, dotted, and dashed lines, respectively. (b) The data of (a), $W_n(L_1, L_2, \eta)$, as a function of $z = (\eta - \eta_c)L^{1/\nu}$. The scaling function for $W_n(L_1, L_2, \eta)$ is denoted by U_n. (c) The universal finite-size scaling functions of W_n with $n = 0, 1, 2, 3, 4$ for soft disks, hard disks, site percolation on square lattice systems. The non-universal metric factors for soft disks, hard disks, and lattice site percolation are $C_1 = 1$, $C_2 = 0.897 \pm 0.029$, and $C_3 = 1.60 \pm 0.07$, respectively.

To check the universality of critical exponents, HW calculated the mean sizes of finite clusters $S(L, \eta_c)$, the mean sizes of percolating clusters $S_p(L, \eta_c)$, and the distributions of cluster sizes, $n(L, s, \eta_c)$, for soft disks, hard disks, and site percolation at their critical points η_c for systems of various linear dimension L. It follows from finite-size scaling and the scaling behavior of

$n(L, s, \eta_c)$ [5] that $S(L, \eta_c) \sim L^{\gamma/\nu}$, $S_p(L, \eta_c) \sim L^{d_f} = L^{d-\beta/\nu}$, $n(L, s, \eta_c) \sim s^{-\tau(L)}$. The critical exponents estimated from these equations are presented in Table 1 which shows that soft disks, hard disks, and percolation on planar lattices are in the same universality class.

Table 1. Universality of $E_p(\eta_c)$ and critical exponents for 2D continuum percolation. For E_p, we consider both free boundary conditions (fbc) and periodic boundary conditions (pbc) in the horizontal direction.

quantities	soft disks	hard disks	LPM	exact
threshold	1.1302 ± 0.0008	0.8503 ± 0.0010	0.5928 ± 0.0010	
$E_p(\eta_c)$ (fbc)	0.50 ± 0.01	0.50 ± 0.03	0.50 ± 0.01	0.5
$E_p(\eta_c)$ (pbc)	0.64 ± 0.02	0.64 ± 0.02	0.63 ± 0.02	
ν	1.39 ± 0.07	1.36 ± 0.04	1.37 ± 0.05	$1.33...$
γ/ν	1.785 ± 0.012	1.790 ± 0.012	1.780 ± 0.020	$1.7916...$
$d - \beta/\nu$	1.889 ± 0.006	1.885 ± 0.008	1.891 ± 0.016	$1.89583...$
τ	2.05 ± 0.02	2.05 ± 0.02	2.04 ± 0.02	$2.0549...$

5 UFSSF in Dynamic Critical Phenomena

In previous sections, we only consider static critical phenomena. Recently, Wang and Hu (WH) considered universality of critical exponent and dynamic finite-size scaling functions (DFSSF) in dynamic critical phenomena of the Ising model on planar lattices [40].

The dynamic critical exponent z [41, 42] which characterizes the critical slowing down near the critical temperature is of much interest. Since there is no exact solution for z, computer simulation plays an important role in the evaluation of z. In the past two decades, estimates of z varied in a large range between 1.7 and 2.3 [43]. Only very recently, several authors reached a consistent value for this exponent with different simulation schemes. ¿From time relaxation of the magnetization and energy of the Ising model, Ito [43] found that $z = 2.165 \pm 0.010$ for the sq lattice Ising model, which was confirmed by other calculations [44, 45, 46]. While the universality of static critical exponents was well established long ago [1], the universality of z [41], in the sense that z does not depend on details of local interactions and lattice structures, is rarely extensively studied in the literature. Almost all simulations are performed on the sq lattice Ising model [47].

The critical exponent z can be evaluated by studying the relaxation of the magnetization M on a lattice with a linear dimension L and N lattice sites, which has following form at the critical temperature T_c [42], $M(T_c, t) \equiv$

$M(L \to \infty, T_c, t)) \sim t^{-\beta/\nu z}$, where β and ν are universal static exponents for M and correlation length, respectively, and are 1/8 and 1 for the two-dimensional Ising model, and t is the number of Monte Carlo steps with the unit of one sweep of all lattice sites. WH [40] used heat bath dynamics with the damage spreading method to study the Ising model on 1000×1000 sq, pt and hc lattices at critical points. From the logarithmic-scaled relaxation curves for $M(T_c, t)$, they estimated z to be 2.166 ± 0.007, 2.164 ± 0.007, 2.170 ± 0.010 for sq, pt, and hc lattices, respectively, which are consistent with each other and also consistent with other calculations [43, 44, 45, 46].

The universality of z provides a good basis to study the universality of DFSSF. Suzuki [42] proposed that when the temperature T is near the critical temperature T_c, the magnetization of a system of linear dimension L at time t, $M(L, T, t)$, may be written as

$$M(L, T, t) = L^{-\beta/\nu} f(L^{1/\nu}(T/T_c - 1), tL^{-z}). \qquad (10)$$

WH [40] first considered the case $T = T_c$ and have

$$M(L, t) \equiv M(L, T_c, t) = L^{-\beta/\nu} f(tL^{-z}), \qquad (11)$$

where $L = \sqrt{N}$. In [23], HLC considered 512×512 sq, 433×500 pt, and 433×250 hc lattices, so that E_p at critical points are identical for all lattices [21], which is a crucial step to obtain UFSSF for E_p and P. To obtain the UFSSF for the Ising model, one should choose the ratios between aspect ratios of different lattices to be approximately equal to those of [23]. Therefore, WH considered the Ising model on 32×51 and 64×102 sq lattices, 27×50 and 54×100 pt lattices, and 27×25 and 81×75 hc lattices. The relaxation of $M(L, t)$ of the Ising model on such lattices are shown in Fig. 4(a); $M(L, t)L^{\beta/\nu}$ as a function of tL^{-z} is presented in Fig. 4(b), which shows that two curves of the same lattice with different L fall onto an identical dynamic scaling function and scaling functions for different lattice structures are different. Following the case of static critical phenomena [7, 23], WH proposed following equation for universal DFSSF $F(x)$

$$D_i M(L, t) = L^{-\beta/\nu} F(C_i tL^{-z}). \qquad (12)$$

Here D_i and C_i for i equal 1, 2, and 3 are the nonuniversal metric factor for sq, pt, and hc lattices, respectively. With $D_i = 1$ ($i = 1, 2, 3$), $C_1 = 1$, $C_2 = 1.222 \pm 0.009$, and $C_3 = 0.693 \pm 0.018$, Fig. 4(c) is obtained, which shows that data for different lattices fall on an universal DFSSF.

Next WH evaluated $M(L, T, t)$ for the Ising model on 32×51 sq lattice, 27×50 pt lattice, and 27×25 hc lattice for $T \neq T_c$ and at some finite scaled times, say $C_i t_i L^{-z} = 1.658g$ with g being 0.5, 1, and 2, which means that $t_1 = 5000g$ MCS for sq lattice, $t_2 = 3332g$ MCS for pt lattice and $t_3 = 2773g$ MCS for hc lattice. Following [23], WH proposed following equation for universal DFSSF F'.

$$D_i M(L,T,t_i)L^{\beta/\nu} = f(E_i L^{1/\nu}(T/T_c-1), C_i t_i L^{-z}) \equiv F'(E_i L^{1/\nu}(T/T_c-1)),$$
$$(13)$$

where D_i and E_i for $1 \leq i \leq 3$ are nonuniversal metric factors. With $D_i = E_i = 1$ and C_i of Fig. 4(c), $D_i M(L,T,t_i)L^{\beta/\nu}$ as a function of $x = E_i L^{1/\nu}(T/T_c - 1)$ is plotted in Fig. 5, which shows that in the critical region and for any value of g, the three lattices have universal DFSSF for $M(L,T,t_i)$.

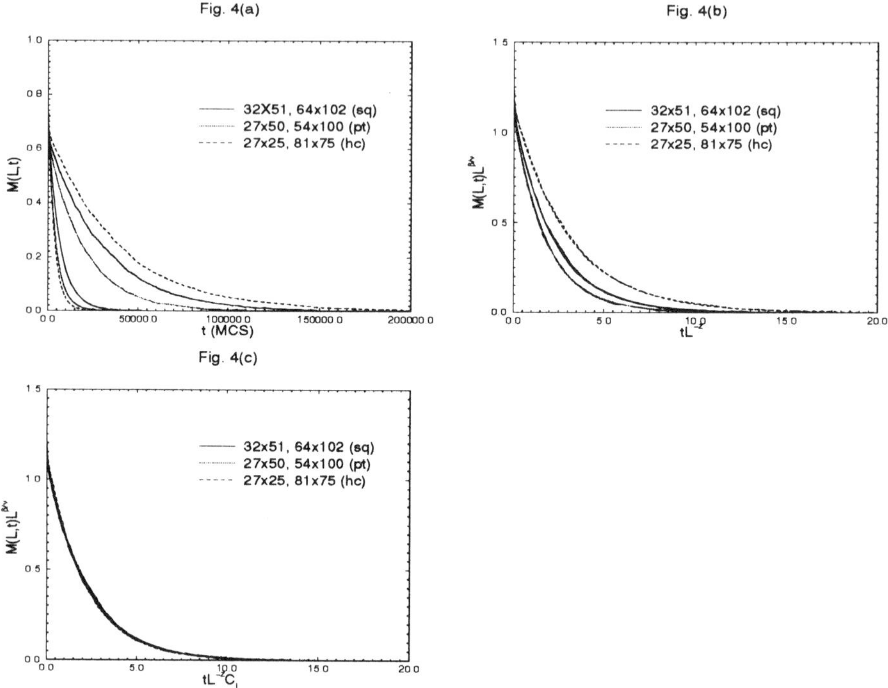

Fig. 4. (a) $M(L,t)$ of Eq.(11) vs. t for the Ising model on sq, pt, and hc lattices. (b) $M(L,t)$ vs. tL^{-z}. (c) $D_i M(L,t)$ vs. $C_i t L^{-z}$ with non-universal scaling factors $D_1 = D_2 = D_3 = 1$, $C_1 = 1$ (sq lattice), $C_2 = 1.222 \pm 0.009$ (pt lattice) and $C_3 = 0.693 \pm 0.018$ (hc lattice).

Figure 5 suggests that as $g \to \infty$, $D_i = E_i = 1$ $(1 \leq i \leq 3)$ still gives universal DFSSF for $M(L,T,t_i)$ and such nonuniversal metric factors should be consistent with nonuniversal metric factors for the static finite-size scaling function (SFSSF) [23, 25]. A cluster Monte Carlo method [48] which can overcome the critical slowing down is used to calculate the equilibrium magnetization M_e of the Ising model on 32×51 sq, 27×50 pt, and 27×25 hc lattices to test this idea. It has been found that $D_i' M_e L^{\beta/\nu}$ as a function of $x = E_i' L^{1/\nu}(T/T_c - 1)$ for three lattices have universal SFSSF [49] with $D_i' = E_i' = 1$ for i=1, 2, and 3. The nonuniversal metric factors for the Ising model obtained by Okabe and Kikuchi [25] correspond to $D'_1 = E'_1 = 1$ for the sq lattice, $D'_2 = 1.02 \pm 0.02$ and $E'_2 = 0.96 \pm 0.03$ for the pt lattice, and $D'_3 = 0.98 \pm 0.02$ and $E'_3 = 1.00 \pm 0.02$ for the hc lattice, which are consistent with $D_i = E_i = 1$, $1 \le i \le 3$, used in Fig. 5 of this paper.

Fig. 5

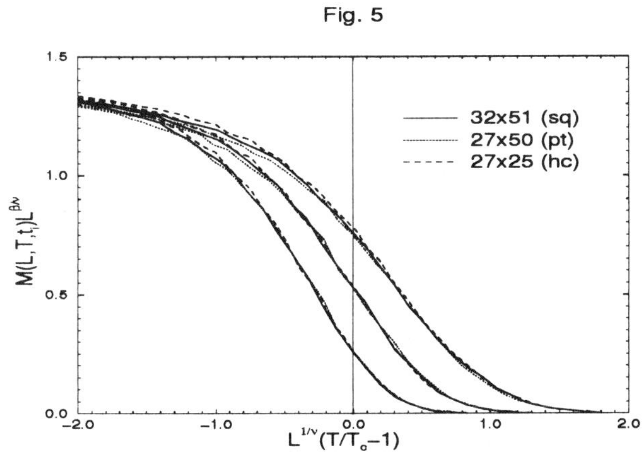

Fig. 5. $D_i M(L,T,t_i)L^{\beta/\nu}$ of (13) vs. $E_i L^{1/\nu}(T/T_c-1)$ with $D_i = E_i = 1$, $1 \le i \le 3$, for the Ising model on sq, pt, and hc lattices near the critical temperature of each lattice and for the scaled times $C_i t_i L^{-z} = 1.658g$ with g being 0.5, 1, and 2. At $T = T_c$, the curves from top to bottom are for g being 0.5, 1.0, and 2.0, respectively.

It is well known that the Ising model and the bond random percolation model (BRPM) correspond to the q-state Potts model [50] with q being 2 and 1, respectively, so that the bond probability p of the percolation model is related to the temperature T of the Potts model by $p = 1 - exp(-2/T)$ [51]. HLC [23] obtained non-universal metric factors for BRPM, using $(p-p_c)L^{1/\nu}$ as a scaling variable. Recalculating their result using $(T/T_c - 1)L^{1/\nu}$ as a scaling variable, we find metric factors of $D'_1 = E'_1 = 1$ for sq lattice, $D'_2 = 1.021 \pm 0.021$ and $E'_2 = 0.996 \pm 0.034$ for pt lattice, and $D'_3 = 0.987 \pm 0.011$ and $E'_3 = 1.011 \pm 0.019$ for hc lattice, which is consistent with $D_i = E_i = 1$, $1 \le i \le 3$, presented in this paper.

Here we present a heuristic argument for the surprising result that $D_i = E_i = 1$, $1 \le i \le 3$, could give good fits to Eqs.(12) and (13). At and near the critical point, two important quantities governing the behavior of a dynamic system are relaxation time τ and correlation length ξ and the details of the lattices and local interactions are not important. The temporal evolutions of the Ising model on sq, pt, and hc lattices in terms of scaled times $C_i t_i L^{-z}$ look similar. Since ratios of aspect ratios for sq, pt, and hc lattices are chosen to be $1:\sqrt{3}/2:\sqrt{3}$, the domains of these lattices and critical behavior of the Ising model on these domains also look similar. Therefore, $D_i = E_i = 1$,

$1 \leq i \leq 3$, can give good fits to Eqs.(12) and (13) and the static Ising model. Similar arguments may be applied to the bond percolation problem.

6 Summary and Final Remarks

Using HMCSM [10, 30] and relative aspect ratios considered by LPPS [21], we found universal finite-size scaling functions (UFSSF) for the existence probability, E_p, the percolation probability, P, and the probability for the appearance of n percolation clusters, W_n, of site and bond percolation on sq, hc, and pt lattices. Using random deposition process, we found universal finite-size scaling functions for E_p and W_n for CPM of soft disks and hard disks and LPM, see Figs. 2 and 3. Table 1 shows that the CPM of soft disks and hard disks are in the same universality class as the lattice percolation models. We have found similar results for the CPM of soft spheres and hard spheres in three dimensional space. Using heat bath dynamics and relative aspect ratios considered by LPPS [21], we found universal dynamic finite-size scaling function for $M(L,T,t)$ without using static nonuniversal metric factors, which implies that C_1 and C_2 of Eqs. (2) and (3) are equal to 1.

We may consider the general case that a disk has a hard core of radius R_1 and a soft shell of radius R_2, where $R_1 < R_2$. The soft disk of Fig. 1(a) corresponds to $R_1 = 0$ and the hard disk of Fig. 1(b) corresponds to $R_1 = R_2/2$. Two disks are in the same cluster if their soft shells overlap. The general case $0 \leq h = R_1/R_2 < 1$ had been considered by Lee [52]. However, he did not reach a definite result about the universality of such general hard disks. Our results show that disks with $h = 0$ and $h = 0.5$ are in the same universality class, which suggests that disks with $0 \leq h < 1$ may be in the same universality class. Further studies in this direction are needed.

Our results suggest that computer simulations can help us to understand the mysteries of nature.

Acknowledgements

We thank B. I. Halperin and R. H. Swendsen for discussions and J. G. Dushoff for a critical reading of the paper. This work was supported by the National Science Council of the Republic of China (Taiwan) under grant numbers NSC 86-2112-M-001-001 and NCHC-86-02-008, the Computing Center of Academia Sinica (Taipei) and National Center for High-Performance Computing (Taiwan).

References

+. Electronic address: huck@phys.sinica.edu.tw.

*. Permanent address: Department of Applied Physics, Shanghai Jiao Tong University, Shanghai 200030, China.

1. H. E. Stanley, *Introduction to Phase Transitions and Critical Phenomena*, (Oxford Univ. Press, New York, 1971).

2. L. P. Kadanoff, Physica A **163**, 1 (1990).

3. C. N. Yang, Phys. Rev. **85**, 808 (1952).

4. C. H. Chang, Phys. Rev. **88**, 1422 (1952).

5. D. Stauffer and A.Aharony, *Introduction to Percolation Theory* 2nd. ed. (Taylor and Francis, London, 1992).

6. M. E. Fisher, in *Proc. 1970 E. Fermi Int. School of Physics*, M. S. Green ed. (Academic, NY, 1971) Vol. 51, p. 1.

7. V. Privman and M. E. Fisher, Phys. Rev. **B30**, 322 (1984).

8. *Finite-Size Scaling*, J. L. Cardy ed. (North-Holland, Amsterdam, 1988).

9. *Finite-size Scaling and Numerical Simulation of Statistical Systems*, V. Privman ed. (World Scientific, Singapore, 1990).

10. C.-K. Hu, Phys. Rev. B **46**, 6592 (1992).

11. C.-K. Hu, Phys. Rev. Lett. **69**, 2739 (1992).

12. C.-K. Hu, Physica A, **189**, 60 (1992).

13. C.-K. Hu and J.-A. Chen, Physica A **199**, 198 (1993).

14. J.-A. Chen and C.-K. Hu, Chin. J. Phys. (Taipei) **32**, 749 (1993).

15. C.-K. Hu, J. Phys. A: Math. Gen. **27**, L813 (1994).

16. C.-K. Hu, Chin. J. Phys. (Taipei) **32**, 519 (1994).

17. C.-K. Hu and J.-A. Chen, J. Phys. A: Math. Gen. **28**, L73 (1995).

18. C.-K. Hu, J.-A. Chen and C.-Y. Lin, Chin. J. Phys. (Taipei) **34**, 727 (1996).

19. P. J. Reynolds, H. E. Stanley, and W. Klein, J. Phys. A **11**, L199 (1978); Phys. Rev. B **21**, 1223 (1980).

20. R. M. Ziff, Phys. Rev. Lett. **69**, 2670 (1992).

21. R. P. Langlands, C. Pichet, Ph. Pouliot, and Y. Saint-Aubin, J. Stat. Phys. **67**, 553 (1992).

22. J. L. Cardy, J. Phys. A: Math. Gen. **25**, L201 (1992).

23. C.-K. Hu, C.-Y. Lin, and J.-A. Chen, Phys. Rev. Lett. **75**, 193 (1995); **75**, 2786(E)(1995).

24. C.-K. Hu, C.-Y. Lin, and J.-A. Chen, Physica A **221**, 80 (1995).

25. Y. Okabe and M. Kikuchi, Int. J. Mod. Phys. C **7**, 287 (1996)

26. J.-P. Hovi and A. Aharony, Phys. Rev. Lett. **76**, 3874 (1996).

27. C.-K. Hu, Phys. Rev. Lett. **76**, 3875 (1996).

28. I. M. Ruzin, N. R. Cooper, B. I. Halperin, Phys. Rev. B. **53**, 1558 (1996).

29. Here "square" means a primitive unit cell of the lattice is square rather than $L_1 = L_2$.

30. C.-K. Hu, J. Korean Physical Soc. (Proc. Suppl.), **29**, S97-101 (1996).

31. C.-K. Hu and C.-Y. Lin, Phys. Rev. Lett. **77**, 8 (1996).

32. C.-K. Hu and B. I. Halperin, Phys. Rev. B **55**, 2705 (1997).

33. N. R. Cooper, B. I. Halperin, C.-K. Hu, I. M. Ruzin, Phys. Rev. B **55**, 4551 (1997).

34. R. Zallen, *The Physics of Amorphous Solid* (John Wiley & Sons Inc. New York, 1983).

35. C.-K. Hu and F.-G. Wang, J. Korean Physical Soc. (Proc. Suppl.), **30**, Sxxx (1997).

36. C.-K. Hu, Chin. J. Phys. (Taipei) **25**, 182 (1987).

37. K. W. Kratky, J. Stat. Phys., **52**, 1413 (1988).

38. J. Hoshen and R. Kopleman, Phys. Rev. **B14**, 3438 (1976).

39. E. T. Gawlinski and H E Stanley, J. Phys. A **14**, 291 (1981),

40. F.-G. Wang and C.-K. Hu, Phys. Rev. E, **56**, (1997).

41. P. C. Hohenberg and B. I. Halperin, Rev. Mod. Phys. **49**, 435 (1977).

42. M. Suzuki, Prog. Theor. Phys. **58** , 1142 (1977).

43. N. Ito, Physica A **196**, 591 (1993).

44. M. P. Nightingale and H. W. J. Blote, Phys. Rev. Lett. **76**, 4548 (1996)

45. F.-G. Wang, N. Hatano and M. Suzuki, J. Phys. A **28**, 4543 (1995).

46. P. Grassberger Physica A **214** , 547 (1995).

47. A coherent anomaly method was used to obtain $z = 2.15(2)$ for the Ising model on a pt lattice. See M. Katori and M. Suzuki, J. Phys. Soc. Jpn. **57**, 807 (1988).

48. R. H. Swendsen and J. S. Wang, Phys. Rev. Lett, **58** 86 (1987).

49. C.-K. Hu, J.-A. Chen and C.-Y. Lin, preprint.

50. F. Y. Wu, Rev. Mod. Phys. **54**, 235 (1982).

51. C.-K. Hu, Phys. Rev. B **29**, 5103 and 5109 (1984) and references therein.

52. S. B. Lee, Phys. Rev. B., **42**, 4877 (1990).

Polymer Surfaces and Interfaces:
A Continuum Simulation Approach

Gary S. Grest[a], *Martin-D. Lacasse*[a], *and Michael Murat*[b]

[a]Corporate Research Science Laboratories, Exxon Research & Engineering Company,
  Annandale, NJ 08801, USA
[b]Soreq Nuclear Research Center, Yavne 81800, Israel* and Max Planck Institut für
  Polymerforschung, Postfach 3148, 55021 Mainz, Germany**

  *permanent address
**present address

Abstract

Results from our molecular dynamics simulations for polymers are reviewed, with emphasis on
surfaces and interfaces. We show that continuum models have certain advantages over the more
traditional lattice model simulation techniques, particularly in dealing with polymers under shear
and polymer-polymer interfaces. As an example of polymers under shear, we present simulations
of end-grafted polymers under shear. We show that results for a simple coarse-grained bead-
spring model are in qualitatively good agreement with recent experiments using the surface force
apparatus. As a second illustrative example, we present results for the interface between immiscible
polymer blends. In this case, continuum models offer a simple way to obtain the surface tension
from the measured pressure tensor.

1 Introduction

As a result of the rapid increase in performance of computers and numerical algorithms, one
is now able to use computer simulations to study the physical phenomena taking place in
more and more complex systems. As such, computer simulations have begun to play a more
critical role in improving our understanding of polymeric systems; this includes not only
single chains in dilute solution but also entangled polymer melts, networks, and polymers at
surfaces and interfaces. Simulations are important not only in testing the basic assumptions
of various theoretical models, but also in interpreting experimental results [1].

Early computer simulations of polymers were mostly carried out on a lattice, using Monte
Carlo methods [1,2]. This approach has lead to significant progress over the past twenty years
and will continue to do so in many areas. In some cases however, e.g. in the study of shear,
lattice models have serious limitations. For this reason, and also due to the availability
of more powerful computers, continuum, off-lattice polymer models have recently become
popular. In this paper, we review some of the recent progress in studying polymers using
continuum models, with a specific emphasis on polymer surfaces and interfaces.

The strength of lattice models stems from the fact that all distances are fixed and discretized and consequently, very efficient algorithms can be written. This constrains the
system to fixed volume and aspect ratios, and therefore, lattice models are best suited for
isochoric ensembles. In contrast, continuum models can easily be designed to study a variety
of thermodynamic ensembles. Although off-lattice methods tend to be slower, the difference
in performance is well compensated for by their flexibility. Moreover, present continuous-
space (CS) algorithms are very efficient so that the trade-off is small. Apart from the
possibility of simulating other ensembles, in particular isobaric, CS models can more easily
represent polymer branches, such as those in star-branched polymers, cross-linked polymer

networks or simple branched polymers. One can also study the effects of volume difference between species, more realistic longer-range potentials, and the effects of shear. For systems with interfaces, CS models have one more advantage: besides their inherent spatial isotropy, they offer a simple way of obtaining the surface tension γ [3],

$$n\gamma = (P_\perp - P_\parallel)L_\perp, \tag{1}$$

from the pressures measured perpendicular $P_\perp$ and parallel $P_\parallel$ to the n interfaces, which are perpendicular to the dimension $L_\perp$ of the simulation cell.

Once a CS approach has been chosen, one has then to decide on an efficient simulation algorithm. At the present time, there principally exist two methods for the simulation of systems at equilibrium, namely Monte Carlo (MC) and molecular dynamics (MD) [1]. Brownian dynamics has also been used but the computational aspects of this method appear to be too inefficient for the simulation of polymers [4]. Although both MC and MD methods give similar results, they are conceptually and computationally very different. However, since MD algorithms include the actual classical, internal dynamics, they also give direct access to the dynamical behavior of the system. Here we mostly concentrate on MD methods.

2 Model and Method

The magnitude of the relevant length and time scales responsible for the macroscopic properties of polymeric systems allows one to use a simplified model for a polymer molecule. For a CS model, a convenient representation of a homopolymer consists in attaching N soft or hard spherical beads, which we refer to as mers, of mass m together to form a chain. The resulting object, representing one chain, has $3N$ spatial coordinates. In practice, a large number of such chains are constructed and put in a virtual system having the desired boundary conditions, i.e. periodic, antiperiodic, or a wall. In an MD simulation, the time evolution of the coordinates of all chains is resolved through Newton's equation of motion [5]. The softness and the stickiness of the beads, is set by the interaction potential.

The motion of the chains is coupled to a heat bath, acting through a weak stochastic force $\mathbf{W}$ and a corresponding viscous damping force with friction coefficient Γ. Besides improving the diffusion of the system in phase space, this coupling has the practical advantage of stabilizing the numerical calculation. Including these terms, the equation of motion of mer i is

$$m\frac{d^2\mathbf{r}_i}{dt^2} = -\nabla_i U - m\Gamma\frac{d\mathbf{r}_i}{dt} + \mathbf{W}_i(t). \tag{2}$$

The last term $\mathbf{W}$ is a white noise having an average strength determined by temperature and the friction coefficient through the fluctuation-dissipation theorem. Thus, for a given interaction potential U, the viscous friction coefficient constant Γ is the only free parameter in the equation. It has to be carefully chosen to avoid overdamping so that the motion of the mers be dominated by inertia. In most of the runs, all of the mers were coupled to the thermal reservoir with $\Gamma = 0.5\tau^{-1}$, where τ is the natural Lennard-Jones unit (see below).

The equations of motion of the mers are integrated using a velocity-Verlet algorithm [5] with a time step Δt. In most of the cases presented here, the time step is chosen between $\Delta t = 0.009$ to 0.012τ, depending on the type of system. However for some runs for polymer brushes under a steady-state shear, only the first ten mers of each end-grafted chain was coupled to the heat bath. In this case, Δt was reduced to 0.006τ to keep the algorithm stable.

The conservative force term derives from a potential energy U which includes: the attractive potential holding adjacent mers along the same chain [6], an interaction potential acting between all the mers, responsible for excluded volume effects, and finally, any potential arising from the presence of walls. The interaction acting between mers is represented by an effective two-body potential, which is often truncated to a short distance r_c, since the number of interacting pairs increases as r_c^3. Such truncated potentials still contain the essential physics required to study collective phenomena which, fortunately, do not strongly depend on the details of the interactions. Thus, the interaction potential between the mers is often modeled as the repulsive core of a central-force Lennard-Jones (LJ) 6:12 potential,

$$U^{\mathrm{LJ}}(r_{ij}) = 4\epsilon \left[\left(\frac{\sigma}{r_{ij}} \right)^{12} - \left(\frac{\sigma}{r_{ij}} \right)^{6} + \frac{1}{4} \right], \tag{3}$$

for $r_{ij} < r_c = 2^{\frac{1}{6}}\sigma$ and zero otherwise. Here r_{ij} is the distance between mers i and j, and ϵ and σ are, respectively, parameters fixing the energy and length scales. All of our results are reported in terms of these natural units, with $\tau = \sigma(m/\epsilon)^{1/2}$ for the time scale.

Using this model, it is possible to simulate a variety of systems. For instance, this model has been shown to be very efficient in studying the properties of entangled chains in polymer melts and networks [6] and for end-grafted chains [7]. The model has also been used in non-equilibrium molecular dynamics to study bulk polymers under shear [8], as well as the properties of polymers confined and sheared by two parallel surfaces [9, 10]. Here, we shall focus on polymers at surfaces and interfaces.

3 Polymers at Surfaces

Surface-polymer interactions are important in many technological applications such as colloidal stabilization [11], adherence and lubrication. Depending on the polymer concentration and on the surface-polymer interaction, a wide variety of interesting phenomena can occur. Polymers at surfaces have been the object of many numerical studies (MC and MD), which span the range from single chains near a surface [12,13] to the interface between a bulk polymer and a substrate [14]. Here we shall focus on polymer brushes and polymer adsorption.

A polymer brush is constructed by end-grafting polymer chains onto a host surface. Computer simulations have been valuable in helping to interpret the experimental data obtained for polymer brushes [7]. Indeed, interesting effects were first predicted by numerical studies and later observed experimentally. This is the case for instance, of the transition occurring from a uniformly stretched brush to a locally, phase-separated state, in which the grafted polymer chains cluster in microdomains as the solvent quality goes from good to poor [15,16]. Israels *et al.* [17] also showed using simulations that by controlling the quality or the pH of the solvent, a small channel covered by end-grafted chains could be used as a microvalve.

Confinement of a polymer chain by a nearby surface leads to configurations which are different from those of free chains. For instance, the end-to-end distance R of free chains scales as $R \sim N^{\nu}$ with ν close to 3/5 in a good solvent. If the chains are end-grafted to a flat surface at a given surface coverage ρ_a, ν changes from 3/5 for values of ρ_a and N for which the attached chains do not overlap, to $\nu = 1$ in the intermediate overlap regime [19–21]. Figure 1 shows a typical configuration of a polymer brush in the overlap regime. If one measures the average brush height h from the host surface in a good solvent, one finds $h \sim N^{3/5}$ for low coverage (the so-called mushroom regime), and $h \sim N\rho_a^{1/3}$ in the intermediate overlap regime.

When two parallel polymer brushes are brought into contact, long-ranged repulsive effects, entropic in origin, act to keep the surfaces apart while maintaining a relatively fluid layer at contact. This gives rise to unusually low friction forces between the two surfaces, as we shall discuss below. The normal [22,23] and shear [24–26] forces acting between the surfaces have been well studied experimentally using the surface force apparatus (SFA), and the atomic force microscope (AFM) [27]. However, neither of these techniques gives a detailed spatial description of the polymer chains. Indeed, because of the small distances involved at contact, it is difficult to extract any configurational information. Computer simulations, however, can provide estimates of the amount of interdigitation between the brushes, while providing calculated forces that compare very well with experiments [7,28]. Here we review some recent simulation results obtained for polymers at surfaces using two different simu-

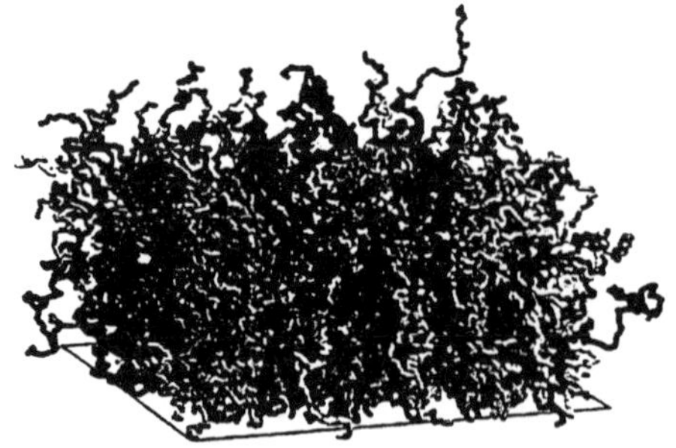
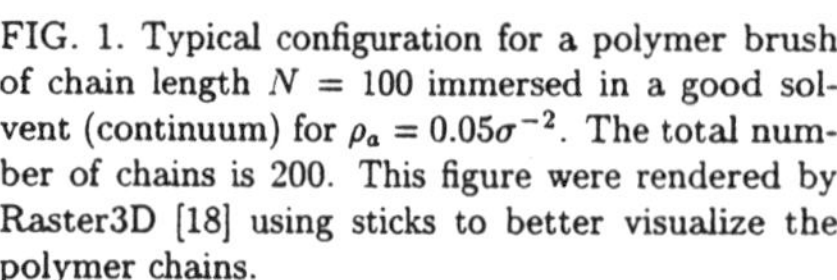

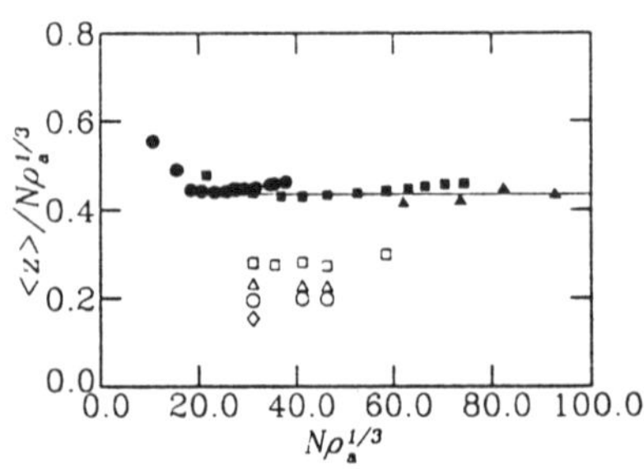

FIG. 1. Typical configuration for a polymer brush of chain length $N = 100$ immersed in a good solvent (continuum) for $\rho_a = 0.05\sigma^{-2}$. The total number of chains is 200. This figure were rendered by Raster3D [18] using sticks to better visualize the polymer chains.

FIG. 2. Average mer distance from the host surface $\langle z \rangle$ versus $N\rho_a^{1/3}$ for a brush in a good solvent at $T = 1.2\epsilon/k_B$. For the continuum-solvent approach, the chain length are $N = 50$ (●), 100 (■), and 200 ($\triangle$). Open symbols are for results of a brush of $N = 100$ immersed in a homopolymer melt of $N_f = 2$ ($\square$), 5 ($\triangle$), 10 ($\bigcirc$), and 40 ($\lozenge$).

lational approaches: treating the solvent as a continuum or including the solvent molecules explicitly.

Continuum Solvent — In this approach, the effects of solvent molecules are included in the effective potential acting between the mers, as well as in the stochastic force and the viscous damping terms of Eq. 2. This is a good approximation since the time scale involved in the motion of small solvent molecules is much smaller than that of the polymer chains, and therefore, the degrees of freedom of the small molecules can be averaged out. In this coarse-grained picture, a theta or poor solvent translates into an additional effective attraction between the mers. Here however, we shall only consider good solvent conditions so that the mer-mer interactions are always purely repulsive.

The simulation results obtained from the continuum-solvent approach support and extend the self-consistent field (SCF) theoretical results [21,29]. For a polymer brush in the overlap regime, this theory predicts a parabolic decay of the mer number density from a maximum at the host surface to zero at h. This parabolic profile was observed in the early simulations of polymer brushes, although at very high coverage, the observed profiles decay more slowly than parabolic [7]. Numerical results for the brush height are presented in Figure 2 for three values of N in the scaling form $\langle z \rangle / N\rho_a^{1/3}$, where the average distance $\langle z \rangle$ of the mers from the host surface is used as a measure of the brush height. For low coverage, $\langle z \rangle$ is nearly independent of ρ_a, as expected in the mushroom regime. For intermediate coverage, the brush height scales as predicted, but at high ρ_a, $\langle z \rangle$ increases faster than $\rho_a^{1/3}$, indicating that 3-body interactions become more dominant than the 2-body ones [30]. The mer number density profile as a function of the distance z from the grafting surface is shown in Figure 3 for $\rho_a\sigma^2 = 0.03$ and 0.10. The profiles are well-described by a parabolic form, in agreement with SCF.

Recently, Overney *et al.* [27] investigated the force between a polymer brush and an AFM tip and compared the results to the force between two surfaces each with end-grafted chains. Murat and Grest [31] extended their simulation of the force between two polymer brushes by modelling the AFM tip as a cylinder with a spherical cap. As the tip is introduced perpendicularly into the brush, the polymer chains oppose its motion. The total force F acting between the tip and the mers of the brush is shown in Figure 4 for $\rho_a = 0.1\sigma^{-2}$ and $0.07\sigma^{-2}$, where F is normalized by the contact area $\mathcal{A}$. The results for the model AFM tip are compared to those obtained for a parallel wall pushed against a brush and those for two parallel brushes pushed against each other. Not surprisingly, $F/\mathcal{A}$ is significantly larger for a wide surface than it is for a tip: the chains in contact with the tip avoid compression by

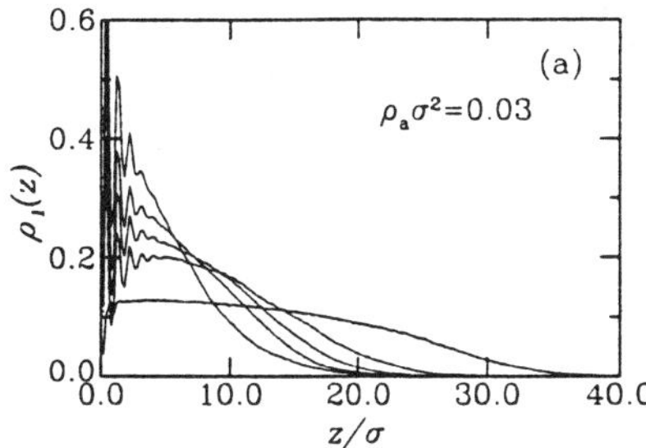
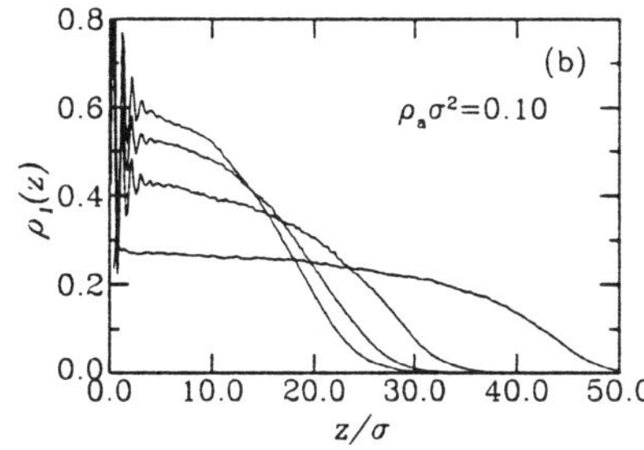

FIG. 3. Brush mer number density $\rho_1(z)$ versus z for a brush of chain length $N = 100$ for (a) $\rho_a\sigma^2 = 0.03$ and (b) 0.10 in a continuum solvent and in a solvent of linear chains of length $N_f = 2, 5$ and 10. In both plots, the flattest curve corresponds to continuum solvent. For $\rho_a\sigma^2 = 0.03$, results for $N_f = 40$ are also shown. The larger N_f, the more the brush collapses as expected from theory.

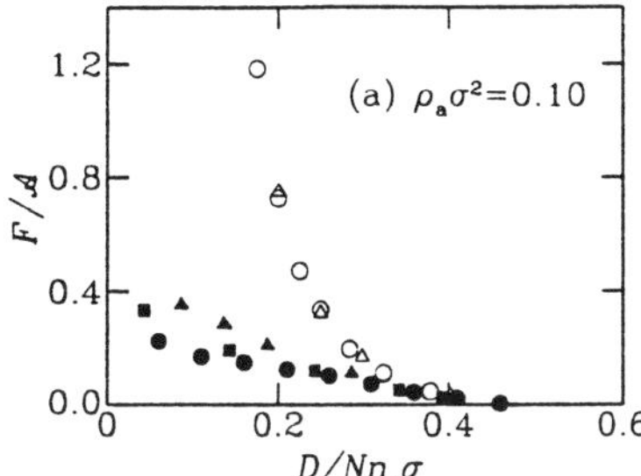
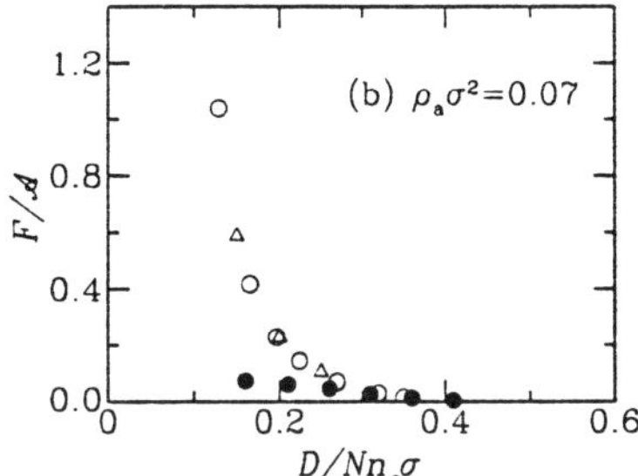

FIG. 4. The force-displacement curves for a polymer brush of 100-mers at a coverage (a) $\rho_a = 0.1\sigma^{-2}$ and (b) $0.07\sigma^{-2}$ for a tip with cylinder radii $r_{\text{cyl}} = 10.0$ ($\bullet$), 14.0 ($\blacksquare$), and 16.0σ ($\triangle$). It is compared with the force between the brush and a bare surface ($\triangle$), and the brush and another identical brush ($\bigcirc$). For comparison, the distance D separating the interacting surfaces has to be divided by two in the two-brush case. n_s is the number of surfaces bearing polymers.

moving away from it, without much additional interactions with neighboring chains. This effect is enhanced in a good solvent where the end-grafted chains are swollen. As expected, when the radius of the tip becomes significantly larger than $\rho_a^{-1/2}$, its compressing effect becomes identical to that of a wide surface. This picture is observed directly by visualization of the chains and from measurements of the mer number density. Another interesting feature is the similarity in results between the compression of a brush by a bare surface and that by another brush. Recent experiments by Pelletier *et al.* [32] have also observed this feature.

We now turn to polymers adsorbed onto a surface. This can be modeled by including attractive interactions between the surface and the polymer chains. This gives rise to a competition between the lowering of the energy resulting from the wall-mer interaction and the loss of configurational entropy from being constrained to a surface. Such a competition is the basic element for a phase transition to take place, and indeed, a transition is obtained by varying either the chain length or the strength of the wall-mer interaction.

Systems of adsorbed polymers are characterized by very long structural relaxation times [33] and hysteresis effects [34]. For this reason, adsorbed chains are more difficult to simulate than end-grafted ones. Consequently, there have been relatively few simulations of these systems, mostly using MC methods. Many of these studies are for a single chain (dilute case) interacting with an adsorbing surface [12, 13]. Only recently have there been numerical results reported for the kinetics of adsorption from semi-dilute solutions [35,36]. Adsorbed polymer layers formed near a surface in contact with semi-dilute solutions are found to exhibit a concentration profile which decays as $z^{-4/3}$, as predicted theoretically by de Gennes [37]. In these simulations, the equilibrium layer thickness is attained quickly. The residence time of the polymer chains in the adsorbed layer depends on how deep the chain is in the layer [35]. This leads to a broad distribution of residence times, the longest of which fixes the time scale for equilibration. Such quasi-equilibrium structures, resulting from the slow motion of chains between the solution and the layer, have also been studied by MD simulations [38]. In these studies, the exchange process is studied by varying the concentration through the pressure using an isobaric model.

While simulations in which the effects of the solvent are included implicitly can contribute significantly to our understanding of polymer chains either adsorbed or attached to a surface, such an approach can only describe the polymer dynamics in what is commonly referred to as the free-draining limit, i.e., all the solvent-mediated hydrodynamic effects are ignored. Technically speaking, to include hydrodynamic effects one also has to turn off the stochastic noise in Eq. 2. Moreover, this approach is not able to predict some more elaborate effects such as local phase separation of mixed solvents near a surface.

Molecular Solvent— We now turn to simulations in which the solvent is modeled directly by the presence of free polymer chains of length N_f acting as solvent molecules. The interactions between a polymer brush and solvent chains are not as well studied using molecular solvents as they are for a continuum solvent. de Gennes [20] was the first to obtain theoretical phase diagrams of such systems. He found that as either ρ_a or N_f increases, the free chains are progressively expelled from the brush. The presence of long solvating chains screens the excluded volume interactions between the brush chains, thus shrinking the brush. In the intermediate overlap regime, Flory-type theory predicts that $h \sim N(\rho_a/N_f)^{1/3}$. This scaling gradually breaks down at high coverage. as the free chains are almost completely expelled from the brush.

Simulating the solvation of a polymer brush by a molecular solvent is very demanding. This is because the number of free chains that must be included is significantly larger than the number of brush chains, even in mer units. Therefore, a larger fraction of the computation time goes to the free molecules, making the computation considerably more costly.

To study the effects of free chains solvating a polymer brush, two parallel surfaces, each bearing end-grafted polymers separated by a distance D are immersed in a solvent of like chains of length N_f at a melt number density $\rho = 0.85\sigma^{-3}$. Systems with values of N_f ranging from single monomers to 40-mers were studied [39]. D is chosen sufficiently large for end-grafted polymers from opposite surfaces not to touch. Results for the average distance from the wall $\langle z \rangle$ of the mers in the grafted chains are shown in Figure 2 for $N = 100$. As expected, the brush height shrinks as N_f increases but the scaling form remains unchanged. For identical potentials, the results obtained with the explicit presence of solvent molecules are naturally different from those obtained with a continuum solvent: the excluded volume effects due to the presence of only a dimer solvent are sufficient to reduce the brush height by almost a factor of two.

As anticipated from the results for the brush height, the structure of the brush is strongly dependent on N_f. This is shown is Figure 3 in which the dependence of $\rho_1(z)$ on N_f is shown for $\rho_a\sigma^2 = 0.03$ and 0.10. For the cases in which both walls are coated with grafted chains, $\rho_1(z)$ is averaged over the two surfaces. Note that, since the overall density of monomers is quite large, $\rho\sigma^3 = 0.85$, the melt chains penetrate into the brush, except very near the wall where the brush monomers dominate. In fact since total monomer density is large everywhere, layering near the wall is evident in both the overall number density and brush monomer number density $\rho_1(z)$. This is in contrast with the continuum solvent case, in which there is a depletion zone near the wall. This difference in the behavior of $\rho_1(z)$ near the wall is due to the osmotic pressure from the solvent not present in the continuum model. Additional results for the properties of a polymer brush in a polymeric matrice, including the dynamics of the chains, can be found in Ref. [39].

With the explicit presence of solvent molecules, it is also possible to study the response of end-grafted polymers to a steady-state shear flow of solvent molecules. Experimental studies by Klein *et al.* [40] reported that, under shear, there is an additional contribution to the normal force for brushes which are closely separated by $D \sim 2h_{\mathrm{max}}$, where h_{max} is the maximum height of the brush. Klein and co-workers [24,25] also found that when compressed, brush-coated surfaces have an anomalously low friction coefficient. This effect was recently studied using a continuum solvent [41]. However, in order to study the penetration of the (simple shear flow) velocity field in the brush and to investigate the normal force for $D \sim 2h_{\mathrm{max}}$, solvent molecules have to be explicitly included [42].

In computer simulations, the shear is imposed by moving two brush-coated plates parallel to each other at a relative velocity v_w. The reported SFA experiments are for polystyrene brushes solvated in toluene: a low molecular weight, good solvent, that is simulated by a dimeric solvent. The relative velocity v_w as well as the separation distance D are held fixed. To avoid any possible biasing of the flow in the shear direction (x), only the y component of the velocity was coupled to the reservoir for $v_w \neq 0$. Thus Langevin noise and frictional terms were added to the equation of motion, Eq. 1, only in the y-direction for nonzero shear velocities. In most of the runs. all of the mers were coupled to the thermal reservoir with a friction coefficient $\Gamma = 0.5\tau^{-1}$. However since introducing a viscous damping screens hydrodynamic interactions [43], some runs were made in which only the first 10 mers of each grafted chain and none of the solvent chains were coupled to the thermal reservoir. In this case, the time step was reduced to 0.006τ to keep the algorithm stable [42]. No significant differences in the two cases were observed.

In order to study the forces between sliding brush-coated plates. a series of simulations were carried out for several values of D. while keeping the overall mer number density at $\rho = 0.85\sigma^{-3}$. In the first series of runs, D was large enough such that the end-grafted chains from the opposing surfaces did not touch. In this case, there is no detectable change in the brush profile or end-to-end distance of the grafted chains for low shear rates [42, 44]. For larger velocities however, some of the brush chains become highly stretched, as the shear rated increased. However this had the effect of slightly decreasing the brush height, in agreement with earlier MC simulations of a brush in a velocity field [45, 46] and the theoretical prediction of Rabin and Alexander [47]. However this results is in disagreement with several later theoretical predictions [48–50] which suggested that the the brush height should increase under shear. In regard to the original experiment by Klein *et al.* [40] which first suggested that the brush height increased under shear, the situation is unclear. Since the experiments were carried out at oscillatory frequencies of a few hundred hertz and not at steady-state shear as in the present simualtions. direct comparison of the two is not possible. Further simulations under oscillatory shear are needed to explore the interplay between the relaxation times of the end-grafted chains and the shear flow.

For $D < 2h_{\mathrm{max}}$, the brushes start to interdigitate as shown in Figure 5. In this figure. the results for the mer number density profiles are presented for three values of D for $v_w = 0$ for $\rho_a \sigma^2 = 0.03$. In this case, the mers from opposite surfaces first begin to overlap at $D \simeq 60\sigma$. In addition to $\rho(z)$, the mer density from only one surface $\rho_1(z)$ is also shown. For very small D, layering occurs near the surface due to the finite volume of each mer. In the inset, the amount of interpenetration $I(D)$, given by the integral of $\rho_1(z)$ from $z/D = 0.5$ to 1.0, is shown. At the highest compression, 30% of the mers connected to one surface are on the opposite half of the midplane. These results are very similar to those obtained previously using a continuum solvent [41].

The shear stress f is obtained from the xz component of the microscopic pressure-stress tensor, $f = -P_{xz}$. The fluctuations in the shear force are much larger than the average force f, requiring long runs. In Figure 6, the dependence of the shear force f on sliding velocity is shown for three values of D. For small v_w and large D, f was found to increase approximately linearly as would be expected for a Newtonian fluid. In this regime there is little change in $\langle R_G^2 \rangle$ (Figure 6) or $\langle R^2 \rangle$ (not shown). There is also no change in $\rho_1(z)$ compared the results shown in Figure 5 for $v_w = 0$. Beyond the Netwonian-like regime, f increases sublinearly for

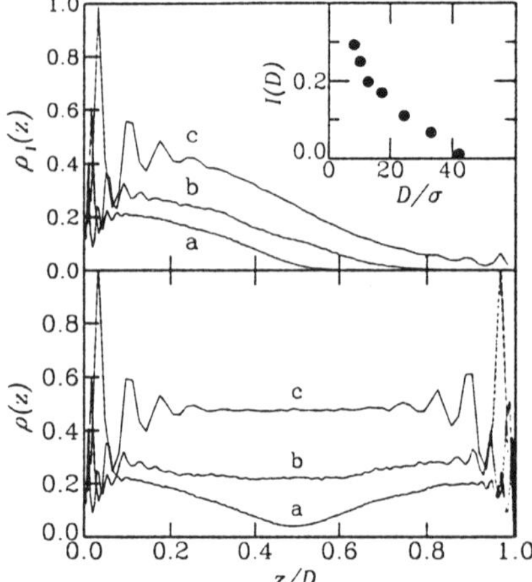

FIG. 5. Brush mer number density profile versus distance from the grafting surface z for polymers of length $N = 100$ and solvent of length $N_f = 2$ at $\rho_a\sigma^2 = 0.03$ for (a) $D = 42.1\sigma$, (b) $D = 24.3\sigma$, and (c) $D = 12.6\sigma$. The total brush mer number density $\rho(z)$ is shown in the lower panel and the mer density $\rho_1(z)$ for one brush is in the upper panel. The inset shows the amount of interpenetration $I(D)$ versus D.

FIG. 6. Shear stress f per unit area and mean squared radius of gyration $\langle R_G^2 \rangle$ as a function of the relative velocity v_w of the two surfaces for $N = 100$ and $N_f = 2$. The data is for $\rho_a\sigma^2 = 0.03$ and $D = 24.2\sigma$ ($\square$), $D = 17.1\sigma$ ($\circ$), and $D = 8.1\sigma$ ($\triangle$).

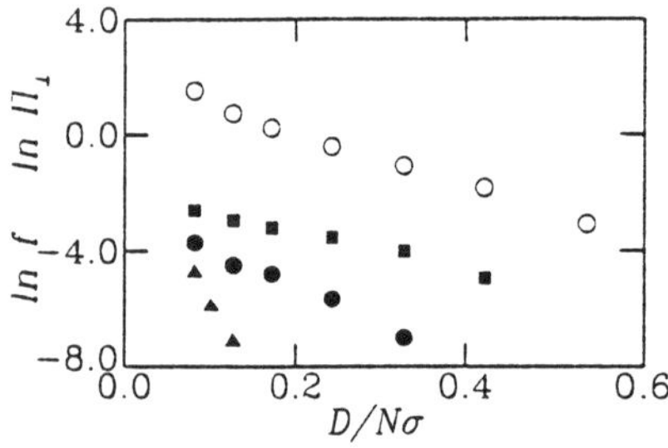

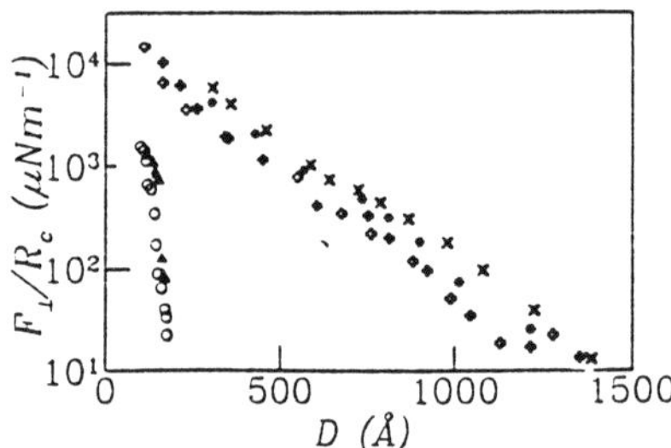

FIG. 7. Semilog plot of the normal osmotic pressure $\Pi_\perp$ ($\circ$) and shear force f versus the plate separation D between polymer brushes of 100-mers at $\rho_a = 0.03\sigma^{-2}$. The shear force is shown for $v_w = 2.0 \times 10^{-4}$ ($\triangle$), 2.0×10^{-3} ($\bullet$), and $2.0 \times 10^{-2}\sigma/\tau$ ($\blacksquare$).

FIG. 8. Experimental results of Klein et al. [24] for the normal (upper set of points) and shear (lower set) forces for mica surfaces bearing polystyrene (140 000 molecular weight) in toluene. The experimental force is normalized by the radii of curvature R_c ($\sim$ 1 cm) of the surface. Results from several different experiments are shown.

large v_w, in analogy to earlier observations [9,51] for small molecule systems confined between two plates. One may call this the non-Newtonian regime. As seen from the mean-squared radius of gyration $\langle R_G^2 \rangle$, the chain stretch in this regime and there is some disentanglement of polymers from the two surfaces. For $D = 8.1\sigma$, $I(D)$ decreases from 0.29 for $v_w = 0$ to 0.27 for $v_w = 0.002\sigma/\tau$ while increasing v_w by a factor of 10 ($v_w = 0.02\sigma/\tau$) decreases $I(D)$

by factor of 2 (0.14). While the amount of interpenetration decreases significantly for large v_w, the normal osmotic pressure actually decreases by about 1%.

Comparison of the normal osmotic pressure and the shear force in this regime for three values of v_w is presented in Figure 7 for $\rho_a \sigma^2 = 0.03$. These results are in qualitatively good agreement with the experiments of Klein et al. [24] shown in Figure 8. The SFA experiments found that f was below their resolution limit for large D and increased rapidly for high compressions. The simulation results do not increase quite as rapidly with decreasing plate separation. As the shear force f is a continuously varying function of velocity even for high shear rates, a true coefficient of friction cannot be defined. However treating the shear-to-normal ratio of the forces as an effective coefficient of friction (as Klein et al. did), $\mu_b \simeq 0.005 - 0.007$ for $D \lesssim 20\sigma$ for $v_w = 0.002\sigma/\tau$ and actually decreases slightly as D decreases in agreement with experiment [25]. This is because the normal force increases more rapidly than the shear force as D decreases. The effect is very strong for large v_w. For example, for $v_w = 0.02\sigma/\tau$, μ_b is largest for large D ($\mu_b = 0.05$) and decreases to 0.014 for $D = 7\sigma$. It is important to remember that since the surface forces experiments are for two crossed cylinders and the simulations are for flat plates, the two cannot be compared directly [52]. In the Derjaguin approximation, the interaction energy per unit area, $E(D) = \int_{2h_{max}}^{D} \Pi_\perp(D')\,dD'$, between two flat surfaces is proportional to the normal force $F_\perp$ divided by the radius of curvature R_c, $F_\perp/R_c$, for two crossed cylinders [52]. The relationship between the shear forces in the two systems is less clear. Note that the range of D over which f is large enough to detect is also larger than that observed experimentally. This is presumably due to the fact that the velocities used here are much larger than those in the experiment. Comparing Figures 7 and 8, the experiments are best fit by the lowest velocity, $v_w = 2.0 \times 10^{-4}\sigma/\tau$. Thus it is likely that the present experimental results on the SFA are only in the low velocity Newtonian-like regime. However since there is presently no experimental data available for the velocity dependence of f for small D, it is not possible to determine for certain which part of the shear force-velocity curve the experiments access.

4 Polymer Interfaces

Another important domain of research is that of interfaces in polymer blends and copolymer systems. In this section, we shall briefly cover some recent results obtained from CS simulations of the bulk properties of binary blends as well as interfaces between two immiscible phases. These systems have also been studied extensively using lattice models [53].

Polymer blends have many technological applications: the main interest in mixing homopolymers resides in the possibility of creating new materials that would combine some of the properties of each component. However, polymer species are in general very difficult to mix, and a great challenge in polymer science is to predict the miscibility of two species from their molecular structures. When the components do not mix, it is often desirable to "glue" two surfaces of immiscible species with the help of a compatibilizing agent [54, 55] acting like a nonionic surfactant at the interface. A standard example is the compatibilization of an interface between a material made of species A and one made of species B by a block copolymer AB [55].

The computer modeling of binary blends is necessarily more elaborate than that of homopolymer melts. This is because the interaction potential U^{LJ} has to take into account the difference in species. A standard approach, which has been used intensively for binary mixtures of simple liquids, is the Lorenz-Berthelot rule [56]. The simpler approach we choose here however, consists in using mers of the same size σ and making the potential amplitude for unlike interacting pairs slightly larger than for like pairs, i.e. to use $(1 + \delta)\epsilon$ instead of the interaction ϵ between like pairs. Simply by varying δ, while leaving the temperature T constant, a phase transition is induced. from a homogeneous blend to a system with two segregated coexisting phases. This effect is reminescent to what is done in recent experiments reported by Gehlsen et al. [57], in which the phase separation of binary mixtures is studied as a function of the difference in deuterium between two otherwise identical polymer species.

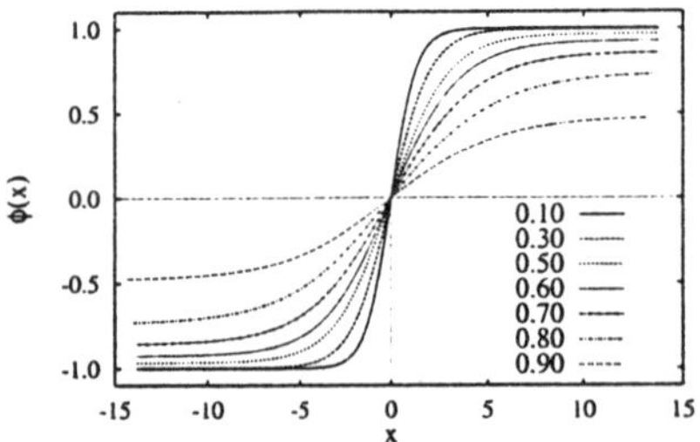 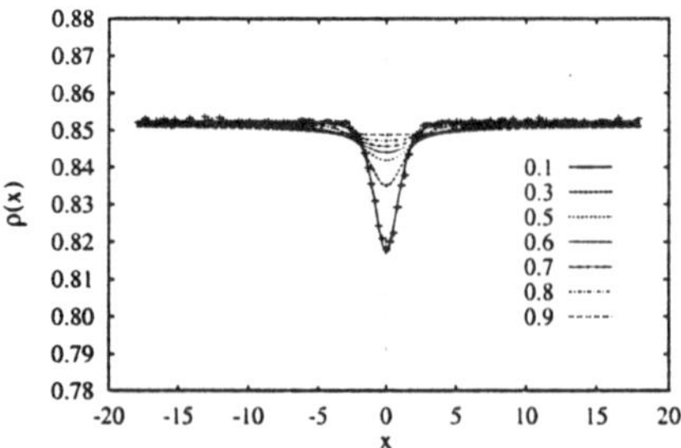

FIG. 9. The interfacial profiles of the yz-averaged order parameter for a system of 600 30-mers with antiperiodic boundary conditions. The values of δ_c/δ are as indicated. The lines are from a fit $\psi(x) = \psi_o \tanh(2x/w)$.

FIG. 10. The number density profile $\rho(x)$ evidencing the density hole at the interface. Systems contain 4000 10-mers. For clarity, data points are only shown for $\delta_c/\delta = 0.1$. The curves are fits to the functional form $\rho(x) = \rho_o[1 - \Delta/\cosh(x/2w_\rho)]$.

In principle, the model and simulation method we described so far would be adequate to study the kinetics of phase separation in binary blends. However, the time scale involved in the unmixing transition —which normally occurs through the diffusive motion of the polymer chains— is extremely long, even more so near T_c. To properly sample the phase space in a reasonable calculation time, we supplement the MD moves with an MC procedure relabeling the type of the homopolymer chains ($A \leftrightarrow B$) according to the Metropolis transition rule. This procedure was first introduced in lattice simulations and has been very successful [58,59]. Our exchange attempt rate is close to M chains per τ, allowing for sufficient relaxation between each of the exchanges. Flory-Huggins theory predicts that the difference required for phase separation δ_c scales as $\delta_c \sim 1/N$; this is directly verified for the present model ($\delta_c = 3.40(5)k_BT/\epsilon N$), for which the phase diagrams of systems of mers ranging from monomers to 50-mers has been calculated [60]. These calculations, which involve finite-size scaling analysis, could determine the full coexistence curve as a function of δ and N. With more detailed interaction potentials, one can in principle refine the present model to improve its predicting power. Meanwhile, a coarse-grained potential yields several interesting results concerning the scaling and the structure of these systems. For more detail see Ref. [60].

We now consider systems in equilibrium in the two-phase region and the interface between the two coexisting A- and B-rich phases. In a virtual system, it is possible to impose so-called antiperiodic boundary conditions, i.e., boundary conditions such that an A mer leaving the simulation box on one side becomes B as it reenters on the opposite side [61]. By doing so in a given direction, say z, one is forcing the existence of an odd number of —most likely one— interfaces normal to that direction. Thus, it becomes possible to study the properties of the interface as a function of the incompatibility δ between the species [62]. Figure 9 shows the interfacial profile of the order parameter $\psi(x) = (\rho_A(x) - \rho_B(x))/\rho$ obtained by spatially averaging $\psi(\mathbf{r})$ in yz slices, perpendicular to the antiperiodic direction x. Each order parameter profile $\psi(x)$ fits very well to a $\psi(x) = \psi_o \tanh(2x/w)$ functional form, thus associating a width w to the interface. The values of ψ_o obtained from the fits are in excellent agreement with values obtained from bulk simulations at the same immiscibility δ.

Similar spatial averages for the number density profile $\rho(x)$ across the interface reveal that the interface between strongly immiscible liquids has a dip in the number density [61,63]. This is shown in Figure 10 where $\rho(x)$ has been fit to a form $\rho_o[1 - \Delta/\cosh(2x/w_\rho)]$, where Δ represents the amplitude of the density dip, and w_ρ is another measure of the interfacial width.

As observed in MC similations [61], the ends of the chains near the interface are more likely (a $\sim$6% effect at $\delta_c/\delta = 0.1$) to be found at the interface rather than in the bulk. For

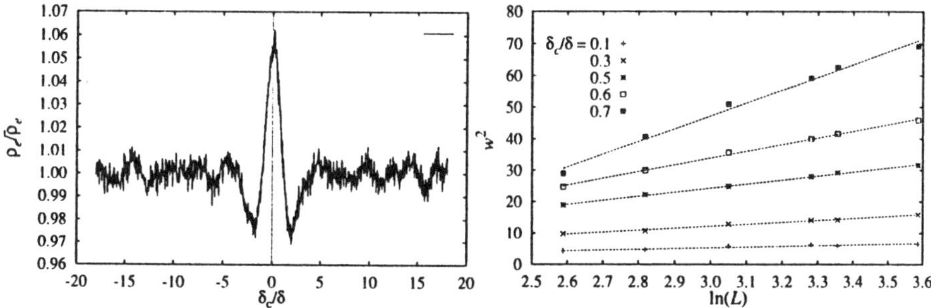

FIG. 11. The normalized end density profile $\rho_e(x)/\bar{\rho}_e$ evidencing a lamellar ordering near the interface. System contain 4000 10-mers and is at $\delta_c/\delta = 0.1$.

FIG. 12. Capillary waves and immiscibility factor dependence of the interfacial width for various systems of 10-mers. The values of δ_c/δ are as indicated. The lines are fits to $w^2 = a_\delta + b_\delta \ln(L)$.

strong immiscibility, this anisotropic ordering of the chains near the interfacial layer induces a small tendency to align the chains perpendicular to the interface in a lamellar structure, at a length scale much larger than w. This is shown in Fig. 11 for a cubic system of 4000 10-mers at $\delta_c/\delta = 0.1$, where the ordering can be observed for a few lamellae and then disappear in the noise.

It is interesting to measure the interfacial width as a function of the immiscibility δ. Our measure of w however, should include capillary-wave effects and a systematic study of w for systems of different sizes should reveal their presence. This is shown in Figure 12, where the dependence of w on both the linear size L of the system and δ is shown, clearly evidencing the presence of capillary waves. To our knowledge, this is the first time that capillary waves are observed in computer simulations of polymers. We fit our data to the theoretical [64] expression $w^2 = a_\delta + b_\delta \ln(L)$, where $b_\delta \sim 1/\gamma_L$, the inverse of the surface tension (the subscript L indicates that γ has been derived from capillary wave effects). Very similar results are obtained from fitting w_ρ with the same functional form. By measuring the surface tension γ_P independently from the difference in the pressures (obtained from the virial [5], hence the subscript P) perpendicular and parallel to the interface [3, 63], one can get another estimate of the surface tension, and compare it with γ_L. In order to compare both measures, extensive calculations of monomeric (Lennard-Jones fluid) systems were performed. Figure 13 shows the two different estimates of γ. Our data show that with $\gamma_L = 1.1/b_\delta$, both measures of γ are consistent. The same factor also reconciles sets of γ_L and γ_P obtained for polymer chains.

We also studied the effects of the chain length on the surface tension for chains of length $N = 5$, 10, 20, and 30. This is shown in Fig. 14 where the scaled surface tension $N\gamma_L$ is plotted as a function of reduced immiscibility. The surface tension has been scaled by a factor N to account for the decrease in mer-mer interactions at fixed reduced temperature $\delta_c/\delta \sim 1/N$. Thus, values of $N\gamma_L$ for two systems at fixed δ_c/δ contain the same enthalpic strength to the surface tension. We see however that at high immiscibility, systems of longer chains exhibit larger $N\gamma_L$, showing that the structure of the interface is strongly influenced by the chain length.

Lastly, we studied the scaling law between the surface tension and the interfacial width. Our data are consistent with the predicted [65] scaling law $\gamma \sim w^{-2}$ for more than a decade in w.

The extension of the present method to study interfaces between block copolymers is a simple matter of bookkeeping. Some of the interesting features of these systems, such as the lamellar order-disorder transition, have been studied with an isobaric version of the present model [60], thus showing the rich capabilities of this approach. A detailed description of these results is beyond the space allocated here.

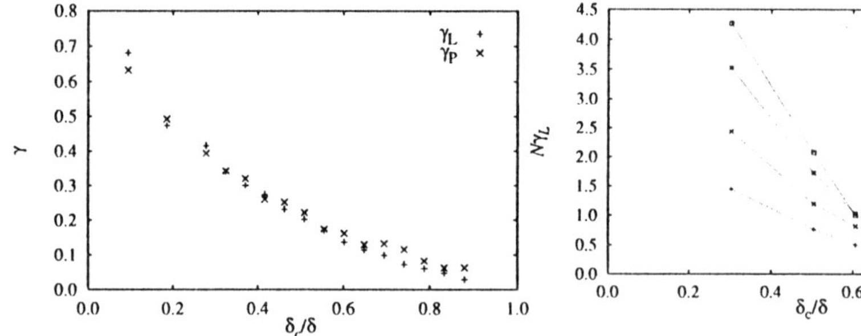

FIG. 13. Two different estimates of the surface tension for a monomeric systems. γ_P is obtained from the difference in pressures parallel and perpendicular to the interface and the surface tension $\gamma_L = 1.1/b_\delta$, where b_δ is obtained as in Fig. 12.

FIG. 14. The behavior of the scaled surface tension $N\gamma_L$ for different chain lengths. The abscissa is in reduced units so that chains of different lengths can be directly compared. Dotted lines are merely guiding the eye.

The inclusion of compatibilizing molecules at the interface between immiscible homopolymer melts can also be studied [66]. In particular, one can study the effect of block copolymers on the surface tension, their configurations at the interface, etc. This work is in progress.

5 Conclusion

Polymers are fascinating materials that can be studied with the use of molecular dynamics simulations. We have shown that a continuous-space model with a rather simple potential allows us to study in detail several important properties that cannot be measured experimentally, especially at the small length scales involved in the interactions of polymer chains with surfaces and interfaces.

The particular approach we presented here has a much broader scope of applications than traditional lattice MC, making it a powerful technique for computational studies of polymer systems. In particular, we used this model to study a variety of systems, including polymers under shear and interfaces between immiscible blends. The natural isotropy of a CS model coupled with a weak mismatch between the polymer chains allowed us to observe strong capillary wave effects at the interface of moderately small systems. One additional advantage of continuum simulations over lattices, which we have not discussed in detail here, is the ability to use isobaric instead of isochoric ensembles. This feature is essential in some cases, such as in the study of the lamellar phase of symmetric diblock copolymers. In such systems, the lamellar spacing depends not only on the chain length but also on the strength of the interaction. Thus, if one is to use a lattice model to study these systems. the lamellar spacing has to be known *a priori* so that the cell dimensions can be commensurately adjusted to the proper value. As the interaction strength is changed, then so does the lamellar spacing, and the simulation cell is very likely to create extraneous effects on the order-disorder transition point. A natural way to study this system is to use an isobaric ensemble, as we demonstrated in Ref. [60].

In a near future. the present work will be extended to larger binary systems and to binary systems containing nonanionic surfactants such as diblock copolymers at the interface. Work is also in progress to study the dynamics of diblock copolymers in both the lamellar and disordered regimes.

34

References

[1] For a collection of recent reviews see *Monte Carlo and Molecular Dynamics Simulations in Polymer Science*, edited by K. Binder (Oxford University Press, New York, 1995).

[2] K. Kremer and K. Binder, Comp. Phys. Rep. **7**, 259 (1988).

[3] T. L. Hill, *An Introduction to Stastistical Thermodynamics* (Dover, New York, 1986) p. 316.

[4] I. M. Neelov and K. Binder, Macromol. Theory Simul. **4**, 119 (1995).

[5] M. P. Allen and D. J. Tildesley, *Computer Simulation of Liquids* (Clarendon, Oxford, 1987).

[6] K. Kremer and G. S. Grest, in ref. [1]; J. Chem. Phys. **92**, 5057 (1990).

[7] G. S. Grest and M. Murat, in ref. [1].

[8] M. Kröger, Rheology **5**, 66 (1995).

[9] P. A. Thompson, G. S. Grest, and M. O. Robbins, Phys. Rev. Lett. **68**, 3448 (1992); P. A. Thompson, M. O. Robbins, and G. S. Grest, Israel J. Chem. **35**, 93 (1995).

[10] R. Khare, J. J. de Pablo, and A. Yethiraj, Macromolecules **29**, 7910 (1996).

[11] D. H. Napper, *Polymeric Stablization of Colloidal Dispersions* (Academic, London, 1983).

[12] A. K. Chakraborty and M. Tirrell, MRS Bulletin **21** (1), 28 (1996).

[13] K. Binder, A. Milchev, and J. Baschnagel, Ann. Rev. Mat. Sci. **26**, 107 (1996).

[14] J. Baschnagel and K. Binder, Macromol. Theory Simul. **5**, 417 (1996).

[15] P.-Y. Lai and K. Binder, J. Chem. Phys. **97**, 586 (1992).

[16] G. S. Grest and M. Murat, Macromolecules **26**, 3108 (1993).

[17] R. Israels, D. Gersappe, M. Fasolka, V. A. Roberts, and A. C. Balazs, Macromolecules **27**, 6679 (1994).

[18] D. J. Bacon and W. F. Anderson, J. Molec. Graphics **6**, 219 (1988); E. A. Merritt and M. E. P. Murphy, Acta Cryst. **D50**, 869 (1994).

[19] S. Alexander, J. Phys. (Paris) **38**, 983 (1977).

[20] P.-G. de Gennes, Macromolecules **13**, 1069 (1980).

[21] S. T. Milner, Science **251**, 905 (1991).

[22] H. J. Taunton, C. Toprakcioglu, L. J. Fetters, and J. Klein, Macromolecules **23**, 571 (1990).

[23] S. S. Patel and M. Tirrell, Ann. Rev. Phys. Chem. **40**, 597 (1989).

[24] J. Klein, E. Kumacheva, D. Mahalu, D. Perahia, and L. J. Fetters, Nature **370**, 634 (1994); J. Klein, E. Kumacheva, D. Perahia, D. Mahalu, and S. Warburg, Faraday Discuss. **98**, 173 (1994).

[25] J. Klein, Ann. Rev. Mat. Sci. **26**, 581 (1996).

[26] S. Granick, A. L. Demirel, L. L. Cai, and J. Peanasky, Israel J. Chem. **35**, 75 (1995); L. L. Cai, J. Peanasky, and S. Granick, Trends Polym. Sci. **4**, 47 (1996).

[27] R. M. Overney, D. P. Leta, C. Pectroski, M. H. Rafailovich, Y. Liu, J. Quinn, J. Sokolov, A. Eisenberg, and G. Overney, Phys. Rev. Lett. **76**, 1272 (1996).

[28] M. Murat and G. S. Grest, Macromolecules **22**, 4054 (1989); Phys. Rev. Lett. **63**, 1074 (1989).

[29] E. B. Zhulina, O. V. Borisov, and V. A. Pryamitsyn, J. Colloid Interface Sci. **137**, 495 (1990).

[30] E. Raphaël, P. Pincus, and G. H. Fredrickson, Macromolecules **26**, 1996 (1993).

[31] M. Murat and G. S. Grest, Macromolecules **29**, 8282 (1996).

[32] E. Pelletier, G. F. Belder, G. Hadziioannouu, and A. Subbotin, J. Phys. II France **7**, 271 (1997).

[33] X. Zheng, B. B. Sauer, J. G. V. Alsten, S. A. Schwarz, M. H. Rafailovich, J. Sokolov, and M. Rubinstein, Phys. Rev. Lett. **74**, 407 (1995).

[34] J. Klein and P. F. Luckham, Macromolecules **17**, 1041 (1984).

[35] R. Zajac and A. Chakrabarti, Phys. Rev. E **52**, 6536 (1995); J. Chem. Phys. **104**, 2418 (1996).

[36] P.-Y. Lai, J. Chem. Phys. **103**, 5742 (1995).

[37] P. G. de Gennes, *Scaling Concepts in Polymer Physics* (Cornell University Press, Ithaca NY, 1979).

[38] M. Murat, Macromolecules **28**, 5928 (1995).

[39] G. S. Grest, J. Chem. Phys. **105**, 5532 (1996).

[40] J. Klein, D. Perahia, and S. Warburg, Nature **352**, 143 (1991).

[41] G. S. Grest, Phys. Rev. Lett. **76**, 4979 (1996).

[42] G. S. Grest, in *Dynamics in Small Confined Systems III*, edited by J. M. Drake, J. Klafter, and R. Kopelman (Materials Research Society, Pittsburgh, 1997), Vol. 464.

[43] B. Dünweg, J. Chem. Phys. **99**, 6977 (1993).

[44] G. S. Grest, M.-D. Lacasse, and M. Murat, MRS Bulletin **22** (1), 27 (1997).

[45] P.-Y. Lai and K. Binder, J. Chem. Phys. **98**, 2366 (1993).

[46] L. Miao, H. Guo, and M. J. Zuckermann, Macromolecules **29**, 2289 (1996).

[47] Y. Rabin and S. Alexander, Europhys. Lett. **13**, 49 (1990).

[48] J.-L. Barrat, Macromolecules **25**, 832 (1992).

[49] V. Kumaran, Macromolecules **26**, 2464 (1993).

[50] J. L. Harden and M. Cates, Phys. Rev. E **53**, 3782 (1996).

[51] S. Granick, Science **253**, 1374 (1992).

[52] J. Israelachvili, *Intermolecular & Surface Forces* (Academic, London, 1994).

[53] K. Binder, Adv. Polym. Sci. **112**, 181 (1994).

[54] J. T. Koberstein, MRS Bulletin **21** (1), 19 (1996).

[55] H. R. Brown, MRS Bulletin **21** (1), 24 (1996).

[56] D. Henderson and P. J. Leonard, in *Liquid Mixtures*, edited by D. Henderson (Academic, New York, 1971), p. 413.

[57] M. D. Gehlsen, J. H. Rosedale, F. S. Bates, G. D. Wignall, L. Hansen, and K. Almdal, Phys. Rev. Lett. **68**, 2452 (1992).

[58] A. Sariban and K. Binder, J. Chem. Phys. **86**, 5859 (1987).

[59] A. Sariban and K. Binder, Macromolecules **21**, 711 (1988).

[60] G. S. Grest, M.-D. Lacasse, K. Kremer, and A. Gupta, J. Chem. Phys. **105**, 10583 (1996).

[61] M. Müller, K. Binder, and W. Oed, J. Chem. Soc. Faraday Trans. **91**, 2369 (1995).

[62] M.-D. Lacasse and G. S. Grest, in *Morphological Control in Multiphase Polymer Mixtures*, edited by R. M. Briber, D. G. Peiffer, and C. C. Han (Materials Research Society, Pittsburgh, 1997), Vol. 461.

[63] J. Stecki and S. Toxvaerd, J. Chem. Phys. **103**, 4352 (1995).

[64] A. N. Semenov, Macromolecules **27**, 2732 (1994).

[65] M. Stamm and D. W. Schuber, Ann. Rev. Mater. Sci. **25**, 325 (1995).

[66] A. Werner, F. Schmid, K. Binder, and M. Müller, Macromolecules **29**, 8241 (1996).

Monte Carlo Simulation of Langmuir Monolayer Models

F. Schmid, C. Stadler, and H. Lange

Institut für Physics, Universität Mainz, D-55099 Mainz, Germany

Abstract. We study a coarse grained, continuum model for Langmuir monolayers, *i.e.*, monolayers of amphiphilic molecules on a polar substrate. The amphiphiles are represented by stiff chains of beads with one end grafted to a planar surface. Monte Carlo Simulations at constant pressure have been performed, using simulation boxes of variable size and variable shape. A number of techniques have been explored in order to obtain an efficient simulation algorithm. We discuss the resulting phase diagram, characterize the different phases, and analyze the conditions, under which they can be found.

1. Introduction

In this contribution, we present Monte Carlo simulations which aim to provide a better understanding of phase transitions in Langmuir monolayers or Langmuir Blodgett films. Langmuir monolayers are monolayers of amphiphilic molecules, with one polar headgroup and one or several nonpolar tails, adsorbed on a liquid or solid substrate. They are of increasing technical importance, since they can be used to design surfaces with well defined structural properties [1]. They are also of biological interest as simple model systems for biological membranes – which are generally pictured as lipid bilayers, consisting of two weakly coupled monolayers, with embedded proteins [2].

In computer simulations, Langmuir monolayers have mostly been studied in atomistic detail using Molecular dynamics [3]–[7]. These studies have been successful in reproducing experimental results and provided useful additional information, *e.g.*, with respect to the conformation of the nonpolar chains. Unfortunately, they are very time consuming, which makes it difficult to obtain an overview over the phase behavior in a wider parameter range. Moreover, the quality of the simulation results depends very much on the quality of the realistic interaction potentials, which have been used as input. The connection between the behavior observed in a simulation, and the details of the potentials, is not always clear.

We will take a different approach here: The study of a highly simplified model for Langmuir monolayers. Such models make rather poor quantitative predictions for monolayer properties. But they allow for an efficient sampling of the phase behavior in parameter space, and they can provide fundamental insight into the underlying physics of the phase transitions, independent of the details of the potentials.

2. Definition of the model

Before defining an actual model, let us discuss briefly the experimental phase diagram of fatty acid monolayers on water, sketched schematically in Figure 1 [8, 9]. The most remarkable transition, which is also present in almost every other Langmuir monolayer system, is the first order transition between the disordered expanded phase (LE) and one of the condensed phases (L_2, Ov or LS). On the condensed side of the phase diagram, a large number of different phases is found, which differ by order in one or more quantities: Quasicrystalline order (CS, L_2''), orientational order of the backbones of the chains (CS, S, L_2', L_2''), and tilt order of chains (L_2 and L_2" are tilted towards nearest neighbors, L_2' and Ov towards next nearest neighbors).

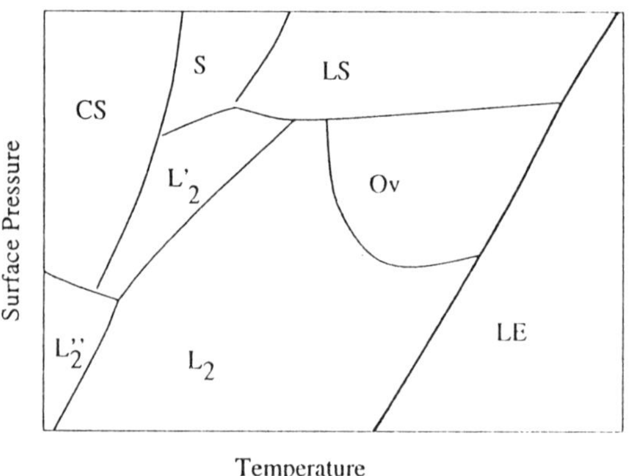

Figure 1: Generic phase diagram for fatty acid monolayers on water. See text for explanation of the phases. Increasing the length of the hydrocarbon chains shifts the phase diagram to higher temperatures (from [9])

Our simplified model does not attempt to preserve all these different possibilities of ordering. We shall only be concerned with two types of transitions: The main transition from the expanded phase into a condensed phase, and the tilting transitions. Theoretical arguments and mean field calculations indicate that the transition into the expanded phase is driven by an increase of the conformational entropy of the chains [10]; hence it is advisable to keep these degrees of freedom, *i.e.*, model the amphiphilic molecules as flexible chains. Since we are not interested in backbone ordering, we give the chains as little internal structure as possible.

This leads to the following "minimal model", which has first been proposed by Haas and Hilfer [11]: Amphiphilic molecules are represented by chains of beads, with one "head" bead confined to a planar surface. The beads are connected by bonds of variable length b, subject to a bond length potential ("FENE" potential)

$$V_{bl}(b) = -\frac{c_{bl}}{2} d_{bl}{}^2 \ln(1 - \frac{(b - b_0)^2}{d_{bl}{}^2}) \qquad (|b - b_0| \le d_{bl}), \quad (1)$$

which is harmonic close to the optimal bond length b_0 and has a logarithmic cutoff at $b_0 \pm d_{bl}$. A very high force constant was chosen, $c_{bl} = 100$ energy units. The stiffness of the chains is represented by a bond angle potential

$$V_{ba}(\theta) = c_{ba}(1 - \cos(\theta)) \qquad (2)$$

for the angle θ between adjacent bonds. Beads which are not direct neighbors within a chain interact *via* a truncated Lennard-Jones potential

$$V_{LJ}(d) = \epsilon \left[(\sigma/d)^{12} - 2(\sigma/d)^6 + 0.031 \right] \qquad (d < 2\sigma). \quad (3)$$

The heads are free to move within the plane, $z = 0$, but cannot move out of it. In a first step, we model heads which are identical to chain monomers, *i.e.*, they interact with the potential (3). In a second step, we will study the effect of making the heads larger than the chains. From a computational point of view, every increase of the range of an interaction slows the algorithm down dramatically. On the other hand, there is no obvious reason why the heads have to attract each other. Hence we can replace the potential between the heads by a short range, purely repulsive soft core potential

$$V_{sc}(d) = \epsilon \left[(\sigma_h/d)^{12} - 2(\sigma_h/d)^6 + 1 \right] \qquad (d < \sigma_h). \quad (4)$$

In the following, we shall put $\epsilon = 1$, $\sigma = 1$, and the Boltzmann factor $k_B = 1$, thereby setting the units of energy, length and temperature. We can attempt a rough translation of these units into standard SI-units when assuming that we model hydrocarbon chains, and that every bead represents two (CH_2) groups. With realistic potential parameters taken from ref. [12], we find that the temperature is given in units of $T_0 = 240$ K, the length in units of $l_0 = 0.35$ nm, and the pressure in units of $\Pi_0 = 25$ mN/m.

The bond length parameters are $b_0 = 0.7$ and $d_{bl} = 0.2$. We have simulated systems of 144 chains containing $N = 7$ beads. A typical disordered high temperature configuration is shown in Figure 2.

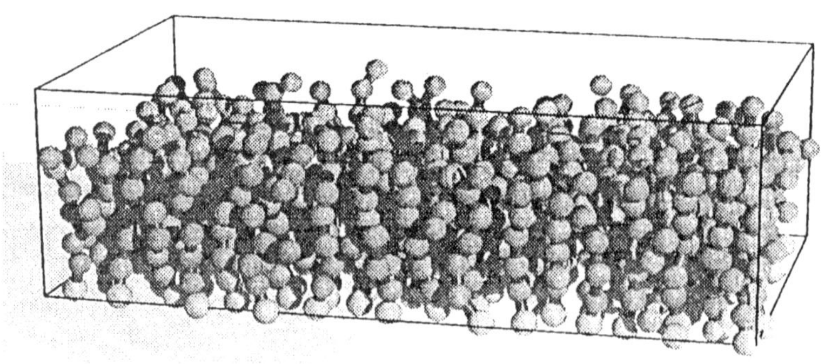

Figure 2: Configuration snapshot in the disordered phase at pressure $\Pi = 40$, temperature $T = 4$, and head size $\sigma_h = 1.1$.

3. Ground state analysis

The classical low temperature behavior is dominated by the internal energy. It is thus useful to determine the ground state, which the system assumes at zero temperature. We suppose that in the ground state, chains are straight, arranged on a distorted hexagonal lattice, and unformly tilted in one direction. With this premise, we minimize the total internal energy E at fixed molecular area a with respect to the parameters tilt angle, bond length and lattice distortion. The surface pressure is obtained using $\Pi = -dE/dA$.

Some important insight into the physics of tilting transitions can already be gained from the study of an even simpler model system, grafted rods. Within our model, the rigid rod limit is described by chains of $N \to \infty$ "beads", which interact *via* a reduced potential $V_{LJ}b_0{}^2$, with $c_{ba} \to \infty$, $c_{bl} \to \infty$, $b_0 = L/N$. Figure 3 shows the ground state phase diagram for rods of length $L = 5$, which are attached to purely repulsive heads of variable head size σ_h (see (4)). One finds that rods with no heads or small heads do not tilt. On increasing the head size, a sequence of tilting transitions is found: A continuous transition from the untilted phase (U) to a phase with tilt towards next nearest neighbors (NNN), and a first order transition to a phase with tilt towards nearest neighbors (NN) [13]. We conclude that in a rigid rod system, tilt is induced by a mismatch of head and tail size, and that the extent of the mismatch determines the direction of tilt. This latter result is in agreement with recent experimental observations [9].

The chains in our simulation model are still straight at zero temperature, but due to the finite number of interaction centers (beads) they have internal structure. As a consequence, tilted chains pack in an energetically more favorable way than untilted chains, and even pure Lennard-Jones chains already tilt towards next nearest neighbors at zero pressure. With increasing

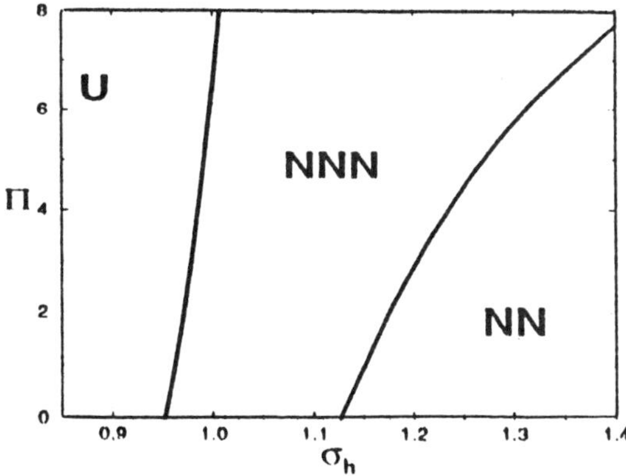

Figure 3: Ground phase diagram for grafted rigid rods with heads of size σ_h, in the plane of σ_h vs surface pressure Π. See text for explanation.

head size, one recovers the swiveling transition to a phase with tilt towards nearest neighbors at $\sigma_h = 1.17$. These calculations yield the true classical ground state, provided that the chains tilt indeed uniformly at zero temperature. Low temperature simulations however indicate that systems of chains with large heads, $\sigma_h = 1.2$, assume modulated structures. We will come back to this point at the end of section 5.

4. Some comments on the simulation technique

We study n chains on a paralellogram with the side lengths L_x and L_y and the angle α (*i.e.*, the total area $A = L_x L_y \sin \alpha$), using periodic boundary conditions in these two dimensions and free boundary conditions in the third. The heads are grafted to the plane $z = 0$. We use the Metropolis algorithm with the three basic Monte Carlo moves

1. Displacement of single monomers (beads).

2. Vary L_x or L_y, *i.e.*, stretch or squeeze the configuration in one direction

3. Shear the box (Change the angle α while keeping A constant).

Since the area A of the simulation box fluctuates, the effective Hamiltonian is given by

$$\mathcal{H} = E + \Pi A - nNT \log A \tag{5}$$

with the internal energy E and the surface pressure Π. The last term accounts for the translational entropy of the particles [14]. We found that the Monte Carlo moves which change the shape of the box, *i.e.*, the aspect ratio L_x/L_y or the angle α, are crucial to maintain constant, *isotropic*, pressure in the condensed phase. The pressure can be checked using the virial theorem. In order to do so, we define the pressure tensor,

$$\Pi_{\alpha\beta} = -\frac{NT}{A}\delta_{\alpha\beta} - \frac{1}{A}\sum_{i=1}^{nN} r_i^\alpha \frac{\partial E}{\partial r_i^\beta}, \tag{6}$$

where α, β denote the x or y component and the sum i runs over all nN monomers. In a system at mechanical equilibrium, the thermal average of the tensor $\Pi_{\alpha\beta}$ has to be diagonal and equal to $\Pi_{\alpha\beta} = \Pi\delta_{\alpha\beta}$. If this is not the case, the system is subject to mechanical stress, caused by the rigid simulation box, which may affect the simulation results. In some cases, the off diagonal elements Π_{xy} reached values almost as large as Π_{xx} and Π_{yy} if the box was not allowed to shear.

We have explored a number of collective Monte Carlo moves in order to obtain an algorithm with improved time correlation properties. In particular, we have implemented and tested continuous configurational biased Monte Carlo moves [16]; unfortunately, it turned out that the improvement in the decorrelation time was more than outweighed by the high increase of actual computer time needed for one Monte Carlo step. We have also implemented collective polymer moves, where whole chains are displaced rather than just single monomers. They did not accelerate the decorrelation. A variant of this idea is the introduction of alternative box moves, in which only the head positions are rescaled along with L_x and L_y, and the relative monomer positions within a chain remain unchanged. Note that the last term in the Hamiltonian (5) then has to be replaced by $-nT\log A$. These moves improved the properties of the algorithm in the highly diluted disordered region of the phase diagram, but not in the condensed region of interest.

The most substantial speedup was achieved using more conventional techniques: The use of Verlet tables in combination with a link cell algorithm, and the introduction of huge arrays to keep track of all interaction energies. The optimized algorithm performs one Monte Carlo step (MCS) in a system of 144 chains, consisting of 1008 attempted monomer moves and one attempt to change L_x, L_y, and the angle α, in 0.14 seconds on a DEC alpha workstation. The characteristic decorrelation time is ~ 200 MCSs. In general, the system was equilibrated for 30.000 to 50.000 MCSs and then sampled for 40.000 – 200.000 MCSs. Data were collected every 100 MCSs.

5. Simulation results

5.1 Lennard-Jones chains

We first discuss the properties of stiff Lennard-Jones chains [15]. Both heads and chain monomers interact with the potential (3), the stiffness constant is taken to be $c_{ba} = 100$. Resulting isobars for various surface pressures Π in area/temperature space are shown in Figure 4. One notices a clear first order transition, which manifests itself in a jump of the area per molecule at the transition temperature $T_0(\Pi)$. Note that the curve for surface pressure $\Pi = 0$ has to describe a metastable state; the true equilibrium state at any nonzero temperature is the infinitely dilute gas state, since the translational entropy diverges logarithmically with the area a. At surface pressures $\Pi \geq 10$, no hysteresis effects were observed.

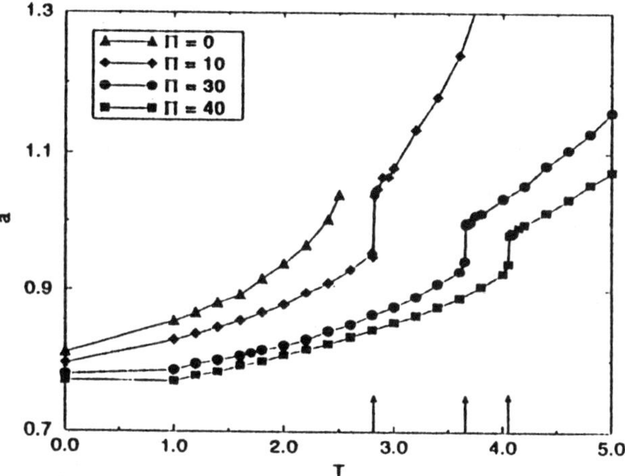

Figure 4: Molecular area a vs temperature T for different pressures Π.

We shall attempt to characterize the transition somewhat further. Figure 5 shows monomer density profiles in the direction z perpendicular to the interface. As can be expected, the profile broadens more and more with increasing temperature. However, no qualitative change is observed right at the transition temperature.

The study of profiles in the (xy) plane is more revealing. Figure 6 shows the two dimensional density profile for the relative position of the six nearest neighbors in the head group plane. It changes quite dramatically at the transition – below the transition, six distinct peaks are found, which disappear above the transition.

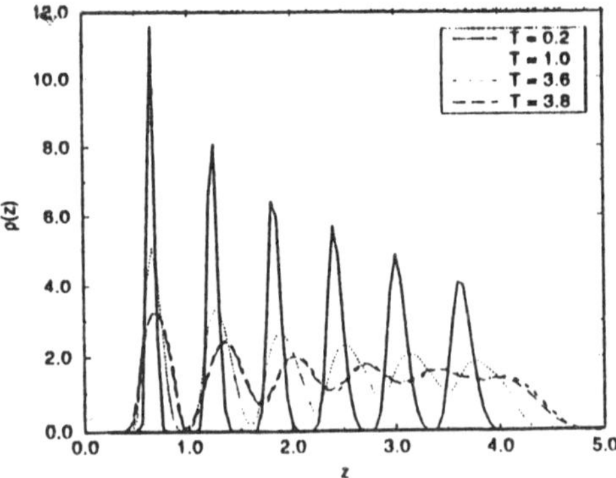

Figure 5: Monomer density profiles in the z direction for different temperatures T at surface pressure $\Pi = 30$.

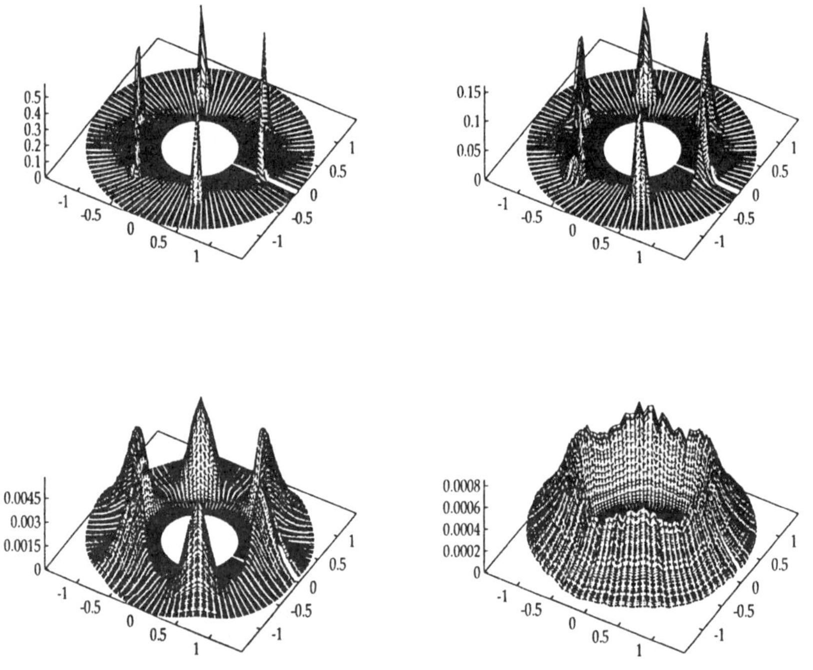

Figure 6: Distribution of relative nearest neighbor positions in the plane of head groups at surface pressure $\Pi = 30$ and temperatures $T = 0.2$ (upper left), $T = 1.0$ (upper right), $T = 3.6$ (lower left) and $T = 3.8$ (lower right).

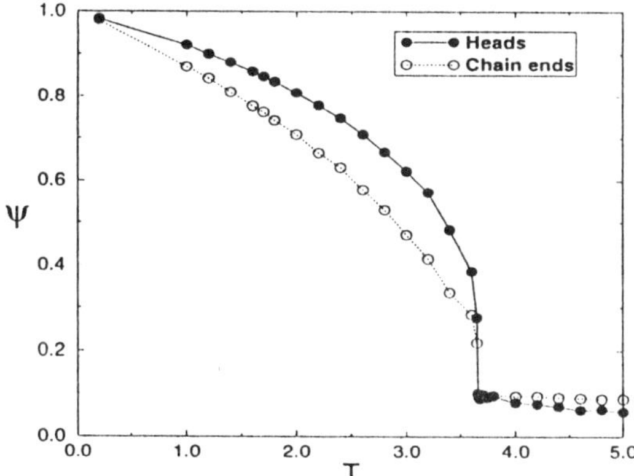

Figure 7: Order parameter Ψ vs temperature T at surface pressure $\Pi = 30$ for head monomers and chain ends.

This observation can be quantified using the "global bond-orientational order parameter" of 2D melting

$$\Psi = |\frac{1}{6n} \sum_{j=1}^{n} \sum_{k=1}^{6} \exp(6i\phi_{jk})|. \tag{7}$$

The sum j runs over all n heads of the system, k over the six nearest neighbors of j, and ϕ_{jk} denotes the the angle between the vector connecting j and k and a fixed reference line, which can be chosen arbitrarily. As shown in Figure 7, the parameter Ψ jumps to almost zero at the transition temperature. For comparison, we have also calculated the corresponding quantity for the end monomers. They are generally less ordered than the head monomers, but their order parameter drops down at the same temperature. We conclude that the transition involves two dimensional melting of the head lattice. On the other hand, it is presumably not solely driven by melting, since it is much more marked than melting transitions in true two dimensional systems. Usually, even deciding whether such a transition is first order or continuous is exceedingly difficult (see ref. [17] for a discussion). Obviously, the chains play a major role for the transition, and most likely drive it.

Apart from the lattice order, we are also interested in the tilting behavior of the chains. It can be characterized by the average tilt angle θ of chains with respect to the surface normal, and by the two dimensional order parameter $\vec{R}_{xy} = (\overline{x}, \overline{y})$, where $\overline{x}$ and $\overline{y}$ are the x and y component of the head-to-end vector of a chain, averaged over the chains in one configuration. In the thermodynamic limit, R_{xy} is zero in a disordered system and nonzero in a system with collective tilt, where the azimuthal symmetry is broken.

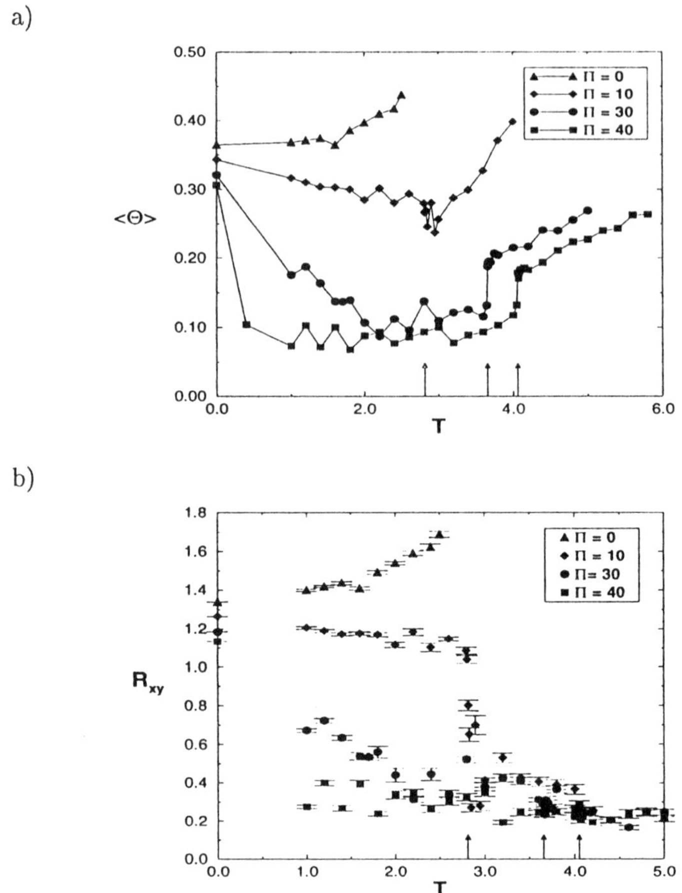

Figure 8: Tilt angle Θ (a) and tilt order parameter R_{xy} (b) vs temperature T for different surface pressures Π.

Figure 8 shows the thermal average $\langle\theta\rangle$ and $R_{xy} = \langle|\vec{R}_{xy}|\rangle$ as a function of temperature at different surface pressures. The arrows indicate the location of the first order "melting" transition discussed above. The tilt angle first drops with increasing temperature and then rises again, in the fluid state. The quantity R_{xy}, on the other hand, drops down and remains close to zero in the disordered state. These results are in agreement with earlier results of Hilfer and coworkers [18, 11]. From Figure 8b), one is lead to suspect that the transition from the tilted state with $R_{xy} \neq 0$ to the untilted state with $R_{xy} \approx 0$ decouples from the melting transition at pressures $\Pi = 30$ and 40 and occurs at a lower transition temperature already. Unfortunately, the data are not of high enough quality to support a definitive statement. Note that

R_{xy}, which is defined as a positive definite quantity, has pronounced finite size tails in the disordered phase. The inspection of the distribution of $\vec{R}_{xy}$ (not shown) reveals that the chains are tilted predominantly towards next nearest neighbors at all pressures and temperatures, in agreement with the ground state analysis of section 3.

5.2 Chains with larger heads

We now turn to the discussion of another system: flexible chains ($c_{ba} = 10.$) with larger heads, which interact *via* the soft core potential (4). We first study chains with heads of size $\sigma_h = 1.1$. A representative set of curves for Ψ and R_{xy} for two different pressures is shown in Figure 9. The decoupling of the tilting and the melting transition at high pressures is very clear here. Three distinct phases are found: a tilted phase with an ordered head lattice, an untilted phase with an ordered head lattice, and an untilted, fluid phase.

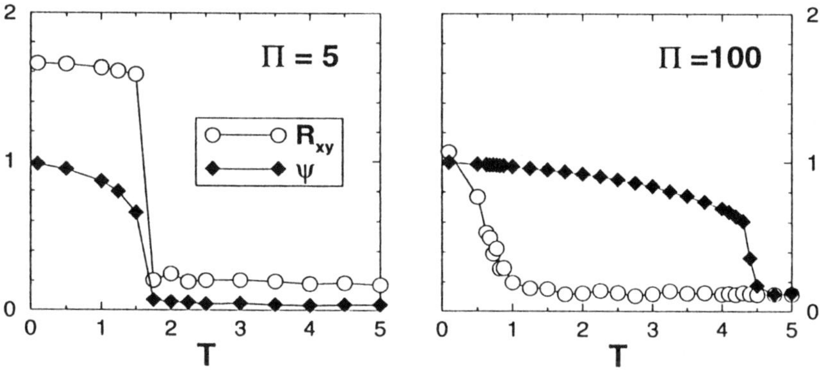

Figure 9: Lattice order parameter Ψ and tilt order parameter R_{xy} vs temperature T for chains with head size $\sigma_h = 1.1$, at surface pressures $\Pi = 5$ and $\Pi = 100$.

If one identifies the fluid phase with the expanded phase LE, and the two ordered phases with the condensed phases LS (untilted) and Ov or L_2 (tilted), the phase diagram has indeed some similarity with the high temperature phases in the experimental phase diagram of Figure 10. Preliminary results for the phase diagram in surface pressure/temperature space are shown in Figure 1 [19]. Note that compared to the results from above, the temper-

atures of the order/disorder transition go down considerably in the present
system of much more flexible chains. This remains true if one compares Figure 10 with the transition temperatures in a directly comparable system of
stiff chains ($c_{ba} = 100.$) with soft core heads of size $\sigma_h = 1.1$ [15]. Hence it is
not an effect of the head interactions, but another indication of the fact that
chain disordering is primarily responsible for the order/disorder transition.

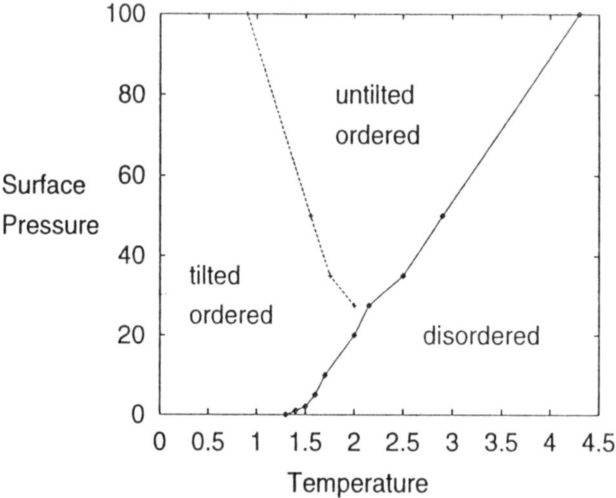

Figure 10: Phase diagram for chains of head size $\sigma_h = 1.1$.

So far the chains were always found to tilt towards next nearest neighbors,
which is presumably a consequence of the fact that the head size was still
chosen too small to favor tilt towards nearest neighbors. This motivated us
to also look at chains of size $\sigma_h = 1.2$. Figure 11 shows the projection of
the head-to-end vectors into the (xy)-plane for a resulting configuration. As
expected, the average tilt direction changes towards nearest neighbors; more
interesting, another striking new feature emerges: The direction of tilt is not
uniform any more, but modulated over several lattice spacings.

The reasons for this unexpected behavior are not yet clear. A number
of different lengths compete in the system – the lattice constant of the head
lattice, the monomer size, the bond length. The modulation presumably
results from a subtle interplay between these three. Whether it is a mere
artefact of the model, or whether it could also be relevant in real systems, is
an issue which will have to be addressed in future work.

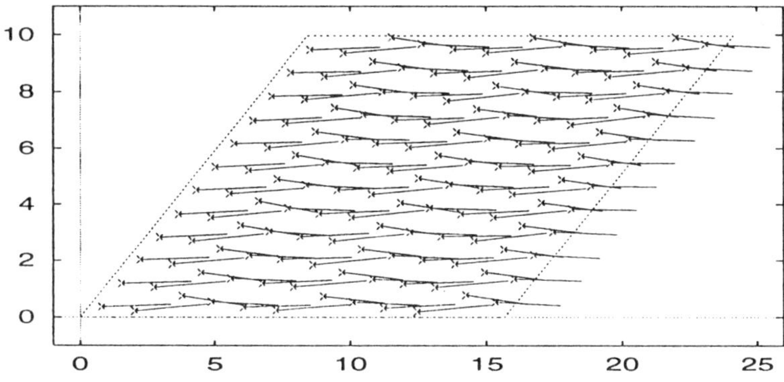

Figure 11: Projection of the head-to-end vector of chains into the (xy)-plane for a configuration of chains with head size $\sigma_h = 1.2$, at surface pressure $\Pi = 1.$, and temperature $T = 0.1$. Crosses mark the positions of the head groups. Also shown are the boundaries of the simulation box.

6. Conclusions

The model presented here reproduces a first order transition between a liquid phase and a tilted or untilted liquid expanded phase. It provides insight into the factors which determine the direction of the tilt, and the possible origin of the swiveling transition between tilt directions. Besides that, our studies have raised a number of new questions, which will have to be resolved in the future.

We have thus demonstrated that Monte Carlo studies of very simplified models for Langmuir monolayers can address and partly answer fundamental questions on the physics of the phase transitions in these systems. While they fail to make good quantitative predictions, they allow for a relatively fast scanning of phase behavior and for systematic variation of specific model parameters, which are less easily accessible to the chemist (like the chain stiffness or the head group size). Hence such studies are useful complements to experiments, and to the traditional Molecular Dynamics simulations in atomistic detail.

Acknowledgements

We thank K. Binder, F. Haas and R. Hilfer for useful discussions. C.S. was supported by the Graduiertenkolleg on supramolecular systems in Mainz.

References

[1] J.D. Swalen, D.L. Allara, J.D. Andrade, E.A. Chandross, S. Garoff, J. Israelachvili, T.J. Carthy, R. Murray, R.F. Pease, J.F. Rabolt, K.J Wynne, H. Yu, Langmuir **3**, 932 (1987).

[2] R.B. Gennis: *Biomembranes*, Springer Verlag (1989).

[3] J.P. Bareman, G. Cardini, M.L. Klein, Phys. Rev. Lett. **60**, 2152 (1988); G. Cardini, J.P. Bareman, M.L. Klein, Chem. Phys. Lett. **145**, 493 (1988); J.P. Bareman, M.L. Klein, J. Phys. Chem. **94**, 5202 (1990); J. Hautman, M.L. Klein, J. Chem. Phys. **91**, 4994 (1989); **93**, 7483 (1990).

[4] J. Harris, S.A. Rice, J. Chem. Phys. **89**, 5898 (1988); M. Li, S.A. Rice, J. Chem. Phys. **104**, 6860 (1996).

[5] M.A. Moller, D.J. Tildesley, K.S. Kim, N. Quirke, J. Chem. Phys. **94**, 8390 (1991).

[6] H.E. Alper, D. Bassolino, T.R. Stouch, J. Chem. Phys. **98**, 9798 (1993).

[7] S. Karaborni, S. Toxvaerd, J. Chem. Phys. **96**, 5505 (1992); **97**, 5876 (1992); S. Karaborni, S. Toxvaerd, O. Olsen, J. Phys. Chem. **96**, 4965 (1992); S. Karaborni, Langmuir **1993**, 1334 (1993). J.I. Siepmann, S. Karaborni, M.L. Klein, J. Phys. Chem. **98**, 6675 (1994); S. Karaborni, G. Verbist, Europhys. Lett. **27**, 467 (1996).

[8] Condensed phases in monolayers are reviewed by C.M. Knobler, R.C. Desai, Ann. Rev. Phys. Chem. **43**, 207 (1992).

[9] B. Fischer, E. Teer, C.M. Knobler, J. Chem. Phys. **103**, 2365 (1995); E. Teer, C.M. Knobler, C. Lautz, S. Wurlitzer, J. Kildae, T.M. Fischer, preprint.

[10] F. Schmid, M. Schick, J. Chem. Phys. **102**, 2080 (1995); F. Schmid, to be published in Phys. Rev. E.

[11] R. Hilfer, F.M. Haas, K. Binder, Nuovo Cimento **16**, 1297 (1994); F.M. Haas, R. Hilfer, K. Binder, J. Chem. Phys. **102**, 2960 (1995). F.M. Haas, R. Hilfer, K. Binder, J. Phys. Chem., **100**, 15290 (1996); F.M. Haas, R. Hilfer, J. Chem. Phys. **105**, 3859 (1996).

[12] J. Baschnagel, K. Binder, W. Paul, M. Laso, U.W. Suter, I Batoulis, W. Jilge, T. Bürger, J. Chem. Phys. **95**, 6014 (1991).

[13] F. Schmid, H. Lange, J. Chem. Phys., in press.

[14] M. Allen, D. Tildesley, *Computer Simulation of Liquids* (Clarendon Press, Oxford, 1987).

[15] H. Lange *et al*, document in preparation; H. Lange, Diplomarbeit Universität Mainz.

[16] D. Frenkel, G. Mooij, in *Monte Carlo and Molecular Dynamics of Condensed Matter Systems*, K. Binder and G. Cicotti eds, p. 163 (1996).

[17] H. Weber, D. Marx, K. Binder, Phys. Rev. B **51**, 14636 (1995).

[18] M. Scheringer, R. Hilfer, K. Binder, J. Chem. Phys. **96**, 2269 (1991).

[19] C. Stadler *et al*, document in preparation.

Cluster Simulation of Lattice Gases

H.W.J. Blöte and J.R. Heringa

Faculty of Applied Physics, Delft University of Technology,
P.O. Box 5046, 2600 GA Delft, The Netherlands

Abstract. Hard-core lattice gases can be simulated by means of cluster Monte Carlo algorithms. In particular we formulate a single-cluster algorithm for lattice gases with nearest neighbour-exclusion, in the absence of further neighbor interactions. In analogy with cluster algorithms for spin systems, this method quite efficiently suppresses critical slowing down. We use this method to derive statistically accurate finite-size data for lattice gases with nearest-neighbor exclusion on the simple-cubic and body-centered-cubic lattices. A finite-size-scaling analysis of these data indicates that these models belong to the Ising universality class. This result contradicts earlier analyses; we explain this disagreement in terms of corrections to scaling, which are quite prominent in these models.

1. Introduction

The renormalization theory has provided a satisfactory explanation of the existence of universality classes of critical phenomena. Each universality class has its own characteristic set of critical exponent and dimensionless amplitude ratios. Furthermore, there is considerable empirical evidence that universality classes are characterized by only a few parameters: the dimensionality, the mathematical nature of the order parameter (e.g., an n-component vector for the $O(n)$ model); and the range and the uniformity of the interactions. Here we focus on relatively simple choices for these parameters: we consider systems with a scalar order parameter (for instance, the Ising model, binary mixtures and simple gas-liquid systems) and with short-range interactions. In particular one may ask what evidence exists for universality of these Isinglike models, because of the non-rigorous nature of the renormalization theory.

In two dimensions, strong evidence for the universality of critical properties of Isinglike systems follows from exact solutions, and from accurate numerical analyses. Also in three dimensions, considerable evidence has been reported in support of universality. For instance, see Ref. [1] for an entry into the existing literature. Critical exponents have been calculated

Springer Proceedings in Physics, Volume 83
Computer Simulation Studies in Condensed-Matter Physics X
Eds.: D. P. Landau, K.K. Mon, H. -B. Schüttler
© Springer-Verlag Berlin Heidelberg 1998

for the ϕ^4 model, for a number of nearest-neighbor lattice models, for lattice models with interactions beyond nearest neighbors, and for a spin-1 lattice model. These investigations use a variety of tools such as ϵ expansion, renormalization, series expansions, Monte Carlo and Monte Carlo renormalization. Typical accuracies of these results are a few times 10^{-3} and, generally speaking, they tend to be in agreement with the universality hypothesis. Moreover, a large body of experimental results exists which tend to cluster about the theoretical results. Deviations exceeding the estimated experimental inaccuracies have been reported; but it is difficult to exclude the possibility that these deviations are due to crossover phenomena and corrections to scaling. This applies especially to those cases where the analysis does not allow for such effects.

However, one may maintain that, from an empirical point of view, the situation in three dimensions is somewhat less clear than in two-dimensions. In the absence of exact solutions, one is restricted to results subject to numerical errors, which thus always allow small deviations from universality. Moreover, recently, Monte Carlo analyses [2, 3] of lattice-gas models with nearest-neighbor exlusion led to results that were inconsistent with Ising universality. These analyses were based on Metropolis-type simulations, and did not include corrections-to-scaling. Here, the problem is that these corrections bring new unknown parameters into the fit formula, so that the errors in the estimated quantities (such as the critical point and the critical exponents) tend to grow considerably, possibly disabling any nontrivial conclusions.

For a meaningful analysis which includes such corrections, it is necessary to obtain accurate numerical data for a sufficiently wide range of system sizes. This is a hard problem in the case of Metropolis-type simulations, because this method suffers considerably from critical-slowing down: the autocorrelation time (expressed in 'sweeps', one update per spin) behaves roughly as L^2 where L is the linear system size. Much faster relaxation occurs in the case of the cluster method originated by Swendsen and Wang [4] and further improved by Wolff [5]. These methods are not well applicable to lattice gases. However, Dress and Krauth [6] have formulated a cluster-simulation method for hard-core gases in continuous space. The underlying ideas are also applicable in the case of lattice gases [7, 8]. We have used a Wolff-like single-cluster method to further investigate the universal properties of the lattice gases on the simple-cubic and the body-centered cubic lattices.

The outline of this paper is as follows. In Section 2, we present a pedagogical description of cluster Monte Carlo for Ising models, in particular of the Wolff algorithm. Elements of this description are applied in Section

3, where we describe the lattice-gas models and the cluster algorithm for their simulation. The numerical analysis and its results are presented in Section 4.

2. Ising Cluster Simulation

The extremely long autocorrelation times associated with standard Metropolis-type simulations near critical points, arise because a change of state of one particle has only a small effect on large-scale structures, such as critical fluctuations.

For a number of model systems, simulation algorithms have been developed that can produce nonlocal changes, affecting many particles at the same time. As a result, critical slowing down is suppressed. We will first describe such a 'cluster' algorithm for spin models on an arbitrary lattice. Although this simulation method is applicable to a variety of models including the Potts, XY and Heisenberg models, here we choose, for simplicity, the language of the Ising model.

In zero magnetic field, its Hamiltonian is

$$\mathcal{H}/kT = -K_\mathrm{I} \sum_{<ij>} s_i s_j \tag{1}$$

where the sum is over all pairs of nearest-neighbor sites (i,j). The spin variables are labeled with the sites they sit on and can assume only values ± 1. The Ising coupling $K_\mathrm{I} = J/kT$ where J is the energy of a pair of antiparallel neighbor spins. We take ferromagnetic interactions: $K_\mathrm{I} > 0$.

2.1 The Random-Cluster Model

Disregarding a constant shift of energy, an alternative way to write the Hamiltonian is

$$\mathcal{H}/kT = -K \sum_{<ij>} \delta_{s_i,s_j} \tag{2}$$

where $K = 2K_\mathrm{I}$. The 'Kronecker delta' δ_{s_i,s_j} is equal to 1 if $s_i = s_j$ and otherwise 0. The partition sum is

$$Z = Z_s = \sum_{s_1,s_2,\cdots,s_N} e^{K \sum_{<ij>} \delta_{s_i,s_j}} = \sum_{s_1,s_2,\cdots,s_N} \prod_{<ij>} e^{K \delta_{s_i,s_j}} \tag{3}$$

where the index s of Z refers to the 'spin' variables. Later we shall express Z in other variables. Expression (3) tells us that the probability of a spin configuration $\{s_i\} \equiv (s_1, s_2, \cdots, s_N)$ is

$$P_s(\{s_i\}) = \frac{\Pi_{<ij>}\, e^{K\delta_{s_i,s_j}}}{Z} \tag{4}$$

Note that the form $e^{K\delta_{s_i,s_j}}$ can assume only two values. These satisfy

$$e^{K\delta_{s_i,s_j}} = 1 + u\delta_{s_i,s_j} \tag{5}$$

where $u = e^K - 1$. Thus (3) becomes

$$Z_s = \sum_{\{s_i\}} \prod_{<ij>} (1 + u\delta_{s_i,s_j}) \tag{6}$$

Next we express (6) in different variables. As a first step we introduce 'bond variables' b_{ij} on the nearest-neighbor links of the lattice, which take values 0 and 1. We rewrite

$$1 + u\delta_{s_i,s_j} = \sum_{b_{ij}=0}^{1} (u\delta_{s_i,s_j})^{b_{ij}} \tag{7}$$

with the convention $0^0 = 1$. Thus we have $Z = Z_s = Z_{sb}$ with

$$Z_{sb} = \sum_{\{s_i\}} \prod_{<ij>} \sum_{b_{ij}=0}^{1} (u\delta_{s_i,s_j})^{b_{ij}} = \sum_{\{b_{ij}\}} \sum_{\{s_i\}} \prod_{<ij>} (u\delta_{s_i,s_j})^{b_{ij}} \tag{8}$$

This expression describes a simultaneous (both in the s and in the b variables) probability distribution [9]

$$P_{sb}(\{s_i\}, \{b_{ij}\}) = \frac{\Pi_{<ij>}(u\delta_{s_i,s_j})^{b_{ij}}}{Z} \tag{9}$$

which satisfies

$$P_s(\{s_i\}) = \sum_{\{b_{ij}\}} P_{sb}(\{s_i\}, \{b_{ij}\}) \tag{10}$$

By construction, Z_{sb} is equal to Z_s; in other words, one can, by summation, eliminate the bond variables in Z_{sb}. As shown by Kasteleyn and Fortuin [10], one can alternatively eliminate the spin variables in Z_{sb}. In other words, for a given configuration of bond variables $\{b_{ij}\}$ one can execute the sum on the spin variables. For simplicity one may first consider bond configurations containing very few nonzero b_{ij}. When all b_{ij} are zero, the factor $(u\delta_{s_i,s_j})^{b_{ij}}$ is equal to one, irrespective of s_i and s_j. Thus the summation on the N spin variables yields a contribution 2^N to Z. A configuration of bond variables with only one nonzero b_{ij} contributes $2^{N-1}u$ to Z. The additional factor u is the weight of a bond variable, and the factor $1/2$ appears because the Kronecker δ consumes the summation over one of the spin variables. Thus one might, for the case of b nonzero bond variables, at first sight expect to find a contribution $2^{N-b}u^b$ to Z. However, this is not correct in general. Suppose that one adds a bond (i.e., a nonzero bond variable) between two spins that were already connected by a chain of bonds. This new bond will add a factor u but not a factor $1/2$ because both spins *are already equal* in all surviving terms in Z. Thus the exponent b has to be replaced by $b - l$ where l is the number of loops closed by the bond variables. Hence we find $Z = Z_s = Z_{sb} = Z_b$ with

$$Z_b = \sum_{\{b_{ij}\}} 2^c u^b \tag{11}$$

where b is the number of bonds and $c = N - b + l$ is the number of clusters: units consisting of lattice sites connected by bonds. Each new bond decreases the number of clusters by one, unless the sites connected by this bond were already in one cluster; in that case, the number of loops l increases by one. The partition sum (11) describes the so called random cluster model [10]. It determines a canonical probability distribution of bond configurations

$$P_b(\{b_{ij}\}) = 2^c u^b / Z_b \tag{12}$$

This distribution is related to the simultaneous probability distribution P_{sb} by

$$P_b(\{b_{ij}\}) = \sum_{\{s_i\}} P_{sb}(\{s_i\}, \{b_{ij}\}) \tag{13}$$

This equation and (10) determine the interrelations between the probability distributions $P_s(\{s_i\})$, $P_{sb}(\{s_i\}, \{b_{ij}\})$ and $P_b(\{b_{ij}\})$. Let us now suppose that we have a stochastic process that generates configurations $\{s_i\}$ with probability $P_s(\{s_i\})$ for instance, by a Metropolis Monte Carlo algorithm. Then, we can also generate mixed configurations $(\{s_i\}, \{b_{ij}\})$ with probability $P_{sb}(\{s_i\}, \{b_{ij}\})$ by using the conditional probabilities

$$\frac{P_{sb}(\{s_i\}, \{b_{ij}\})}{P_s(\{s_i\})} = \prod_{<ij>} \frac{(u\delta_{s_i,s_j})^{b_{ij}}}{1 + u\delta_{s_i,s_j}} =$$

$$\prod_{<ij>} [\delta_{s_i,s_j} \frac{u^{b_{ij}}}{1 + u} + (1 - b_{ij})(1 - \delta_{s_i,s_j})] \tag{14}$$

The form between square brackets says that, if $s_i = s_j$, we have to choose $b_{ij} = 1$ with probability $\frac{u}{1+u} = 1 - e^{-K}$ (or $b_{ij} = 0$ with probability $\frac{1}{1+u} = e^{-K}$); and if $s_i = -s_j$, we always choose $b_{ij} = 0$. Note that for each given spin configuration, the bond probabilities are independent: the bond variables can thus be chosen by a simple Monte Carlo procedure.

From the probability distribution $P_{sb}(\{s_i\}, \{b_{ij}\})$ one finds, by summing on the spin configurations (eq. 13), the probability distribution $P_b(\{b_{ij}\})$. In the mixed ensemble that we just generated, this simply means that we may 'forget' the spin variables. This provides a Monte Carlo method that generates bond configurations $\{b_{ij}\}$) with probability $P_b(\{b_{ij}\})$.

One can also follow the reverse path and generate mixed configurations by inserting spin variables into bond configurations. From (9) and (12) it follows that

$$\frac{P_{sb}(\{s_i\}, \{b_{ij}\})}{P_b(\{b_{ij}\})} = 2^{-c} \prod_{<ij>} (\delta_{s_i,s_j})^{b_{ij}} \tag{15}$$

The bond variables divide the spins in $c = N - b + l$ clusters; the spins in one cluster must have the same sign because of the Kronecker δ's. But, apart from this restriction, plus and minus signs are equally probable; the signs of the clusters are independent random variables. The signs can thus be generated by a simple Monte Carlo procedure. Using c random variables ± 1 one thus generates mixed configurations with probability $P_{sb}(\{s_i\}, \{b_{ij}\})$. One may now forget the bond variables: thus one generates spin configurations $\{s_i\}$ with the Boltzmann probability distribution $P_s(\{s_i\})$, as required. The Monte Carlo process has meanwhile introduced nonlocal changes, as needed for fast relaxation. Repetition of these steps produces a Monte Carlo process that is called the Swendsen-Wang method [4]; at the Ising critical point, its efficiency may exceed Metropolis type simulations by orders of magnitude. This efficiency is related to the fact that the cluster formation process takes place at the percolation threshold. This follows directly from the known properties of the random-cluster model. Typical clusters are large in comparison with a single spin, but considerably smaller than the whole system. The latter point is significant: if typical clusters occupy most of the system, the algorithm does little except changing the signs of all spins and loses its efficiency.

2.2 The Wolff Cluster Algorithm

A simplification of the Swendsen-Wang algorithm can be realized by forming and flipping just one cluster. The advantage of this approach is that only one cluster need be formed; the rest of the lattice needs no work. This cluster is built around a randomly chosen site and formed by introducing bond variables with the appropriate probabilities. This stochastic process produces a cluster that may be large or small. If it is small, the change of the spin configuration is insignificant, but this applies as well to the work involved to form the cluster. One cluster flip in the Wolff algorithm involves the following steps:

1. choose a random lattice site; denote it i.

2. $s'_i = -s_i$ (flip spin i)

3. for all neighbor sites k of i do the following:

   (a) if $s_k = -s'_i$ do the following with probability $1 - e^{-K}$:
       i. $s'_k = -s_k$ (spin k included in cluster);
       ii. write k in a list of addresses (called stack).
   (b) if $s_k = s'_i$, do nothing.

4. read an address l from the stack;

5. execute the steps listed under 3 for the neighbor sites k of l;

6. erase the address l from the stack;

7. repeat steps 4-6 until the stack is empty.

When the stack is empty, the cluster is completed and flipped. This algorithm is even more efficient than the Swendsen-Wang method. For critical systems of size L^d, it is roughly L^2 times more efficient than the Metropolis method.

2.3 Detailed Balance for Wolff Clusters

In order to prove that these cluster algorithms reproduce the Boltzmann distribution, it is sufficient that the transition probability matrix is irreducible and satisfies detailed balance. The condition of irreducibility poses here no problem; that of detailed balance was proved by Wolff [5].

We consider a Wolff step that transforms a spin configuration $\{s_i\}$ into $\{s'_i\}$ by flipping the spins in a region $\mathcal{C}$ wich is a subset of the lattice $\mathcal{L}$. Note that all spins located on $\mathcal{C}$ have the same sign. We denote the probability that the Wolff cluster formation process leads to a bond configuration whose bonds within $\mathcal{C}$ connect all sites of $\mathcal{C}$, by $W(\mathcal{C}, \{s_i\})$. Consider those nearest-neighbor pairs having one member within $\mathcal{C}$ and one outside $\mathcal{C}$. Let the boundary line surrounding $\mathcal{C}$ intersect n_p such pairs of parallel neighbor spins of $\{s_i\}$, and n_a such pairs of antiparallel neighbors. The probability that the Wolff algorithm does *not* connect these parallel pairs is $(e^{-K})^{n_p}$. It thus follows that the probability of a Wolff step from $\{s_i\}$ to $\{s'_i\}$ is

$$T(\{s'_i\}, \{s_i\}) = e^{-n_p K} W(\mathcal{C}, \{s_i\}) \tag{16}$$

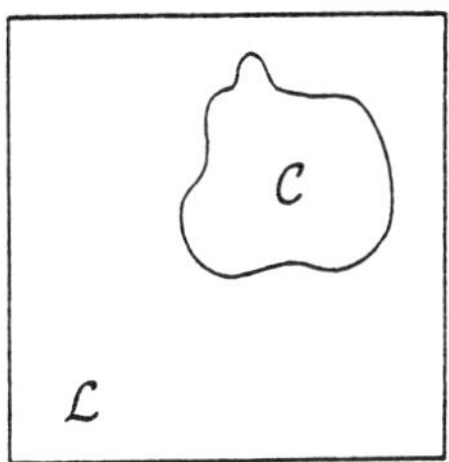

Figure 1. A Wolff cluster step inverts the spins inside a region $\mathcal{C}$ of the lattice $\mathcal{L}$.

Next, we consider the probability of a Wolff step that transforms $\{s_i'\}$ into $\{s_i\}$, thus involving an inversion of the spins in the same region $\mathcal{C}$. It is equal to

$$T(\{s_i\}, \{s_i'\}) = e^{-n_a K} W(\mathcal{C}, \{s_i'\}) \tag{17}$$

The spin pairs divided by the boundary line of $\mathcal{C}$ that were antiparallel for state $\{s_i\}$, are parallel in $\{s_i'\}$. Therefore we now have n_a instead of n_p. The spins in the region $\mathcal{C}$ of $\{s_i\}$ are inverted with respect of $\{s_i'\}$. Since a global inversion does not affect the probabilities used in the cluster formation process, $W(\mathcal{C}, \{s_i\}) = W(\mathcal{C}, \{s_i'\})$. Thus, (16) and (17) combine into

$$T(\{s_i'\}, \{s_i\})/T(\{s_i\}, \{s_i'\}) = e^{(n_a - n_p)K} \tag{18}$$

This is precisely the condition of detailed balance, namely

$$T(\{s_i'\}, \{s_i\})P_{eq}(\{s_i\}) = T(\{s_i\}, \{s_i'\})P_{eq}(\{s_i'\}) \tag{19}$$

where $P_{eq}(\{s_i\})$ is the Boltzmann distribution, because

$$P_{eq}(\{s_i'\})/P_{eq}(\{s_i\}) = \exp[-\frac{E(\{s_i'\}) - E(\{s_i\})}{kT}] =$$

$$e^{2(n_a - n_p)K_I} = e^{(n_a - n_p)K} \tag{20}$$

Thus the Wolff method is guaranteed to reproduce the Boltzmann distribution once the initialization effects have decreased to negligible levels.

2.4 Wolff with Demons

In Monte Carlo methods, demons tend to behave decently and follow the accepted rules of ensemble statistics, in contrast with their forefather called Maxwell's demon, whose behavior was in strong violation of such rules. The Wolff cluster formation probabilities can be reformulated using demons

[11, 12]. These demons are energy reservoirs with, for simplicity, a discrete spectrum, bounded from below but not from above, and with an energy spacing equal to KkT. These demons are first brought in equilibrium with a heat bath at temperature T: their energy distribution thus is the canonical one. Then, the demons are disconnected from the heat bath, and one is placed on each nearest-neighbor bond of our Ising lattice. A microcanonical step follows:

1. choose a random lattice site; denote it i.

2. $s_i' = -s_i$ (flip spin)

3. for all neighbor sites k of i do the following:

    (a) if the demon on the bond (i, k) cannot absorb the change of energy associated with the inversion of spin i with respect to spin k, do the following:
        i. $s_k' = -s_k$ (spin k included in cluster);
        ii. write k in a list of addresses (called stack).

    (b) if the demon on the bond (i, k) can absorb the change of energy associated with the inversion of spin i with respect to spin k, do nothing.

4. read an address l from the stack;

5. execute the steps listed under 3, reading l instead of i, for the neighbor sites k of l;

6. erase the address l from the stack;

7. repeat steps 4-6 until the stack is empty.

A demon whose spectrum is not bounded from above will always be able to absorb energy in step 3. But if it is in its ground state, it is unable to give off any energy so that the neighbor spin s_k may have to be included in the cluster.

This formulation can easily be generalized to the case where antiferromagnetic couplings, or couplings of both signs, are present.

3. The Lattice Gas Cluster Algorithm

In order to describe lattice gases we use variables σ_k that take the value 0 or 1. When a particle occupies lattice site k, $\sigma_k = 1$; if this site is empty,

$\sigma_k = 0$. We restrict ourselves to the case of nearest-neighbor exclusion. We formulate this exclusion by terms $K_{\mathrm{nn}}\sigma_i\sigma_k$ in the Hamiltonian and considering the limit $K_{\mathrm{nn}} \to -\infty$. Further interactions between the particles are absent, but we include a magnetic-field-like term in order to control the particle density. The Hamiltonian is thus

$$-\mathcal{H}/kT = K_{\mathrm{nn}} \sum_{<i,j>} \sigma_i\sigma_j + H \sum_k \sigma_k \qquad (21)$$

The first sum is over all nearest-neighbor pairs. Periodic boundary conditions are imposed. The field H represents the chemical potential of the lattice-gas particles. This formulation of lattice-gas models is still quite similar to the Ising model; by a mapping of the variables $\sigma_i = (s_i + 1)/2$ on Ising variables $s_i = \pm 1$, we get an antiferromagnetic Ising model in a field. Both the nearest-neighbor coupling and the field are infinite in magnitude, but their effects on a spin surrounded by minus spins almost cancel: such a spin has a nonzero probability to be in the $+1$ state as well as to be in the -1 state. However, two neighboring plus-spins cost an infinite energy: this corrresponds with the nearest-neighbor exclusion in the lattice-gas representation.

In this work, we consider lattice gases on the simple-cubic and on the body-centered cubic lattice. Both lattices can be divided into two equivalent sublattices. This is an essential property that determines the character of the ordering transition. As long as the lattice gas is sufficiently dilute, the particles are evenly distributed over both sublattices. When the density increases, correlated regions grow where the particles tend to be on the same sublattice. At the critical point, such a region may grow to an infinite size, and at even higher densities, the majority of the particles "chooses" one of the sublattices: the sublattice symmetry of the system as a whole is spontaneously broken. The associated order parameter is the staggered density: the difference in occupation between the two sublattices. It assumes a nonzero value in the ordered phase and may be compared with the spontaneous magnetization of the Ising model.

The cluster simulation technique described above for the Ising model uses the global plus-minus symmetry of the Hamiltonian; this symmetry is one of the conditions that enables one to flip clusters as a whole. The proof of detailed balance uses the fact that this symmetry operation, namely a change of sign, is its own inverse. However, the plus-minus symmetry of the Ising model is absent in the lattice-gas: there is a nearest-neighbor exclusion of particles, but no such thing as a nearest-neighbor exclusion of vacancies. Thus, cluster steps that interchange particles and vacancies do, in general, not satisfy the condition of detailed balance.

It is fortunate that there exist other symmetry operations that meet this condition. In their study of hard-core gases in continuous space, Dress and Krauth [6] introduced a cluster method for such systems. This method uses spatial symmetry operations that move a lattice gas particle to a new location. In two dimensions, such a symmetry operation may, for instance, be a mirror reflection of a plane with respect to a line, or a rotation of π radians about a point. A particle (a hard disk in two dimensions) serves as the first member of the cluster and is subjected to the symmetry operation. If it arrives at a position that does not overlap the hard-core of any other particle, the cluster is finished. If it does, these other particle(s) is (are) included in the cluster. This means that they are, just as in the Ising case, entered into the stack memory, and subjected to the symmetry operation. This procedure is in analogy with the picture sketched above for Wolff cluster formation using demons. The energy change involved in moving a particle is either zero or infinite; the probability that the demon can absorb this energy change is 1 or 0 respectively. Thus, after choosing the symmetry operation and the first particle of a cluster, the cluster formation continues in a deterministic way.

Unfortunately, the percolation threshold of this cluster formation process was not found to coincide with the critical point [6]. On this basis one may expect that cluster algorithms for continuous hard-core gases suppress critical slowing down less effectively than those for the Ising model do.

However, the cluster simulation of the present lattice gas models allows certain differences with respect to that of Ref. [6] for continuous models. One may take advantage of the fact that the simple-cubic and body-centered cubic lattices consist of two sublattices. This can de done by taking symmetry operations that map the sites of one sublattice onto the other. As we shall see, this leads to clusters that are restricted to one sublattice. This is very similar to the situation in the ferromagnetic Ising model, where the cluster formation process accepts only spins of one sign, and does take place at the percolation threshold. Note that, as a consequence of the nature of the order parameters in the respective models, the sign of an Ising spin corresponds with the sublattice number of a lattice-gas particle. Hence, the possibility to construct an efficient algorithm looks more promising than in the case of continuous lattice gases. In the simple cubic lattice we restrict ourselves to $L \times L \times L$ lattices, with L even. Labeling the lattice sites by integers x, y, z one may choose the symmetry operation as a mirror inversion with respect to a randomly chosen lattice axis. For instance, for inversion with respect to the x direction, one may write $x' = x_0 - x \bmod L$, $y' = y$, $z' = z$, where x_0 is odd.

These operations do not interchange the sublattices of the body-centered cubic lattice. For this lattice we use coordinates x, y, z that are either all integers or all half odd integers. For $L \times L \times L$ lattices, we may use an inversion such as $x' = x_0 - x \bmod L$, $y' = y_0 - y \bmod L$, $z' = z_0 - z \bmod L$, where x_0, y_0, and z_0 all are half-odd integers.

Let us now follow the cluster formation in detail. The algorithm chooses an arbitrary particle to be the first member of the cluster. The symmetry operation moves this particle to a new position that is one of three distinct types:

1. an empty site whose neighbor sites are also empty;

2. an empty site of which one or more neighbor sites are occupied;

3. an occupied site.

In case 1, the cluster formation process stops after moving the first particle. The cluster consists of precisely 1 particle. In case 2, the cluster formation process continues: the neighbor particles are recorded in the stack memory and are included in the cluster. In case 3, let us say that a particle on site k is moved to site l that is already occupied. Therefore the latter particle is moved to site k; particles l and k are interchanged. Since the lattice-gas particles are identical, there is no net change of configuration and we assign a size 0 to this cluster. It is important to note that this situation can only occur right at the beginning of a new cluster. Any occurrence at a later stage would imply a violation of the nearest-neighbor exclusion. Thus each cluster is restricted to sites on one sublattice.

After completion of the cluster, one may proceed to form clusters using particles that were not yet included in a cluster, until all particles are member of a cluster. Then, we may decide for each cluster separately whether or not it will be subjected to the geometric symmetry operation. Random decisions lead to an algorithm similar to that of Swendsen and Wang [4]. On the other hand, we may alternatively define a Monte Carlo step as the formation of one cluster which is then moved according to the geometric symmetry operation. This leads to an algorithm analogous to the single-cluster method described by Wolff [5].

In spite of these analogies with the cluster simulation of the Ising model, there are important differences. In the first place, the procedure described above conserves the number of particles. This could be compared with conservation of energy in the Ising model: a microcanonical simulation. Another difference is that at high particle densities, the condition of irreducibility of the transition probability matrix is violated. Unfortunately, we have no exact information about the density where this violation first occurs. We chose a safe way out, namely to alternate one or more single-cluster steps by one Metropolis sweep, so that any particle configuration can be reached with a nonzero probability.

Although this procedure introduces critical energy fluctuations, which may be expected to slow down the rate of relaxation, we found that the statistical accuracies are approximately the same as for Wolff simulations of Ising models, using the same lattice size and the same number of clusters. Moreover, the averaging on the number of particles, i.e. the grand canonical ensemble in the lattice gas language, enables a more direct comparison with Ising simulations using the canonical ensemble.

4. Simulations and Results

The simulations of the simple-cubic gas used linear lattice sizes $L = 6, 8, 10, \cdots, 32, 48$ and 64. For the body-centered cubic model, even lattice sizes $L = 6, 8, 10, \cdots, 32, 36$, and 40 were used. For each system size, data were taken about one million times. These data were taken at intervals separated by one Metropolis sweep and a number of single-cluster steps, varying from one for small systems to 20 for the largest systems. The simulations took place at chemical potentials within a margin of a few times 10^{-3} of the respective critical points. These data include those used in Refs. [7, 8] as well as new simulations for larger system sizes.

The data sampled during the simulations include the order parameter, i.e., the staggered density. For the simple cubic model it is defined as

$$\rho_s = \frac{1}{N} \left\langle \sum_{x+y+z=\mathrm{odd}} \sigma_{x,y,z} - \sum_{x+y+z=\mathrm{even}} \sigma_{x,y,z} \right\rangle \tag{22}$$

where N is the number of sites in the lattice. On this basis we could also sample the dimensionless amplitude ratio

$$Q_L = \frac{\left\langle \rho_{s,L}^2 \right\rangle^2}{\left\langle \rho_{s,L}^4 \right\rangle} \tag{23}$$

at criticality. It is simply related to the Binder cumulant [13]. An index L has been appended in order to indicate the finite-size dependence. Its value in the limit $L \to \infty$ is a universal constant. Its Ising analogon has recently been accurately determined as $Q = 0.6233(4)$ [1], using periodic systems with cubic symmetry. This allows a first test of universality: if it is satisfied, the present lattice gases must have the same Q. In analogy with Ref. [1] we expect the following scaling behavior near the critical point:

$$Q_L = Q + a_1(H - H_c)L^{y_t} + a_2(H - H_c)^2 L^{2y_t} + a_3(H - H_c)^3 L^{3y_t} \quad (24)$$
$$+ b_1 L^{y_i} + b_2 L^{y_2}$$

where Q, a_i, b_i, c_i, p_0, q_i, r_1, s_0 and H_c are in principle unknown parameters. In the fits for Q, we have assumed the Ising temperature exponent $y_t = 1.587$, the irrelevant exponent $y_i = -0.83$ and $y_2 = d - 2y_h$. For further details, see Ref. [1]. The fitting procedure indicated that Eq. (24) describes the numerical data well for $L \geq 6$: χ^2 agrees well with the number of degrees of freedom. For both lattices we find values of Q close to what expected on the basis of the universality hypothesis (see Table 1). After fixing the value of this parameter to the $Q = 0.6233$ in the fit formula, we obtain the critical points somewhat more accurately (also in Table 1).

Table 1. Numerical results for the critical points and several universal constants obtained from simulations of the simple cubic and body-centered cubic lattice gases. The errors in the last decimal places are shown between parentheses, and correspond with one standard deviation.

quantity	simple cubic		b.c. cubic	
Q	0.622	(2)	0.621	(3)
H_c (Q fixed)	0.05449	(3)	-0.32822	(2)
y_h	2.482	(4)	2.479	(5)
y_t	1.61	(3)	1.60	(4)

Another quantity of interest is the staggered compressibility

$$\chi_L = N(\langle \rho_s^2 \rangle - \langle \rho_s \rangle^2) \quad (25)$$

which has been sampled as well. It is similar to the susceptibility in the Ising model; we thus expect the following scaling behavior:

$$\chi_L = c_0 + c_1(H - H_c) +$$
$$L^{2y_h - d}[a_0 + a_1(H - H_c)L^{y_t} + a_2(H - H_c)^2 L^{2y_t} + b_1 L^{y_i}] \quad (26)$$

We have fitted this expression to the data for χ_L. Results for the magnetic exponent y_h are included in Table 1. The estimated errors in these results include the uncertainties due to the errors in H_c.

Finally, we have sampled and analyzed the particle density. From its fluctuations we determined its derivative to H, a quantity that is analogous to the specific heat of the Ising model. The analysis is similar to that in Refs. [1, 7, 8] and leads to results for the temperature exponent y_t that are also displayed in Table 1.

These results agree well with Ising universality; in addition to the universal numbers given above we quote $y_h = 2.4815(15)$; see e.g. Refs. [1, 14] and references therein.

It seems thus reasonable to conclude that there is no numerical evidence that the present lattice gases violate Ising universality. The analyses indicating such a violation did not take into account corrections to scaling [2, 3]. We observe that, for the two models investigated, the finite-size correction-to-scaling amplitude b_1 (see above) assumes a value twice that for the simple-cubic Ising model. This magnitude is undoubtedly one of the reasons behind the large discrepancies reported [2, 3]. When we remove the term with amplitude b_1 from our fit formula, we obtain results that deviate strongly from the Ising universal ones. This is, however, accompanied by an unacceptable increase of the least-squares criterion.

Finally, we remark that the magnitude of the latter increase is linked to the statistical accuracy of our Monte Carlo data. The present scaling analysis, which includes two different finite-size corrections for Q, was only possible because an efficient cluster algorithm was available: thus we could compute data of a sufficient quality.

References.

[1] H.W.J. Blöte, E. Luijten and J.R. Heringa, J. Phys. A **28**, 6289 (1995).

[2] A. Yamagata, Physica A **222**, 119 (1995).

[3] A. Yamagata, Physica A **231**, 495 (1996).

[4] R.H. Swendsen and J.-S. Wang, Phys. Rev. Lett. **58**, 86 (1987).

[5] U. Wolff, Phys. Rev. Lett. **62**, 361 (1989).

[6] C. Dress and W. Krauth, J. Phys. A **28**, L597 (1995).

[7] J.R. Heringa and H.W.J. Blöte, Physica A **232**, 369 (1996).

[8] J.R. Heringa and H.W.J. Blöte, submitted Physica A (1997).

[9] R.G. Edwards and A.D. Sokal, Phys. Rev. D **38**, 2009 (1989).

[10] P.W. Kasteleyn and C.M. Fortuin, J. Phys. Soc. Jpn. Suppl. **26**, 11 (1969).

[11] M. Creutz, Phys. Rev. Lett. **69**, 1002 (1992).

[12] J.R. Heringa and H.W.J. Blöte, Phys. Rev. E **49**, 1827 (1994).

[13] K. Binder, Z. Phys. B **43**, 119 (1981).

[14] A.L. Talapov and H.W.J. Blöte, J. Phys. A **29**, 5727 (1996).

Computer Simulations of Fracture in Disordered Visco-elastic Systems

K. Kaski and P. Heino

Helsinki University of Technology, Laboratory of Computational Engineering, Miestentie 3, FIN-02150 Espoo, Finland

Abstract. Dynamics of fracture and its instabilities have been studied using two-dimensional visco-elastic models. Two models have been developed to describe disordered systems, in which the disorder appears either as topological disorder or as non-uniform mass distribution at mesoscopic length scale. The first model is based on a network of dissipative Born springs and the second model is based on finite element method with a similar dissipative force relaxation mechanism as in the first model. Results of computer simulations in a topologically disordered system show a very similar crack branching, branch bending and velocity oscillation behaviours as found in recent experiments. Also the conjectured scaling of branch bending has been tested favourably. In systems with non-uniform mass-distribution, crack curving and crack arrest were found to occur especially for strongly correlated disorder.

1. Introduction

Fracture phenomena in disordered materials have attracted a lot of interest due to their intrinsic technological importance. Examples of interesting and complex materials are certain polymers and fibrous compounds like paper. Typically their mechanical response show also a time dependent dissipative phenomenon, i.e. visco-elasticity. Thus from the physics point of view fracture in disordered materials involves processes on a wide range of length and time scales.

Over the last few years instability in dynamic fracture has become to the keen focus of experimental [1, 2, 3, 4], analytical [8, 9] and computer simulation [5, 6, 7] studies. This instability manifests itself such that when the velocity of the propagating crack reaches a critical value it starts to change in magnitude, or direction or both. The most spectacular experimental observation of instability is crack branching in a brittle amorphous material, studied first by Sharon *et al.* [1]. In this study the propagating main crack was found to branch to a set of two fairly symmetrically appearing daughter cracks, which propagated short distances and carried away some energy from the crack tip before stopping. The formation of these millimeter scale daughter cracks occurred periodically in time such that the velocity of the main crack showed oscillations. Although the instability in dynamic fracture has been known for some time, theoretical advances have been lacking since the pioneering work by Yoffe [10].

Springer Proceedings in Physics, Volume 83
Computer Simulation Studies in Condensed-Matter Physics X
Eds.: D. P. Landau, K.K. Mon, H. -B. Schüttler

For these complex phenomena at millimeter scale and for compounds like paper also with the basic length scale up to millimeters we have developed computer simulation models. These mesoscopic models are considered dynamic not only because of local elasticity in the system but also because we include local dissipation. The first model, we consider here is the Born-Maxwell (BM) model. It is a discrete network system of interacting mass sites, connected with Born-springs, that have two degrees of freedom, namely tensile and bending stiffness. The second model is based on discretizing the continuum to a mesh, which is solved with finite element method (FEM). In order to mimic plasticity or visco-elasticity we have included a Maxwell-type force-dependent dissipative mechanism to both of these models.

In what follows we shall introduce these models only briefly, since they have been discussed in detail elsewhere [11, 12]. This section is devided into the subsections of BM-model, FEM-model, crack formation and propagation, disorder, and stability and algorithm. In the third section computer simulation results of dynamic fracture for a homogeneous system, a topologically disordered system and a system with correlated disorder are discussed. Finally we draw conclusions.

2. Description of the Models

In order to make the comparison between the BM-model and FEM-model easier we consider them under the same footing by discretizing them similarly. Thus in both models the system under consideration is divided into mesoscale triangles. The corners of these triangles are considered as the initial locations of the mass sites in the Born-Maxwell model and nodes in the FEM-model. In the BM-model the two mass sites that belong to the same triangle are considered nearest neighbours and the sides of these triangles are the bonds describing interactions between the mass sites. In the FEM-model the triangles are called elements and their free energy is determined using linear elastic theory and linear interpolation within the element[1]. Previously the BM-model has also been studied in a square lattice by Rautiainen *et al.* [13].

When the system is subjected to a load, it deforms, which results in displacements of the mass sites or nodes in the BM-model or FEM-model, respectively. These displacements change the total free energy of the system, and since the velocities of the mass sites or nodes change, the time derivatives of the displacements determine the total kinetic energy of the system. Thus we may write the equations of motion for the mass sites or nodes [11, 12].

2.1 Born-Maxwell Model

In the Born-Maxwell model we consider two nearest neighbour mass sites, i and j. These mass sites do not correspond atoms or molecules in a specific crystaline structure, but they form a network in which each site is considered to represent

---

[1] With different mesh geometries other interpolation schemes have to be used.

a certain piece of material in its immediate vicinity at the mesoscopic length scale, see Fig. 1a. The mass sites are connected with a Born spring of tensile and bending stiffness, as described with the following potential energy

$$H_{ij} = \frac{\alpha}{2}[(\vec{u}_{e,i} - \vec{u}_{e,j}) \cdot \vec{d}_{\parallel,ij}]^2 + \frac{\beta}{2}[(\vec{u}_{e,i} - \vec{u}_{e,j}) \cdot \vec{d}_{\perp,ij}]^2, \tag{1}$$

where $\vec{u}_{e,i}$ is the elastic displacement of site i, $\vec{d}_{\parallel,ij}$ is the unit vector from site i to site j corresponding to the undeformed system, $\vec{d}_{\perp,ij}$ is the unit vector perpendicular to $\vec{d}_{\parallel,ij}$, and α and β are the parameters of tensile and bending stiffness of a bond, respectively. As a representation of plasticity or visco-elasticity the local elastic displacements are allowed to relax according to the following equation

$$\frac{\partial \vec{u}_{e,i}}{\partial t} = \frac{\partial \vec{u}_i}{\partial t} - \frac{1}{\tau}\vec{u}_{e,i}, \tag{2}$$

where $\vec{u}_i$ is the total displacement (i.e. the sum of elastic and plastic displacements) of mass site i and τ is a phenomenological relaxation coefficient. As the mass site is considered to represent a certain piece of material, we may calculate its mass by integrating the density over this piece. Since the forces acting on a mass site are known, the Newton's equations of motion for the site can be written.

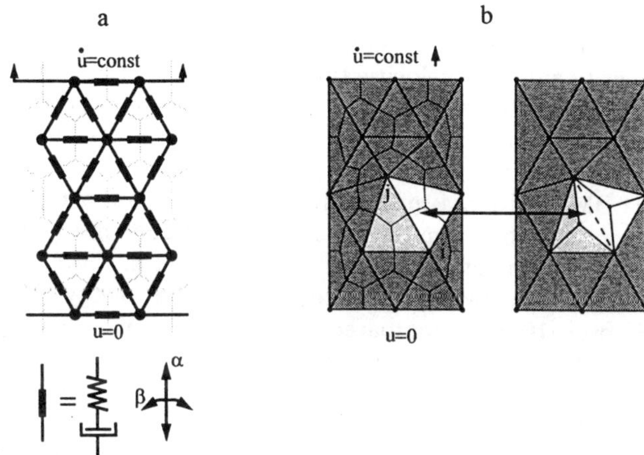

FIG. 1. Models of fracture. (a) Born-Maxwell model, (b) Finite element model. The shaded lines represent the piece of material that is included to a mass site or node. Nearest neighbour mass sites are coupled with a dissipative spring or dissipative linear elastic element. When a crack propagates across a side of a triangle, the interaction between the ends of this side, i.e. mass sites or nodes ceases.

2.2 Finite Element Model

In the finite element model the displacement field is assumed to have a known functional form of the displacements of the nodes. For this we use simply a linear function of the location (x, y) in each element:

$$\vec{u}(x,y) = \vec{u}_0 + x\vec{u}_x + y\vec{u}_y, \tag{3}$$

where the three unknown vectors are functions of the displacements of the nodes of a triangular element and vary from element to element. This assumption is made concerning both the elastic and total displacements giving us the strain field in each element. The material in each element is assumed linear, homogeneous and isotropic such that the Hooke's law describes the stress field within each element. Thus the free energy of each element can be obtained by integrating the free energy density, which is given by the sum of componentwise products of the stress and strain tensors [14]. This integration over the element i results in the following free energy

$$H^i = \frac{1}{2}(\vec{\delta}_e^{\,i})^T K^i \vec{\delta}_e^{\,i}, \tag{4}$$

where K^i is the stiffness matrix of the element i and $\vec{\delta}_e^{\,i}$ is a vector of the elastic displacements of the nodes of this element. Again the displacements are assumed to relax according to Eq. (2). The total force acting on a node i is obtained by summing the contribution of all elements containing the node i.

On the other hand the kinetic energy of an element can be obtained from the known mass density and local velocity, that is determined as the time derivative of Eq. (3). For computational efficiency we have used a lumped mass approximation [15], in which $\frac{1}{3}$ of the mass of each element round the node is assigned to the node as depicted in Fig. 1b.

In both the BM- and FEM-model the force acting on a mass site or node depends on the location of the mass site or node and all its nearest neighbours. In the BM-model the potential is a pair potential and in the FEM-model all three nearest neighbours are used to determine the free energy of an element. Ashurst and Hoover [16] have compared the free energy of a FEM-model with that of a regular spring model in a triangular lattice and found them to be identical. In addition it is noted that the mass of a site or a node is obtained similarly in both models.

2.3 Crack Formation and Crack Propagation

We now consider how a mesoscopic crack forms in the BM- or FEM-model. In both cases we assume that when the distance between two nearest neighbour mass sites or nodes of an element exceeds a given threshold a small rupture occurs. As the equations of motion for the mass sites or nodes are solved until a macroscopic breakdown, this rupture criterion determines also the conditions for crack propagation. When such a crack forms, the interaction between the mass sites or nodes i and j ceases and the crack propagates across the line joining them, which we call a bond. In this study we have used 1% and 2% as the values of rupture thresholds for the BM- and FEM-model, respectively.

For the reason that many real materials - especially ductile - have larger macroscopic fracture strains, some tests were also done with the rupture threshold of up to 5%. As a result of this increase in the threshold value we found that in crack branching the tendency to form daughter cracks and the frequency of crack velocity changes were reduced [11]. For comparison it is noted that crack branching has not been observed in ductile materials. Therefore, we expect our models with small rupture thresholds to describe the salient physical features of dynamic fracture in brittle materials. In addition they are computationally more ameanable.

2.4 Disorder

Solid materials show various types of disorder, which can be visualized differently at different length scales. In order to characterize disorder it is natural to start from the length scale of the basic constituents of the material. In most solids the basic length scale is truely microscopic, at which disorder is often topological. On the other in case fibrous composites such as paper the basic length scale ranges from tens of micrometers to few millimeters, i.e. it is mesoscopic. In case of paper disorder is caused by the fibers being spatially and orientationally distributed. At larger length scales than the basic one disorder in solids can be characterized with mass density and its variations, which are measurable with various methods.

A very interesting example of topologically disordered material is brittle amorphous polymethyl metacrylate (PMMA), which Sharon *et al.* [1] used in their experimental studies of dynamic fracture instability. Since they observed crack propagation and branching at about millimeter scale, we have introduced mesoscale disorder to our computer simulation models. In order to characterize topogical disorder, we choose randomly the initial locations for the mass sites or nodes of the Born-Maxwell or finite element model, respectively. Then the nearest-neighbour-connected network of mass sites or nodes and corresponding triangles are found with the help of Voronoi polygons by performing the Delaunay's triangulation [19].

As mentioned above in paper there is a random spatial and orientational distribution of fibers, which manifest a strong disorder in the mass density. This can be determined by the beta-radiogram measurements at the millimeter scale resolution. In addition it is quite commonly known that in paper the elastic modulus depends linearly on the mass density. Thus the basic parameters of the BM- and FEM-model can be determined by first calculating the mass of a site or node in the manner described above and accordingly varying the elastic parameters. It is also known that disorder in paper shows correlations which appear as strong elliptic areas called flocs and weak spots. This can be modelled by placing elliptical unit mass flocs of given size at random locations on a two dimensional model sheet [17]. Then the mass distribution is the result of overlapping flocs. Here we allowed anisotropy in the mass distribution by choosing the principal axis of ellipses to be either in X- or Y-direction. Recently Salminen *et al.* [18] found that in paper the disorder should be taken into account in the rupture threshold rather than in the elastic parameters. Thus in our simulations for the purpose of mimicing paper we have introduced disorder as correlated mass density distribution and made the rupture threshold linearly dependent on the mass density by setting the mean threshold to be 1%.

2.5 Issues of Stability and Algorithm

Intuitively both our models are stable, since Eq. (2) implies that the energy of a bond or an element in the BM- or FEM-model, respectively, tends to 0 as time goes to ∞. However, we have tested the stability also analytically by solving the equation of motion for a system, in which a spring and a dashpot are in series, i.e. for a Maxwell-element. The results show, that the solution is stable and produces the harmonic oscillator solution when $\tau \to \infty$.

In order to solve trajectories of mass sites in the BM-model we have used the original Verlet's algorithm [20], known to be sufficiently stable numerically. In the case of FEM-model we have used the Adams-Moulton predictor-corrector scheme [21], which yields a good accuracy ($\propto O^5$) and stability. Since the determination of crack formation requires knowledge of nodal displacements, the corrector is only approximative when a small crack forms, which in the simulations takes place after updating the nodal displacements. However, the results did not show change when the time step of the algorithm was decreased, thus indicating a stable solution. In both algorithms (Verlet and Adams-Moulton) we used an adaptive time step to improve the accuracy of the solution near the forming crack. Next we describe how this is done. Since the strain of each bond is calculated each time step, it is easy to find the maximum strain of yet unruptured bond. When this strain exceeds 97.5% of the rupture threshold, the time step of trajectory calculation is reduced by a factor of 20 and kept small until the bond breaks and for 300 time steps thereafter. However, it is noted that the change in time step needs to be compensated by modifying the integration scheme slightly.

3. Results and Discussion

We have studied three kinds of systems. In the first case a small cut was introduced to the left free boundary of a homogeneous system to serve as a seed for crack propagation. Secondly we have studied systems with topological disorder. For both of these cases we used the Born-Maxwell and the finite element model. In the third case disorder was introduced to the density and rupture threshold either in a correlated or uncorrelated manner. These situations have been studied using only the BM-model.

For both models the loading is introduced in two stages by moving the upper-most row of mass sites or nodes while the lowermost row of mass sites or nodes stay fixed. First for the time interval $0 \leq t \leq T_r$ the displacement rate of the uppermost row of mass sites or nodes increases gradually as $\frac{1}{2}v_s[1 - cos(\pi \frac{t}{T_r})]$, and then for $t > T_r$ the rate is kept constant, v_s. For the constant strain rate we use $v_s/l = 10^{-4}$, where l is the height of the system.

3.1 Homogeneous System with an Initial Crack

In Fig. 2(a)-(b) we show fracture paths simulated with the BM-model for the purely elastic and dissipative systems. They resemble the experimentally ob-served fracture paths [1]. First of all there is a main crack which appears straight and leads eventually to the macroscopic breakdown of the system. The most striking feature, however, is the appearance of several daughter cracks more or less symmetrically on both sides of the propagating main crack. These daugh-ter cracks appear periodically and they proceed short distances before stopping. Although not shown here the inclusion of disorder to the system makes the ap-pearance of daughter cracks less periodic and the fracture pattern more irregular. It has also been found that dissipation damps the daughter crack propagation, in effect making them shorter.

In Fig. 2(c) we show the simulated fracture path obtained with the FEM-model. One can see quite clearly a *mirror-mist-hackle* behaviour [2], where the terms mirror, mist and hackle refer to the type of surface left behind by the propagating crack. First, when the crack propagates straight, the surface it creates is mirror-like, but as the crack speed increases the crack surface becomes increasingly rough. In the FEM-model the crack branching does not appear as strongly as in the BM-model.

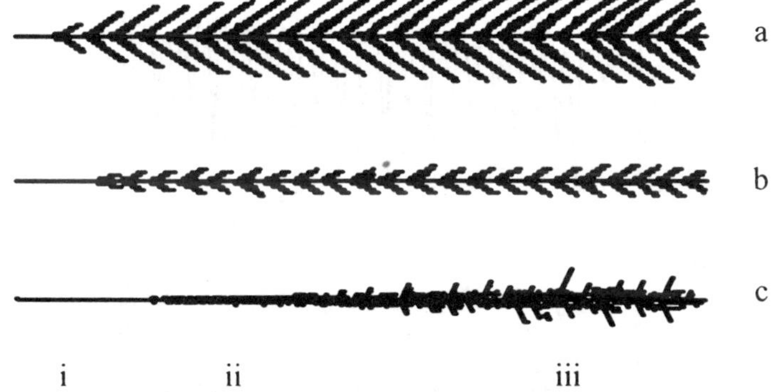

FIG. 2. Fracture paths obtained by using the Born-Maxwell model for (a) $\tau = \infty$ (purely elastic system) and (b) $\tau = 25$ (dissipative system). The main crack is seen to produce several daughter cracks. Fracture path obtained by using the finite element model (c) contains the mirror (i), mist (ii) and hackle (iii) regions.

73

Earlier studies with the Born-Maxwell model using a square mesh and loading it parallel to the bonds, did not reveal crack branching [13]. On the other hand when loading is along the diagonal of the square mesh, crack bifurcations have been observed. This directional dependence can be understood because a crack must propagate perpendicular to the bonds. However, crack bifurcations can be observed in a square mesh even when loading is parallel to the bonds, as shown by Åström *et al.* [22]. In their simulations an initial crack was introduced to a previously strained system.

In Fig. 3 we show the velocity of the main crack (a) in X-direction and the boundary velocity (b) of the moving main crack tip in Y-direction as a function of the crack width. After an initial rapid increase in these velocities - while the crack appears straight - an oscillating behaviour sets in at certain threshold speeds, which coincides with the emergence of periodic crack branching. These velocities decrease every time a pair of daughter cracks forms and increase again until the formation of a new pair of daughter cracks. Though not shown here the oscillatory behaviour are seen also as functions of time in agreement with experimental observation of the crack velocity, see Sharon *et al.* [1]. However, for the BM-model in the oscillatory regime we do not find any increase in the mean crack speed, while in our FEM simulations the crack is found to accelerate slowly in agreement with Sharon *et al.* The local potential energy (c) and kinetic energy (d) densities near the crack tip are also found to oscillate with the same frequency as the crack velocity. The local kinetic energy density at the crack tip comes from the moving boundaries and the potential energy is concentrated just ahead the crack tip. On the other hand the local potential energy at the tip of the main crack is found to decrease shortly after a pair of daughter cracks formed. An interesting observation is that the local kinetic energy has

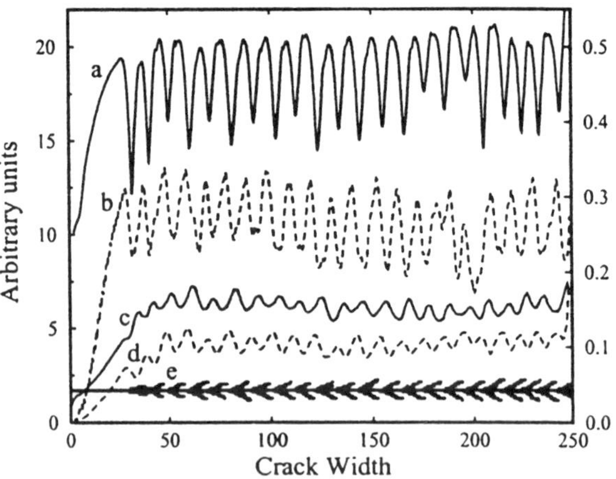

FIG. 3. (a) Crack velocity in the Born-Maxwell model, (b) the velocity of a mass site of crack boundary near the crack tip, local (c) potential and (d) kinetic energies near the crack tip. (e) Fracture path.

74

its maximum when this pair forms and peaks in the potential energy follow the maxima of the kinetic energy with a phase shift. This shows a conversion of kinetic energy of the moving crack boundaries to potential energy in the region ahead the main crack tip.

3.2 Topological Disorder with an Initial Crack

In case of topological disorder we chose the initial locations of mass sites at random and set the elastic parameter values inversely proportional to the initial length of the bond. The elastic parameters determine the macroscopic features of the system, i.e. the Poisson ratio and the Young's modulus. These properties were solved by applying the strain of 10% to the system and solving the stable configuration.

The dependence of the Poisson ratio on the ratio $r = log_e(\alpha/\beta)$ is depicted in Fig. 4. Also shown is a fit to an S-shaped sigmoid function of the form $y = a_1 + b_1/(1 + exp(c_1(r - d_1)))$. In a topologically disordered system the Young's modulus should be related to the average values of stiffnesses α and β and goes to zero as both of them go to 0. Thus we were led to try an estimate for the Young's modulus of the form $E = a_2\alpha + b_2\beta + c_2\alpha^{d_2}\beta^{1-d_2}$. In Fig. 4 the Young's modulus of the system is plotted versus the estimate. Now having got the values for the fitting parameters a_2, b_2, c_2 and d_2 we can use Fig. 4 to choose the average values of α and β, consistent with the Young's modulus and Poisson ratio.

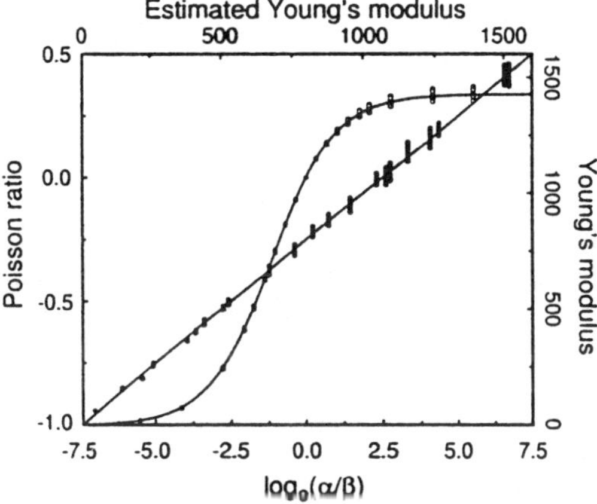

FIG. 4. The Poisson ratio (open circles) of the system as a function of the ratio between the tensile and bending stiffness, $log_e(\alpha/\beta)$, and a fit to an S-shaped sigmoid function. Young's modulus (full circles) as a function of the estimated Young's modulus (shown on the top horizontal axis). Since these points fall on a straight line, average stiffness values can be consistently related to the Young's modulus and Poisson ratio of the system, cf. text.

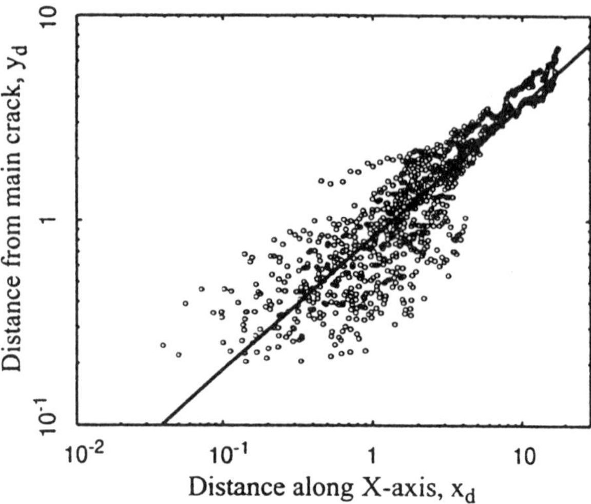

FIG. 5. The distance of daughter crack away from the main crack vs. the distance of daughter along X-axis and a power-law fit to this data. Long daughter cracks show a branching angle of about 15°.

Contrary to a triangular lattice, a random network of sites does not have prescribed directions for easy crack porpagation. In a triangular lattice the branches have to propagate perpendicularly to the bonds which prevents daughter cracks from curving. However, in random networks the daughter cracks are seen to curve during propagation such that the branching angle depends on the length of the daughter crack. This is illustrated in Fig. 5, where we show the distance of a daughter crack from the main crack as a function of the length of the daughter crack along the X-axis. The data obeys roughly a power-law: $y_d = a x_d^b$. For small x_d the distance from the main crack is not of functional form, since these small branches are probably responsible for the mist-behavior of the fracture surface. For larger x_d the given functional form is clearly seen and these large branches are seen as the hackle-behavior. Uncertainties in Fig. 5 arrise since it is a quite subjective issue to determine, which cut bonds belong to the mist region and which are classified as branches. Here the bonds not on the main crack are treated as branches, and for this reason the very short branches complicate the determination of curving behavior. The power law exponent b was found to be 0.65 ± 0.02. Results obtained with the FEM-model are very close to those obtained with the BM-model [12] and both of them are in good agreement with the experimental value 0.7 [1]. In addition we observed that the longest daughter cracks appear with an angle of about 15°, which is in excellent agreement with experiments in brittle amorphous materials [23].

From Fig. 3 it is evident that in a triangular network crack speed oscillation is observed. However, it is not obvious that one should observe crack speed oscillations in a topologically disordered system. Nevertheless this is the case. In Fig. 6c we show the behaviour of a typical crack speed obtained from a topo-

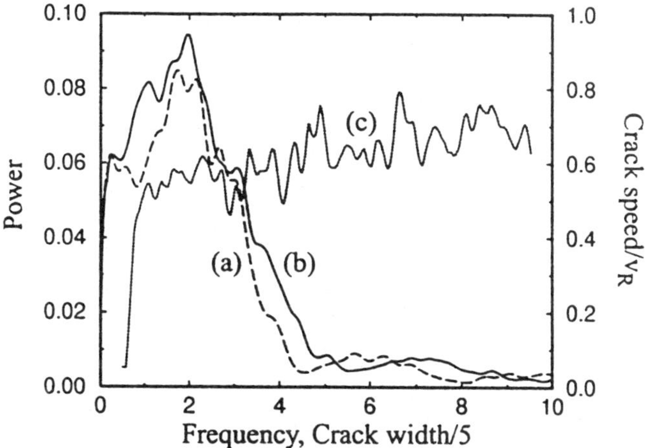

FIG. 6. Spectra of $(v_c - v_{c,mean})$ as averages over five dissipative ($\tau = 100$) (a) and as averages over eight non-dissipative FEM systems (b). Here v_c is the crack speed, $v_{c,mean}$ its short time average and v_R is the Raleigh speed. The spectra indicate that the crack speed oscillates rather than fluctuates randomly. This is contrary to what one might have expected from crack propagation in a random network. The mean crack speed (c) is found to increase slightly in the course of crack propagation.

logically disordered FEM simulations. Here the crack speed seems to increase on the long-time average, but on the short time average also decreases in speed are seen. In order to study whether these changes are periodic or not, we calculated the spectra of crack speed minus its short time average for several systems. These results are shown in Fig. 6 for dissipative and non-dissipative systems. In both cases the spectrum with a clear but broad peak indicates, that the changes in speed are periodic.

3.3 Correlated Disorder without Initial Crack

In relation to how crack propagation could happen in paper, that has correlated mass distributions and distributions of local rupture threshold, we show simulation results of fracture paths in the BM-model. As can be seen in Fig. 7 the crack branches tend to be shorter for a strongly disordered system than for a weakly disordered system. In Fig. 7a it is evident how the initial location for crack formation can be determined. The location for initial crack formation must lie between daughter cracks that have propagated in opposite directions, showing less than 90° angle with the main crack. In addition it is seen that in systems with strongly correlated disorder there are areas that arrest the daughter crack propagation (see Fig. 7b) or cause the main crack to change its direction (see Fig. 7c).

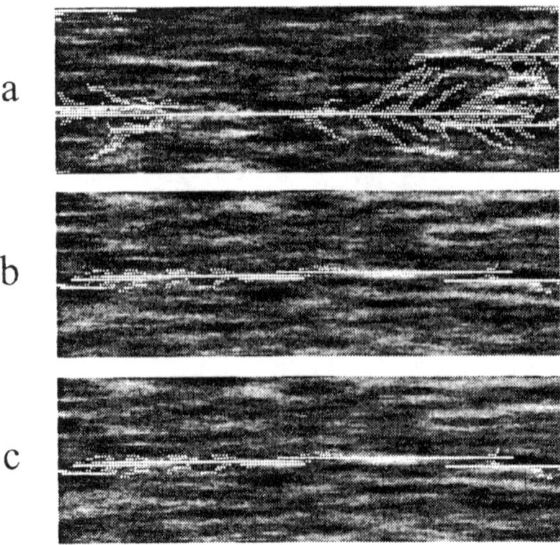

FIG. 7. Fracture paths (black dots) in the Born-Maxwell model with distributions of mass density and rupture threshold. Dark areas correspond large mass densities and thus large rupture threshold. Mass density distributions are generated using unit mass elliptic flocs with radius 5 in X-direction and radius 1 in Y-direction using either 5% mass variation (a), or 20% mass variation (two samples b and c).

In systems with uncorrelated disorder a crack propagates differently since the systems do not include strong areas, which might effectively arrest crack propagation. Therefore, the daughter cracks are longer in these systems than in systems with correlated disorder. On the other hand single exceptionally weak bonds can serve as seeds for initial crack formation. However, with uncorrelated disorder the system is not on the average weak near bonds, that can prevent small cracks from propagating. Therefore, systems with uncorrelated disorder are on the average stronger (measured in terms of maximum load) than systems with correlated disorder.

4. Conclusion

We have studied dynamic fracture of disordered visco-elastic systems using two different models and two types of disorder. In both of the models (Born-Maxwell and finite element model) crack branching and speed oscillation seem to rise from the dynamics of the system. Both these phenomena are seen in ordered and disordered, elastic and visco-elastic systems. In addition visco-elasticity is seen as a mechanism for cracks arrest, with the effect of reducing the length of daughter cracks. Futhermore, in topologically disordered systems like amorphous materials crack curving and branch bending behaviour were found. The branch bending was observed to obey the same scaling behaviour as conjectured by Sharon *et al.*. In systems with strongly correlated disorder crack arrest was found to occur and cause the daughter cracks being short. In this case also the crack curving behaviour was present.

Acknowledgments

We would like to thank Prof. J. Fineberg for very useful discussions and suggestions during the course of this study. This work is in part supported by the Academy of Finland through the MATRA programme.

References

[1] E. Sharon, S.P. Gross and J. Fineberg, Phys. Rev. Lett. **74**, 5096 (1995).

[2] J. Fineberg, S.P. Gross, M. Marder, and H.L. Swinney, Phys. Rev. B **45**, 5146 (1992) and Phys. Rev. Lett. **67**, 457 (1991).

[3] S.P. Gross, J. Fineberg, M. Marder, W.D. McCormick and H.L. Swinney, Phys. Rev. Lett. **71**, 3162 (1993).

[4] A. Yuse and M. Sano, Nature (London) **362**, 329 (1993). Cf. M. Marder, Nature (London) **362**, 295 (1993).

[5] F.F. Abraham, D. Brodbrck, R.A. Rafey, and W.E. Rudge, Phys. Rev. Lett. **73**, 272 (1994).

[6] M. Marder, and X. Liu, Phys. Rev. Lett. **71**, 2417 (1993).

[7] H. Furukawa, Progr. Theor. Phys. **90**, 949 (1993)

[8] K. Runde, Phys. Rev. E **49**, 2597 (1994).

[9] J.S. Langer, Phys. Rev. Lett. **70**, 3592 (1993).

[10] E.H. Yoffe, Philos. Mag. **42**, 739 (1951).

[11] P. Heino and K. Kaski, Phys. Rev. B **54** 6150 (1996)

[12] P. Heino and K. Kaski, *Mesoscopic Maxwell-dissipative Finite Element Model for Crack Propagation*, Accepted for publication in Int.J.Mod.Phys.C (1997).

[13] T.T. Rautiainen, M.J. Alava and K. Kaski, Phys. Rev. E **51** R2727 (1995).

[14] L.D. Landau and E.M. Lifshitz, *Theory of Elasticity*, 3rd revised english ed., Pergamon Press (1986).

[15] O.C. Zienkiewicz and R.L. Taylor, *The Finite Element Method*, Fourth Edition, Vol 1 and 2, McGraw-Hill Book Company (1994).

[16] W.T. Ashurst and W.G. Hoover, Phys. Rev. B, 14 1465 (1976).

[17] M. Korteoja, A. Lukkarinen, K. Niskanen, and K. Kaski, Proc. Int. paper physics conference, Niagara (1995) (cf. also the work by R.R. Farnood and C.T.J. Dodson).

[18] L. Salminen, M. Korteoja, M. Alava and K. Niskanen, *Paperin lujuuden vaihtelusta*, KCL Paper Science Centre communications 89. (In finnish, to apper in english.)

[19] A.V. Potapov, M.A. Hopkins, and C.S. Campbell, Int. J. Mod. Phys. C **6** 3 371-425 (1995).

[20] M.P. Allen and D.J. Tildesley, *Computer simulation of liquids*, Oxford university press (1990).

[21] G. Sewell, *The Numerical Solution of Ordinary and Partial Differential Equations*, Academic Press Inc (1988).

[22] J. Åström, M. Kellomäki and J. Timonen, *Crack bifurcations in a Disorderless System*, preprint (1996).

[23] E. Sharon and J. Fineberg, *The Micro-Branching Instability and the Dynamic Fracture of Brittle Materials*, preprint (1996).

Part II
Quantum Systems

Quantum Phase Transitions in Random Magnets

R.N. Bhatt

Department of Electrical Engineering, Princeton University, Princeton,
NJ 08544-5263, USA

Abstract. This article presents a summary of recent studies of quantum phase transitions in random magnets using Monte Carlo methods. The quantum Ising model with random interactions in the presence of a transverse magnetic field is used as a primary example of a quantum phase transition driven by the strength of the applied field. After a brief discussion of experimental results and analytical results in one-dimension, results of numerical Monte Carlo simulations on the quantum critical behavior of the nearest neighbor Ising spin glass in two and three dimensions, as well as the random Ising ferromagnet in two dimensions are described in detail. Rare fluctuations of the quenched random bond distribution are found to have an important effect on the response of the system in the vicinity of the quantum phase transition, and are found to lead to divergent response in the paramagnetic phase itself. Such nonuniversal behavior is exhibited by many quantum models with quenched randomness; their effect on the description of the quantum phase transition is briefly discussed.

1 Introduction

Continuous (i.e second order) thermal phase transitions have been studied intensely over the past forty years since Landau's pioneering work and the subsequent development of the Renormalization Group method. Many seminal ideas have sprung from this effort including the concepts of broken symmetry[1], scaling and universality[2]. Transitions in seemingly unrelated systems have been brought together under the rubrik of universality classes, and a systematic scheme of classification by dimensionality of the system, and of the order parameter characterizing the broken symmetry has been put into place.

Quantum phase transitions, on the other hand, while also having a long history, have had a more convoluted and meandering path to their understanding. By quantum phase transitions, we are referring to transitions that result from changing of some control parameter (such as a coupling strength, or an externally applied field) at absolute zero ($T = 0$). Thus we are dealing in this case not with thermal fluctuations, but instead with quantum fluctuations, and while there is some similarity, there are also fundamental differences, because of which the latter offer, in the author's view, a much richer array of possibilities. (Of course, one can turn that around, and say that they exhibit, in some sense, less universal behavior !)

The most celebrated example of a quantum phase transition is the metal-insulator transition in electronic systems, on which Mott[3] wrote his classic paper in 1949, and on which experimental work goes back even further[4]. This was not looked upon from the point of view of scaling and universality till the late 1970s, when the work of Wegner[5] and of the gang of four[6] brought out the essential similarity to conventional phase transitions. Over the past decade, however, research on quantum phase transitions has intensified[7], and a greater awareness and understanding has developed of the similarities and differences with thermal phase transitions. Numerous models have been investigated applicable to *e.g.*, superfluid-insulator transitions, quantum Hall transitions, magnetic transitions, etc., and a good understanding has been obtained in many cases at a qualitative level, with quite a few quantitiative details as well. Ironically, the original fermionic metal-insulator transition problem in the presence of electron-electron interactions[8] has remained theoretically unsolved to this day !

In this short article, we review some of these points by taking as our example a simple magnetic model with randomness. While the randomness is a complicating feature, when compared with uniform systems that possess various additional symmetries (translational as well as rotational), it also adds to richness of the problem, when compared with the corresponding counterpart in thermal phase transitions. We consider in particular the Ising random magnet, either in the spin glass regime, or in the ferromagnet regime, in the presence of an applied magnetic field, which is transverse to the Ising axis.

In the next section, we describe briefly the model, and various motivations for studying it, including an experimental system in three dimensions to which the model potentially applies[9]. We then summarize some of the results in one dimension where analytic approaches exist[10, 11, 12] , concentrating on the unusual aspects of the one-dimensional model. The following section 3 is devoted to a description of the critical behavior of the Ising spin glass in a transverse field in two and three dimensions, obtained using Monte Carlo methods coupled with a novel finite size scaling approach[13, 14, 15, 16]. The Suzuki-Trotter[17] formalism is used to map the d-dimensional quantum problem into an anisotropic $(d+1)$ dimensional classical model. A new shape-scaling method is described which allows one to determine, without bias, the dynamical exponent characterizing the scaling of the space and imaginary time directions of the anisotropic classical model. This, in turn, enables one to perform an anisotropic, but one parameter, size scaling to determine various critical exponents characterizing the quantum phase transition.

In section 4, we concentrate on the effect of rare fluctuations (known as Griffiths singularities) in the paramagnetic phase, because of which various linear (or nonlinear) susceptibilities diverge in the paramagnetic phase *before the transition to the ordered phase*. This behavior, which is quite common in many quantum phase transitions of systems with quenched randomness, is shown to be not confined to one-dimensional models, though the magnitude decreases with increasing dimensionality. This is followed by a brief discussion of recent results on the random quantum Ising *ferromagnet* in a transverse field in two dimensions[18, 19] in section 5. Finally, in the concluding section, we summarize the main results and put the model in the context of other random magnetic models which exhibit quantum phase transitions.

2   The Ising Model in a Transverse Field

The quantum Ising model in a transverse field is described by the Hamiltonian:

$$H = -\sum_{\langle ij \rangle} J_{ij}\sigma_i^z\sigma_j^z - \Gamma\sum_i \sigma_i^x, \tag{1}$$

where σ_i^α are Pauli spin matrices on a d-dimensional lattice, and the interactions J_{ij} (of the Ising form coupling the z-components of the spins) are quenched random variables taken from a probability distribution $P(J)$. The second term in the Hamiltonian represents the coupling to a transverse applied magnetic field along the x-direction. (In some studies, the applied fields are also taken to be random; this does not alter the model in a fundamental way, and we shall restrict ourselves to the case of uniform field in this article). We further restrict ourselves to the case where the couplings J_{ij} are of the nearest neighbor type on a simple hypercubic lattice in d-dimensions with d = 1, 2 or 3. For a discussion of the mean-field limit represented by the infinte range model, on which most of the early work was concentrated, the reader is referred to the original literature[20].

The phase diagram of the Hamiltonian of Equation (1) in the transverse-field-temperature $(\Gamma - T)$ plane is shown in Figure 1. At $T = 0$, the model in all dimensions is in an ordered phase (spin-glass or ferromagnetic, depending on the distribution $P(J)$) at small transverse fields, Γ, and has a transition to the paramagnetic state at large Γ. The ordered phase extends to nonzero temperature if the system is above the lower critical dimensionality, d_l, of the classical model represented by the first term in Equation (1). This is shown in the lower part of Figure 1, while the upper part corresponds to the phase diagram for dimensions below d_l. Thus, the upper part applies to the ferromagnetic case for $d = 1$, and to the spin glass case for $d = 1$ or 2. In this article, we concentrate on the $T = 0$ quantum phase transition; the transition for nonzero T is expected to be asymptotically governed by the classical model, though the asymptotic region shrinks to zero as $T \to 0$.

For the case of one dimension $(d = 1)$, an equivalent model, the two dimensional classical Ising model with disordered bonds that are perfectly correlated in one direction but independent in the other, was first studied analytically by McCoy and Wu[10]. More recently, the quantum model was studied by Shankar and Murthy[11], and in greater detail by Fisher[12], who

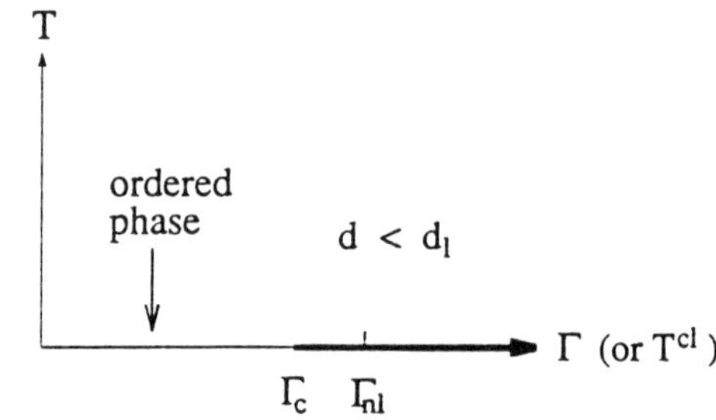

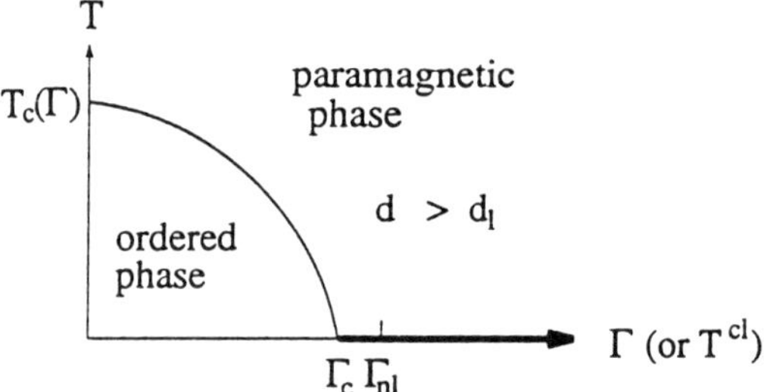

Fig. 1. Phase diagram of a quantum Ising spin glass in a transverse field. Top (bottom) figure corresponds to dimensions below (above) the lower critical dimension of the classical model (from reference 16).

made a number of interesting predictions that have subsequently been confirmed and extended by Young and Rieger using numerical approaches[21]. We describe some of the salient features in the next paragraph.

In one dimension, one may perform a gauge transformation which makes all the couplings J_{ij} and fields Γ_i (taken to be random for the moment) to be positive. Consequently there is no difference between the ferromagnet and the spin glass problem - in effect, there is no frustration corresponding to the spin glass in one dimension with nearest neighbor bonds. This is of course not true in higher dimensions, and correspondingly, one cannot simply "gauge away" the sign of the bonds and fields. However, because the interaction is of the Ising form, involving only the z-component of the spin, whereas the field is a in a transverse direction, one may still choose to keep the fields positive by suitable redefinitions of the spin variables. Nevertheless, the ferromagnet and the spin glass in transverse fields are two distinct problems with different universal exponents, as we shall see below.

An experimental system which is thought to be described by the transverse field Ising Hamiltonian in three dimensions, is the dipolar diluted antiferromagnet, $LiHo_xY_{1-x}F_4$, in which the magnetic ions, Ho occupy randomly the sites occupied by Y in the parent pure compound $LiYF_4$, and interact via dipolar interactions. The magnetic ion has a doubly degenerate ground

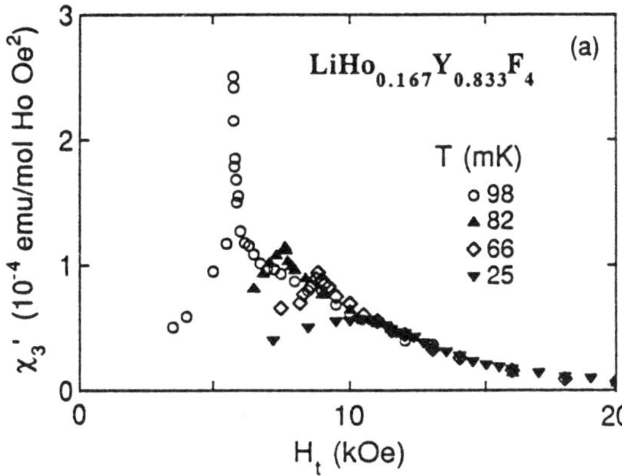

Fig. 2. Plot of the nonlinear susceptibility of $LiHo_{.167}Y_{.833}F_4$ as a function of transverse field strength $H_t = \Gamma$ for various temperatures, T. Note the decrease in the singular behavior as $T \to 0$. From Wu *et al.* [9].

state which is split by application of a field along the c axis of the tetragonal crystal. Experiments on the diluted compound with $x = 0.167$, show evidence of a spin glass transition at very low temperature which gets reduced as a transverse field is applied. By studying the field- dependence of the nonlinear susceptibility at various fixed temperatures (starting with high field, and approaching the transition from the paramagnetic phase), Wu *et al.*[9] found that the divergence of the nonlinear susceptibility became weaker, the lower the temperature (see Figure 2). They concluded that their results were consistent with there not being *any* divergence of the nonlinear susceptibility (the hallmark of a spin glass transition) in the $T = 0$ limit (i.e for the quantum phase transition). Both this result, and the linear phase boundary in the magnetic-field-temperature ($\Gamma - T$) plane (see Figure 3) are in contrast with

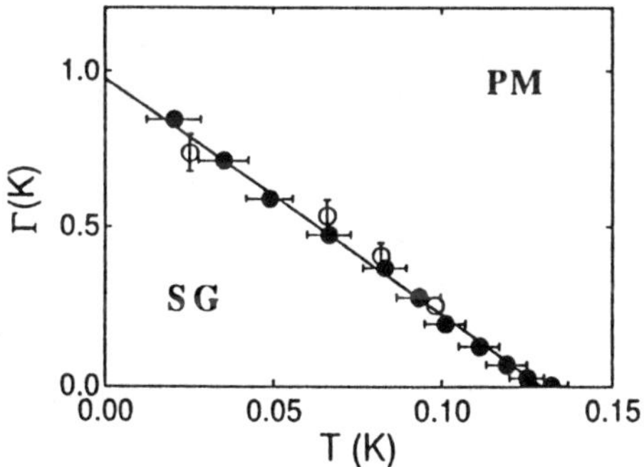

Fig. 3. Experimental phase diagram of $LiHo_{.167}Y_{.833}F_4$ in the $\Gamma - T$ plane. From Wu *et al.* [9].

the results on the infinite range model[20]. However, the differences with the one-dimensional model are even more striking, as we discuss below.

Among the interesting results that emerge from the one-dimensional analytical studies are the following: (i) the model undergoes a phase transition from an ordered state characterized by long range spin correlations to a paramagnetic state at a finite, nonzero value of the magnetic field; (ii) the transition has unusual dynamic scaling, in that the correlation time diverges exponentially, while the correlation length diverges in a power law fashion, so that the effective dynamic exponent z characterizing the scaling between the characteristic relaxation time and correlation length, $\tau \sim \xi^z$, is infinite; (iii) because of the inherent randomness in the Hamiltonian, one obtains distributions for various quantities, but the distributions become so broad in the thermodynamic limit that different moments scale differently ! (Thus, for example, the average and the typical correlation lengths, defined in the usual manner through the average and typical values of the correlation function as a function of distance, diverge with *different exponents* at the transition); (iv) another consequence of the randomness is the presence of Griffiths singularities due to rare configurations of the random variable. For thermal phase transitions, these rare fluctuations are found to have a very weak effect[22]. For the McCoy-Wu model, the classical counterpart of the quantum Ising chain a transverse field, however, McCoy's work showed over 25 years ago, that it led to a divergent susceptibility in the *paramagnetic phase* as the field strength was reduced, well before the transition to the ordered phase. This is to be contrasted with the infinite range model, which being a mean field description, has no such Griffiths phase effects. It is also in contrast with the experimental result, where even the nonlinear susceptibility barely shows a divergence.

The study of the one-dimensional model raises a number of questions, all of which can be put into one common format - are the results obtained of a general nature, or are they specific to one dimension, or even more, specific to this model ? In particular, one would like to know if (i) the non-power-law dynamic scaling is special or general, (ii) if the diparity between the average and typical correlation functions/lengths is special or generic, and (iii) if the effects of rare fluctuations (Griffiths singularities) is restricted to one dimension or is a more general phenomenon. To investigate these issues, as well as see if the effect of higher dimensionality brings one closer to the equally surprising experimental result, that of no divergence in the nonlinear susceptibility, it becomes necessary to use simulational methods.

3   The Transverse Field Ising Spin Glass in Two and Three Dimensions

In higher dimensions, exact analytical results do not exist. While approximate treatments using Migdal-Kadanoff decimation schemes have been carried out[23], we concentrate on the results of Monte Carlo simulations performed in the past three years. To simulate the Hamiltonian in Equation (1), for for $d = 2$ and 3, the Suzuki-Trotter method[17] is employed to transform the quantum Hamiltonian to an anisotropic classical Hamiltonian by using the path integral representation of the quantum problem. In this way we obtain, for a d-dimensional quantum Hamiltonian given by Eq. (1), a classical Hamiltonian an action corresponding to a classical Hamiltonian for a $(d+1)$ dimensional anisotropic model, given by:

$$H_{cl} = - \sum_{\langle ij \rangle, \tau} J_{ij} S_i(\tau) S_j(\tau) - \sum_{i, \tau} S_i(\tau) S_i(\tau + 1). \qquad (2)$$

In the above equation S_i are classical Ising spins (which take on the values $+1$ and -1), coupled by the same coupling constants J_{ij} in the d dimensions as the quantum problem, but with nearest neighbor *ferromagnetic* couplings in the additional (imaginary time) dimension. A rescaling of the couplings has been done; for details, the reader is referred to the original papers[13, 18]. The important point is that the couplings in the spatial directions, J_{ij} depend only on the the spatial coordinates (ij), and are independent of the time coordinate, τ. Thus, not only is the model anisotropic, but the randomness is *perfectly correlated* in the extra dimension. We can therefore already see that fluctuation effects, which have point defect like effects in typical random classical models with uncorrelated (or short-range-correlated) randomness, will be much more severe here, because the perfect correlation in the imaginary time dimension makes them act like line defects which are known to be stronger perturbations on the uniform (perfectly crystalline) models.

The thermodynamic limit of the quantum model corresponds to the d spatial dimensions of the classical model going to infinity, while the limit of the time dimension going to infinity corresponds to the $T \rightarrow 0$ limit of the quantum model. The temperature of the classical model corresponds to the transverse field of the quantum problem, with high temperature representing the high field paramagnetic phase, and low temperature corresponding to low transverse fields, where the system is in the ordered phase. Thus, to obtain properties of the quantum model in the vicinity of the critical point, where universal behavior may be expected, we need to employ Monte Carlo simulation methods on an *anisotropic* classical model with correlated randomness and one added dimension. These we have performed[13, 16] on a SIMD massively parallel machine at Princeton University (MASPAR 1104 with 4096 processors) and at Bell Laboratrories (MASPAR 1216 with 16,384 processors). Results reported in this work took approximately 6 months of continuous running time on the former, and an additional six weeks on the latter, with four times the speed/capability. Simulations were also carried out concurrently by Rieger and Young, who concentrated on the two dimensional Ising spin glass in a transverse field[14, 15].

We simulate the classical anisotropic Hamiltonian on a sequence of finite sized lattices of dimension L in the d space-like dimensions and L_τ in the $(d+1)$th time-like dimension. To extrapolate to the thermodynamic limit ($L \to \infty$), and the quantum model's zero temperature limit ($L_\tau \to \infty$), we use finite size scaling techniques, which have been found to be very successful in dealing with random and frustrated classical models, where the slow relaxation for prohibits equilibration of large sized systems in realistic computer times[24].

In these previous studies, it was found that rather than directly performing a multi-parameter scaling fit to correlation functions, or susceptibilities, which involve at least three adjustable parameters, it is more useful to consider dimensionless quantities, which have a simpler scaling form with fewer adjustable parameters. One such quantity is :

$$g(L, L_\tau, T) = \frac{1}{2}\left[3 - \frac{\langle q^4 \rangle}{\langle q^2 \rangle^2}\right],\tag{3}$$

where q is the overlap of spins between two independently equilibrated copies (1,2) of a finite sized ($L^d \times L_\tau$) system with the same bond configuration (replicas), given by:

$$q = \frac{1}{N}\sum_{i,\tau} S_i^{(1)}(\tau)S_i^{(2)}(\tau).\tag{4}$$

In the above equations, the angular brackets $\langle \cdots \rangle$ denote a thermal (time) average for a given realization of the (random) bonds, and the square brackets $[\cdots]$ denote an average over the bond (J_{ij}) ensemble for a fixed size ($N = L^d \times L_\tau$).

The behavior of g with the temperature and size in the vicinity of the paramagnetic to ordered phase may be obtained from the following general argument. In the paramagnetic phase, as the system dimensions becomes larger than the corresponding correlation lengths, the distributions of the overlap tend towards a gaussian form centered around zero, and we thus have $< q^4 > = 3 < q^2 >^2$, so g tends towards zero in the thermodynamic limit. On the other hand, in the ordered phase, the distributions tend to delta functions with a nonzero mean at $+q(T)$ and $-q(T)$, so g approaches 1 in the thermodynamic limit. For finite but large sizes (L, L_τ) much greater than microscopic lengths but having an arbitrary ratio with the correlation lengths (a requirement that we will assume to hold for the rest of this article) one expects, by finite size scaling, that g will have the scaling form:

$$g = \tilde{g}\left(\frac{L}{\xi}, \frac{L_\tau}{\xi_\tau}\right) = \bar{g}\left(\frac{L}{\xi}, \frac{L_\tau}{L^z}\right).\tag{5}$$

To obtain the second equality, we have used the standard dynamic scaling hypothesis relating the time and space correlation lengths near the critical point:

$$\xi_\tau \sim \xi^z.\tag{6}$$

In this anisotropic classical analog of the quantum Hamiltonian, with its manifest difference between the d space and $(d + 1)$th time direction in the classical analog of the quantum model, the universal scaling function g involves *two* variables. This prevents a direct single parameter size scaling from being directly performed (as is done in isotropic classical models) without knowledge of the dynamical exponent z. To obtain z, we employ the following technique: consider the behavior of g for a fixed but large size L in the space dimensions, as a function of L_τ, the size in the time-like dimension, near T_c, the critical temperature of the anisotropic classical model. (It should be remembered that the critical temperature of the classical problem corresponds to the critical value of the transverse field of the original quantum problem. The real temperature of the quantum model is set to zero by taking the $L_\tau \to \infty$ limit). For small L_τ, one has a d-dimensional system which is in its paramagnetic phase (since the ordering temperature of the d-dimensional system, should be lower than that of the $(d + 1)$ dimensional system); consequently, if the size L is large, the value of g should be small. At the other extreme, for L_τ large and increasing, one has effectively a one-dimensional system, which has a finite correlation length at any nonzero temperature; consequently, $g \to 0$ as $L_\tau \to \infty$. In between the two extremes, we expect g to exhibit a maximum for some optimal shape, (see top part of Figure 4). This maximum should be independent of size L precisely at T_c, since then ξ is infinite and the first argument of Eq. (5) is zero; the maximum corresponds to a fixed (optimal) value of the second argument for each L. For T below (above) T_c, the curves of g versus L_τ should show maxima that increase (decrease) with increasing size L (Figure 4 top panel).

The technique described in the former paragraph was used by Guo, Bhatt and Huse[13] for the Ising spin glass in three space dimensions, taking a simple cubic lattice with nearest neighbor (only) interactions, distributed according to a Gaussian with zero mean and unit variance [i.e. $P(J) = (2\pi)^{-1/2}exp(-J^2/2)$]. They found that the model exhibited a phase transition at a temperature $T_c = 4.32 \pm 0.03$. Furthermore, by scaling the data for g at T_c, where ξ is infinite, so by equation (5) the curves for different linear dimensions L should scale if plotted versus L_τ/L^z (see Figure 4 bottom panel), they obtained numerically for the first time, in a bias-free way, the dynamical exponent $z = 1.3$. At the same time, simulations performed by Rieger and Young[14] on the corresoponding $(2+1)$ dimensional model gave $T_c = 3.275 \pm 0.025$, and $z = 1.5$. Thus the data in higher dimensions $d = 2$ and 3 are consistent with conventional power-law dynamic scaling with a finite z, unlike the one-dimensional case.

With z determined, a standard one-parameter finite size scaling can be performed, using a fixed value of the aspect ratio L_τ/L^z, and the static exponents determined. Scaling of the dimensionless parameter g can be used to determine the correlation length exponent ν, and for $d = 3$, gives[13] $1/\nu = 1.3$ (see Fig. 5), while a scaling of the spin glass susceptibility at the

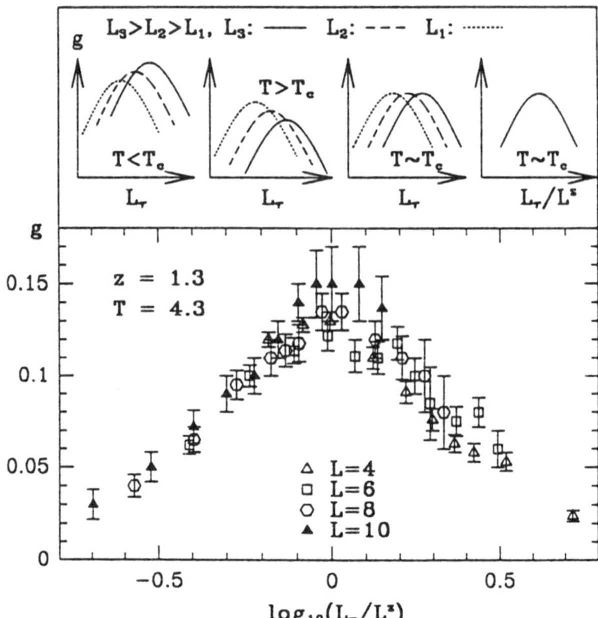

Fig. 4. Top panel shows a schematic plot of the variation of g (Eq. 5) with L_τ for different L for temperatures on either side of the critical point T_c. Bottom panel shows the scaled data for the three-dimensional Ising spin glass, which enables a determination of the dynamical exponent z. (From Guo, Bhatt and Huse [13]).

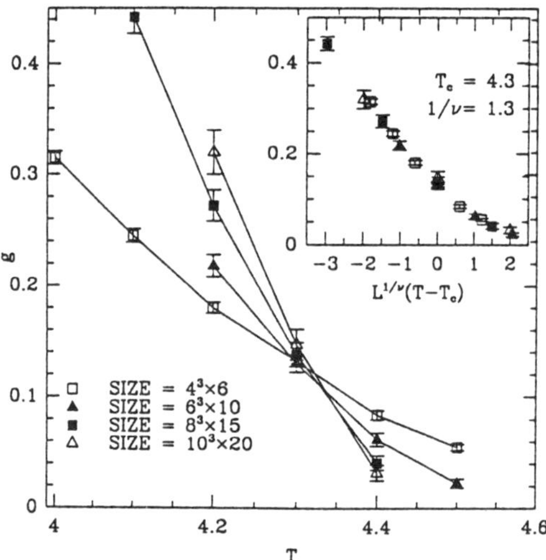

Fig. 5. Dependence of the dimensionless parameter g with T in three space dimensions, for various sizes ($L \times L_\tau$) with a fixed shape L_τ / L^z. Data cross at the critical point and collapse onto a single curve (inset) when the temperature is scaled as $L^{1/\nu}(T - T_c)$, which allows $1/\nu$ to be determined. From Guo *et al.*[13]

transition point, $\chi_{sg} = N[<q^2>] \sim L^{2-\eta}$, gives $\eta = 1.1$. For $d = 2$, a similar analysis[14] yields $\nu = 1.0$ and $\eta = 0.5$. Thus the model appears to behave in qualitatively the same way in two and three dimensions, though the critical exponents are somewhat different, as expected. One difference is that the *linear* susceptibility is found to diverge at the critical point in $d = 2$, while it appears to remain finite in $d = 3$. This difference not withstanding, the *nonlinear* susceptibility is found to diverge with a large exponent at the transition in both cases, in contrast to the experimental results on $LiHo_xY_{1-x}F_4$. (For $d = 3$, one obtains $\chi_{nl} \sim (\Gamma - \Gamma_c)^{\nu(\eta-2-2z)}$, which diverges with an exponent 2.7, not very different from the result for the *classical* model in three dimensions[25]). While there does not appear to be a complete explanation of this big difference, we discuss in the following section the complicating effects of rare fluctuations in the random bond distribution, known as Griffiths singularities.

4   Singularities due to Rare Regions

Rare configurations of the quenched random variable are known to lead to unusual effects in the paramagnetic phase itself; in classical models the effects on the static thermodynamic properties are rather weak[22], though they are in principle expected to dominate the long-time dynamics[26]. In quantum models, where statics and dynamics are coupled, they can be quite strong[12, 27, 28] and the reader is referred to a recent review article[29] for a discussion of this point.

In the context of the Ising spin glass in a transverse field, the strong effect of rare clusters in the paramagnetic phase can be understood from the following argument[30, 16]. The probability of anomalously strong bonds in a region of spatial size L *decreases* exponentially with the volume, $P(L) \sim \exp(-aL^d)$. On the other hand, from the classical mapping of the quantum problem, this propagates unabated along the time-like direction, so we have effectively a one dimensional chain, with a correlation length along the time axis that *increases* exponentially with the coupling, which in turn is proportional to the volume in the spatial dimensions, *i.e.* $\xi_\tau \sim \exp(bL^d)$. The contribution of such rare clusters to various susceptibilities, which are proportional to various powers of the length scale, depends on the relative magnitudes of the parameters a and b, which in turn depend on the distance to the critical point. Since $P(L)$ scales as a power of ξ_τ, as does the susceptibility, these effects give rise to power law tails in the distribution of *local* susceptibilities. This was found in detailed simulations in $d = 2$ by Rieger and Young,[15] and by Guo, Bhatt and Huse [16] in both $d = 2$ and $d = 3$.

As an example, we show in Figure 6 the integrated histograms of the of the *local, nonlinear* susceptibility, obtained [16] for the $d = 2$ case, at two temperatures in the paramagnetic phase, for a sequence of sizes. The linear behavior for large sizes in the double logarithmic plot implies power law tails

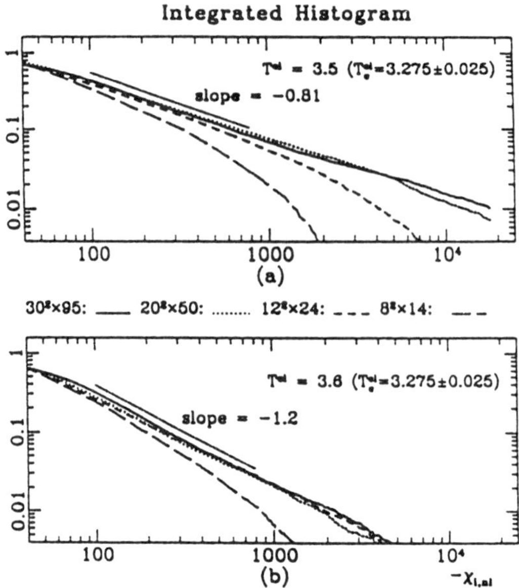

Fig. 6. Integrated histogram of the local nonlinear susceptibility for the model in two space dimensions, on a log-log plot, for two temperatures in the paramagnetic phase. A divergent average nonlinear susceptibility is implied by slopes less than unity. (From Guo, Bhatt and Huse [16]).

for large local susceptibilities with an exponent given by the slope. A slope less than unity leads to a divergent average nonlinear susceptibility, and this is clearly seen to be the case at the lower temperature. From a detailed analysis at a sequence of temperatures[15, 16], the onset temperature for this divergence is found to be $T = 3.55$, well above the critical temperature $T_c = 3.275 \pm 0.025$. For $d = 3$, the data also support the existence of such a region in the paramagnetic phase[16], though the extent is much smaller. This shows that the strong effect of rare fluctuations in the transverse field Ising model is not an effect ascribable to one-dimension, but continues to exist in higher dimensions as well. Griffiths phase effects with nonuniversal, continuously varying, singular behavior is seen in many quantum spin models including the dimerized spin-1/2 Heisenberg chain with quenched disorder, as well as the spin-1 chain with disorder[28]. Thus, it appears to be a phenomenon of more general relevance in quantum models with quenched disorder than in classical thermal phase transitions.

The existence of a divergent average nonlinear susceptibility in three dimensions clearly complicates the interpretation of experimental results, though at this stage it is unclear if it is capable of explaining the large discrepancy in the observed behavior of the nonlinear susceptibility in the experimental system visavis the simple model. Another difference is that the experimental system has dipolar couplings, whereas the model has short ranged (nearest neighbor) interactions with no directional dependence. The effects of these differences are also not clear at present.

The strong (nonuniversal) effects of rare fluctuations do result in some individuality of quantum models with quenched disorder, as contrasted with the tame "herd-mentality" of classical models governed by the overriding principle of universal behavior near continuous phase transitions. The experiences with the varied behavior seen in the studies of the quantum Ising spin glass in a transverse field as well as other models[29] suggests that greater care must be taken in lumping models into universality classes in the case of quantum systems than was the case for classical models.

5 Random Ising Ferromagnet in a Transverse Field

We now briefly allude to some recent work on the ferromagnetic counterpart of the transverse field Ising spin glass, with random bonds picked out of a ferromagnetic distribution. For $d = 1$, with no frustration, the sign of the interactions can be gauged away; however, that is not possible in higher dimensions. One of the issues that is of interest is that for uniform couplings the $d = 2$ quantum model corresponds to the $d = 3$ classical model, for which the correlation length exponent is $\nu = 0.64$. In the presence of quenched disorder in the two spatial dimensions, following the general arguments of Chayes *et al.*[31], one would expect the exponent of the random quantum model to obey the inequality $\nu \geq 2/d = 1$, implying that the randomness is relevant. A similar argument also works for the effects of quenched randomness in the three dimensional classical Ising ferromagnet; however, numerical simulations on the model[32] were unable to demonstrate this relevance, because of the long length scale needed for the crossover to the random fixed point, presumably because the uniform system *almost* satisfies the CCFS inequality $\nu \geq 2/3$ in that case.

The violation of the CCFS inequality is much clearer for the uniform quantum ferromagnet in $d = 2$, implying that randomness is clearly relevant. Nevertheless, upon addition of bond randomness in the spatial directions, it was found[18, 19] that small sizes continue to scale with an apparent correlation length exponent that violates the CCFS inequality, and is often close to the three-dimensional uniform Ising ferromagnet. This is demonstrated in Fig. 7 for a binary bond distribution $[P(J) = 0.8\delta(J - 0.625) + 0.2\delta(J - 2.5)]$. Using the shape scaling to determine the dynamical exponent, one obtains from small sizes ($L \leq 12$), an apparent critical point $T_c = 4.43$, and a dynamical exponent $z = 1.15$. However, performing the standard size scaling for fixed aspect ratio L_τ / L^z, leads one to a correlation length exponent (see Figure 7) $\nu = 0.8$, which violates the CCFS inequality, and data for larger sizes do not fall on the scaling curve (Figure 7). An analysis of the larger

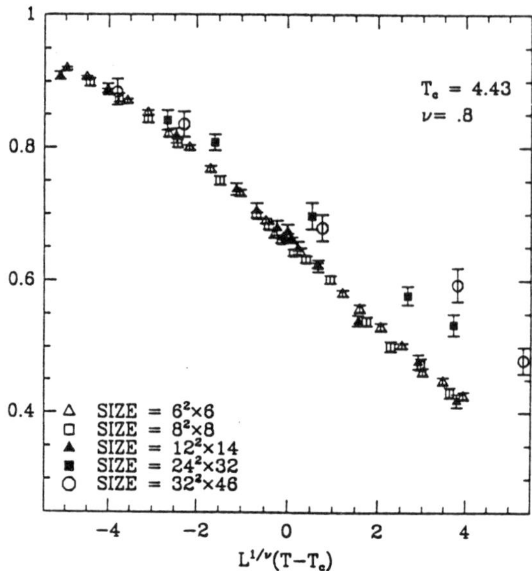

Fig. 7. Scaling plot of the dimensionless parameter g (along the y-axis) for the random ferromagnet. The correlation length exponent, ν obtained from the best fit for small sizes ($L \leq 12$) violates the CCFS inequality. (From Guo and Bhatt [18, 19]).

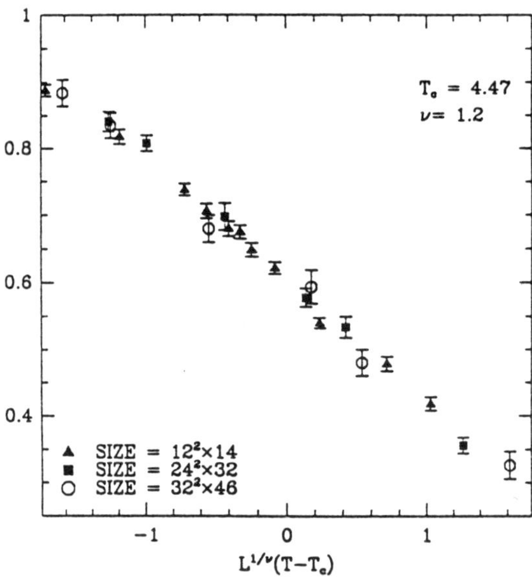

Fig. 8. As in Figure 7, but now with larger sizes. Note that the best fit is obtained with a ν which no longer violates the CCFS inequality. (From Guo and Bhatt [18, 19]).

96

size data (L between 12 and 32) yields a slightly different $T_c = 4.47$, and a correlation length exponent $\nu = 1.2$ which satisfies the CCFS inequality (see Figure 8). Though it is not clear that the data are quite in the asymptotic regime, (which can only be verified by simulations on still larger sizes), the relevance of randomness is clearly demonstrated.

6 Concluding Remarks

This article has provided a brief summary of the role of computer simulations in the recent advances made in the understanding of random magnetic systems undergoing a continuous quantum ($T = 0$) phase transition. Extensive Monte Carlo studies over the last four years relevant for the two and three dimensional short range quantum Ising spin glass in a field transverse to the Ising axis have established that the model undergoes a quantum phase transition at $T = 0$ from a spin glass ordered phase to a paramagnetic phase as a function of the applied transverse field which is characterized by a set of exponents that is distinct from the finite temperature thermal transition. The quantum transition from the paramagnetic phase appears to have conventional dynamic scaling, *i.e.* the correlation time τ scales as a power law of the correlation length ξ, as $\tau \sim \xi^z$, with a finite, nontrivial dynamic exponent z. The same is found for the random ferromagnet in $d = 2$. In this regard, it appears that the two and three dimensional models are more similar to the mean field model than the model in one dimension, where $z = \infty$, because the correlation time scales exponentially with the square root of the correlation length. However, a careful study of the distribution of local nonlinear susceptibility for the quantum spin glass shows that it develops a power law tail in the paramagnetic phase due to rare fluctuations in the quenched random bond distribution, because of which the average susceptibility *diverges in the paramagnetic phase* before the transition to the ordered spin glass phase in both two and three dimensions. In this respect, the models are akin to the one-dimensional case, showing strong Griffiths singularity effect, which are absent in the infinite range mean field model.

The dominance of atypical couplings in random quantum models has been seen in many cases besides the quantum Ising model in a transverse field, such as disordered spin-1/2 and spin-1 Heisenberg chains[33, 27, 28] as well as highly disordered quantum spin-1/2 Heisenberg antiferromagnets in higher dimensions[34]. These are a unique stamp of quantum models, and in many instances leads to a nonuniversality not found in classical models[27, 29]. How ubiquitous these effects are will have to await further investigation; they are clearly not limited to one specific problem or dimension, and the number of examples keeps growing.

One other feature which deserves further study is the possibility that average and typical (median) behavior may be different in the models in higher dimension, like the one-dimensional random quantum Ising model in a transverse field. In that case, the average and typical correlation lengths have very different critical exponents. Further, one of them, the average satisfies the CCFS inequality, while the other violates it. It would be of great interest to see if any of these features carry over to higher dimensions. It should be remarked that in one dimension, the exponent of the average correlation length satisfies the extreme case of the CCFS inequality, *i.e.*, when it becomes an equality $\nu = 2/d$. While in the models we have studied, the best estimates are sometimes not quite at the limit, they are never very far, and it cannot be ruled out that corrections due to finite size effects could be responsible for the small difference. This issue deserves further investigation.

Finally, and most important, is the issue of the rather different behavior seen in experiments on the diluted dipolar Ising magnet, $LiHo_xY_{1-x}F_4$, where the nonlinear susceptibility divergence characteristic of a spin glass appears to get weaker as the quantum $(T = 0)$ boundary is approached, a behavior not exhibited by any of the theoretical models. Whether this is because of the nature of the couplings (long range and/or angular dependence), or the complicating aspect of Griffiths singularity effects, or some other heretofore unknown reason, only future research will tell.

Acknowledgments

The author would like to thank David Huse and Peter Young for many fruitful discussions, and especially Muyu Guo, whose dedicated efforts led to the results discussed in Sections 3, 4 and 5. D. Bitko provided copies of Figures 2 and 3. This work was supported by NSF Grant DMR-9400362. The grant of a Fellowship by the J. S. Guggenheim Foundation, during which some of the research discussed was undertaken, is gratefully acknowledged.

References

1. P. W. Anderson, *Basic Notions of Condensed Matter Physics*, (Benjamin, 1984).
2. M. E. Fisher, *Rev. Mod. Phys.* **46**, 597 (1974).
3. N. F. Mott, *Proc. Phy. Soc. (Lond.)* **A62**, 416 (1949).
4. J. H. de Boer and E. J. W. Verwey, *Proc. Phy. Soc. (Lond)*. **A49**, 59 (1937).
5. F. J. Wegner, *Z. Phys.* **B25**, 327 (1976).
6. E. Abrahams, P. W. Anderson, D. C. Licciardello, and T. V. Ramakrishnan, *Phys. Rev. Lett* **42**, 673 (1979).
7. As evidenced by many workshops and schools on the subject in recent years, *e.g., Quantum Phase Transitions in Condensed Matter*, R. N. Bhatt and M. P. A., Fisher (coordinators), Institute of Theoretical Physics (Santa Barbara, 1992); *Quantum Phases and Transitions*, R. N. Bhatt, A. Georges and S. Sachdev (coordinators), ICTP Spring College, (Trieste, 1994).

8. P. A. Lee and T. V. Ramakrishnan, *Rev. Mod. Phys.* **57**, 287 (1985); D. Belitz and T. R. Kirkpatrick, *Rev. Mod. Phys.* **66**, 261 (1994).

9. W. Wu, B. Ellman, T. F. Rosenbaum, G. Aeppli and D. H. Reich, *Phys. Rev. Lett.* **67**, 2076 (1991); W. Wu, D. Bitko, T. F. Rosenbaum and G. Aeppli, *Phys. Rev. Lett.* **71**, 1919 (1993).

10. B. M. McCoy and T. T. Wu, *Phys. Rev.* **176B**, 631 (1968); *Phys. Rev.* **188**, 982 (1969); B. M. McCoy, *Phys. Rev. Lett.* **23**, 383 (1969); *Phys. Rev.* **188**, 1014 (1969).

11. R. Shankar and G. Murthy, *Phys. Rev.* **B 36**, 536 (1987).

12. D. S. Fisher, *Phys. Rev. Lett.* **69**, 534 (1992); *Phys. Rev.* **B51**, 6411 (1995).

13. M. Guo, R. N. Bhatt and D. A. Huse, *Phys. Rev. Lett.* **72**, 4137 (1994).

14. H. Rieger and A. P. Young, *Phys. Rev. Lett.* **72**, 4141 (1994).

15. H. Rieger and A. P. Young, *Phys. Rev.* **B54**, 3328 (1996).

16. M. Guo, R. N. Bhatt and D. A. Huse, *Phys. Rev.* **B54**, 3336 (1996).

17. M. Suzuki, *Prog. Theo. Phys.* **56**, 1454 (1976); H. F. Trotter, *Proc. Am. Math. Soc.* **10**, 545 (1959).

18. Muyu Guo, *Ph.D. Thesis, Princeton University* (1995, unpublished).

19. M. Guo and R. N. Bhatt, *Preprint* (1997).

20. Y. Y. Goldschmidt and P.-Y, Lai, *Phys. Rev. Lett.* **64**, 2467 (1990) and references therein. See also B. K. Chakrabarti, A. Dutta and P. Sen, *Quantum Ising Phases and Transitions in Transverse Ising Models* (Springer, Berlin, 1996).

21. A. P. Young and H. Rieger, *Phys. Rev.* **B53**, 8486 (1996).

22. R. B. Griffiths, *Phys. Rev. Lett.* **23**, 17 (1969).

23. B. Boechat, R. R. dos Santos and M. A. Continentino, *Phys. Rev.* **B49**, 6404 (1994).

24. R. N. Bhatt and A. P. Young, *Phys. Rev.* **B37**, 5606 (1988).

25. R. N. Bhatt and A. P. Young, in *Heidelberg Colloquium on Glassy Dynamics*, L. van Hemmen and I. Morgenstern, editors (Springer-Verlag, Berlin, 1987).

26. M. Randeria, J. P. Sethna and R. J. Palmer, *Phys. Rev. Lett.* **54**, 1321 (1985).

27. D. S. Fisher, *Phys. Rev.* **B50**, 3799 (1994).

28. R. A. Hyman, K. Yang, R. N. Bhatt and S. Girvin, *Phys. Rev. Lett.* **76**, 839 (1996); R. A. Hyman and K. Yang, *Phys. Rev. Lett.* **78**, 1783 (1997).

29. R. N. Bhatt, in *Phase Transitions and Random Fields*, A. P. Young, editor (World Scientific, 1997).

30. M. J. Thill and D. A. Huse, *Physica A* **214**, 321 (1995).

31. J. Chayes, L. Chayes, D. S. Fisher and T. Spencer, *Phys. Rev. Lett.* **57**, 2999 (1986).

32. D. P. Landau, *Phys. Rev.* **B22**, 2450 (1980).

33. C. Dasgupta and S. K. Ma, *Phys. Rev.* **B22**, 1305 (1980).

34. R. N. Bhatt and P. A. Lee, *J. Appl. Phys.* **52**, 1703 (1981); R. N. Bhatt and P. A. Lee, *Phys. Rev. Lett.* **48**, 344 (1982).

Metal–Insulator Transitions in Strongly Correlated Systems

M. Imada

Institute for Solid State Physics, University of Tokyo, Roppongi 7-22-1,
Tokyo 106, Japan

We review recent progress in understanding the nature of metal-
insulator transitions in strongly correlated electron systems. In particular, using improved algorithms of quantum Monte Carlo simulations for strongly correlated electrons, the universality class of the
transition between the Mott insulator and metals in two dimensions
has been clarified and analyzed numerically and compared with the
prediction from the scaling theory. It shows the existence of new
universality class of the metal-insulator transition characterized by
unusual suppression of coherence in the metallic side due to a large
dynamical exponent.

1. Introduction

Algorithms of numerical computation for strongly correlated systems have
recently been developed and applied extensively. The quantum Monte Carlo
method provides a way of calculating strong correlation effects. A frequently
used algorithm of quantum Monte Carlo calculation is based on the path
integral formalism combined with the Stratonovich-Hubbard transformation.
There exist various types of algorithms for quantum Monte Carlo simulations.
In the main part of this review, we present results obtained from the projector Monte Carlo method where the $T = 0$ ground state is studied. To reach
the ground state quickly in the path integral formalism, we employ an optimized trial wavefunction [1]. This has made it possible to investigate the
metal-insulator transition (MIT) in terms of a quantum phase transition at
$T = 0$.

2. Quantum Monte Carlo Results in Two Dimensions

Near the continuous MIT point, various quantities show critical fluctuations
of the transition. Here we discuss calculated results for the single-band Hubbard model defined by the Hamiltonian

$$\mathcal{H}_H = \mathcal{H}_t + \mathcal{H}_U - \mu N \tag{1a}$$

$$\mathcal{H}_t = -t \sum_{\langle ij \rangle} (c_{i\sigma}^\dagger c_{j\sigma} + h.c.) \tag{1b}$$

$$\mathcal{H}_U = U \sum_i (n_{i\uparrow} - \frac{1}{2})(n_{i\downarrow} - \frac{1}{2}) \tag{1c}$$

Springer Proceedings in Physics, Volume 83
Computer Simulation Studies in Condensed-Matter Physics X
Eds.: D. P. Landau, K.K. Mon, H. -B. Schüttler
© Springer-Verlag Berlin Heidelberg 1998

and

$$N \equiv \sum_{i\sigma} n_{i\sigma}, \tag{1d}$$

where the creation (annihilation) of the single-band electron at site i with spin σ is denoted by $c_{i\sigma}^{\dagger}(c_{i\sigma})$ with $n_{i\sigma}$ being the number operator $n_{i\sigma} \equiv c_{i\sigma}^{\dagger} c_{i\sigma}$.

2.1. Charge Compressibility

In the Mott insulating state of the Hubbard model at half filling, a charge excitation gap opens. Direct estimation of the charge gap is possible from the relation

$$\Delta_c \equiv \frac{1}{2}(E_g(\frac{N}{2}, \frac{N}{2}) - E_g(\frac{N}{2} - 1, \frac{N}{2} - 1)), \tag{2}$$

where $E_g(n_1, n_2)$ is the ground state energy of the system with n_1 up-spin and n_2 down-spin electrons in a N-site system.

Quantum Monte Carlo method of the finite temperature algorithm applied by Moreo et al [2] as well as of the projector algorithm for $T = 0$ applied by Furukawa and Imada [3] to the nearest-neighbor Hubbard model at $U/t = 4$ show the charge gap $\Delta_c \sim 0.6$. Recently a more precise way to estimate the charge gap at $T = 0$ has been developed by Assaad and Imada [4] using an elaborated algorithm for the Mott insulating phase using stabilized matrix calculations. The application of this algorithm used for the Hubbard model at $U/t = 4$ shows $\Delta_c/t = 0.66 \pm 0.015$ [5].

The charge compressibility κ or charge susceptibility χ_c are defined as

$$\chi_c = n^2 \kappa = \partial n / \partial \mu \tag{3}$$

where n is the electron density and μ is the chemical potential. The charge susceptibility vanishes in the Mott insulating phase because of the incompressibility. However, quantum Monte Carlo results on two-dimensional systems show that doped systems become more and more compressible near the MIT in the metallic side [3], [6]. In two-dimensional systems, the charge susceptibility χ_c does not gradually and continuously vanish when one controls the doping concentration from the metallic phase to the MIT point $\delta = 0$. The scaling plot of χ_c even shows singular divergence in the form

$$\chi_c \propto 1/|\delta| \tag{4}$$

for $\delta \neq 0$ [3], [6]. Similar behavior is also observed in the 2D t-J model [7], [8].

When the charge susceptibility is singularly divergent as a power law of δ for $\delta \to +0$, the single-particle description of low-energy excitations has to show the divergence of the effective mass of relevant particle. This is in contrast with the usual MIT between metals and the band insulator, where the number of carriers vanishes.

2.2. Magnetic Correlations

Effects of spin fluctuations are another intensively studied issue near the MIT in 2D. In low dimensional systems, the antiferromagnetic correlation is under a strong influence of quantum fluctuations. Quantum Monte Carlo results in the two-dimensional Hubbard model suggest the existence of the antiferromagnetic long-range order in the Mott insulating phase, $\delta = 0$ [9]. In the two-dimensional Hubbard model, the long-range order appears to be lost immediately upon doping [3]. The Fourier transform of the equal-time spin-spin correlation function

$$S(\mathbf{k}) = \frac{1}{3} \int e^{-i\mathbf{k}\cdot\mathbf{r}} \langle \mathbf{S}(\mathbf{r}) \cdot \mathbf{S}(0) \rangle d\mathbf{r} \tag{5}$$

shows short-ranged incommensurate peak at $\mathbf{k} = \mathbf{Q}$ away from the antiferromagnetic Bragg point $\mathbf{k} = (\pi, \pi)$ when the doping proceeds. The peak value $S(\mathbf{Q})$ becomes finite for $\delta \neq 0$ which means the disappearance of the long-range order in the whole metallic region, $\delta \neq 0$. The scaling of $S(\mathbf{Q})$ appears to follow

$$S(\mathbf{Q}) \propto 1/\delta, \tag{6}$$

which has been interpreted by the antiferromagnetic correlation length ξ_m scaled by $\xi_m \propto \delta^{-1/2}$ [3]. Here, it should be noted that the existence of the antiferromagnetic correlation length is a nontrivial fact in the metallic phase because the correlation at asymptotically long distance should follow power-law decay at $T = 0$ in usual spin gapless metals and naively, it leads to divergence of correlation length. The Monte Carlo results in two dimensions suggest that the spin correlation of the period given by the wave number $\mathbf{Q}$ is constant as in the Heisenberg model for the distance $1 \ll r \ll \xi_m$ while it decays as a power law $\propto 1/r^\gamma$ with $\gamma > 2$ in the region $r \gg \xi_m$. In the Fermi liquid, it has to follow $\langle \mathbf{S}(0) \cdot \mathbf{S}(\mathbf{r}) \rangle \propto r^{-\gamma}$ with $\gamma = 3$ at $r > \xi_m$ in 2D. In the one-dimensional doped Hubbard model, it has been established that the incommensurate spin correlation decays as $1/r$ for $1 \ll r \ll \xi_m$ while as $1/r^{1+K_\rho}$ for $r \gg \xi_m$ with $\xi_m \propto \delta^{-1}$ [10], [11]. K_ρ is the exponent of the Tomonaga-Luttinger liquid.

Temperature dependence of the antiferromagnetic correlation length ξ_m has also been studied extensively by numerical approaches. In the undoped case, ξ_m continuously and rapidly increases until $T \to 0$ because the antiferromagnetic transition takes place at $T = 0$ in 2D. Chakraverty, Halperin and Nelson [12] have pointed out the relation of the quantum Heisenberg model in 2D to the O(3) nonlinear sigma model in 3D and suggested the existence of two regions, namely, $\xi_m \propto 1/T$ at higher temperatures (quantum critical regime) while ξ_m increases exponentially at lower temperatures in the renormalized classical regime. This basic feature has been more or less numerically confirmed (for example, [13]).

At finite doping, $\xi_m(T)$ has also been computed for the two-dimensional Hubbard model [14]. It shows a sharp contrast with the undoped case. ξ_m as well as peak value of the equal-time spin structure factor $S(Q)$ grows at high temperatures, $T > J$. With decreasing temperature, however, they show rapid saturation at rather high temperatures. For example, at $\delta \sim 0.15$, ξ_m more or less saturates at $T \sim J$ and does not grow below it. This means ξ_m quickly approaches the intrinsic correlation length $\sim 1/\sqrt{\delta}$ given from (6). The temperature below which ξ_m becomes more or less temperature independent is denoted as T_{scr}. T_{scr} rapidly increases from 0 at $\delta = 0$ to $T_{\mathrm{scr}} \sim J$ at $\delta \sim 0.15 \sim 0.2$. Similar and consistent result has also been obtained in the two-dimensional t-J model by Jaklič and Prelovšek [7]. This temperature-insensitive peak structure is associated with the incoherent part of the spin correlation while the coherent part has a small weight at higher frequencies when the doping concentration becomes low. This coherent part may give rise to the Fermi surface effect at low temperatures associated with the power-law decay of the spin correlation at long distance in metals.

These numerical results are consistently interpreted by the following picture as discussed by Imada [14] and Jaklič and Prelovšek [15]. It is useful to analyze the spin correlation as a sum of incoherent and coherent parts. As the doping concentration decreases, the weight of the coherent part decreases to vanish and the incoherent weight becomes dominant. The coherent part comes from asymptotically long-time and long-distance part beyond $\tau_m (\equiv \xi_m^z)$ and ξ_m where the correlation decays as power-law at $T = 0$. Here z is the dynamical exponent. The incoherent part arises from the short-time and short-distance part ($\tau < \xi_m^z$ and $r < \xi_m$). As $\xi_m \to \infty$ with decreasing δ, it is clear that the incoherent part dominates and the coherent weight vanishes. The spin structure factor

$$S(\mathbf{q},\omega) \equiv \int_{-\infty}^{\infty} \mathrm{d}r\mathrm{d}t \ e^{-i\omega t} e^{i\mathbf{q}\cdot\mathbf{r}} \langle \mathbf{S}(\mathbf{r},t) \cdot \mathbf{S}(0,0) \rangle \tag{7}$$

may be given as the sum of the coherent and incoherent contributions as

$$S(\mathbf{q},\omega) \simeq S_{\mathrm{inc}}(\mathbf{q},\omega) + S_{\mathrm{coh}}(\mathbf{q},\omega). \tag{8}$$

The incoherent part S_{inc} has typical widths ξ_m^{-1} in the q-space and ξ_m^{-z} in the ω-space. Essential feature of numerical results may be approximately and phenomenologically written as

$$S_{\mathrm{inc}}(\mathbf{q},\omega) \sim \frac{C}{-i\omega + D(\mathbf{q}^2 + K^2)} \tag{9}$$

with $K \equiv \xi_m^{-1}$, $D_s \propto \xi_m^{2-z}$, $C \propto \xi_m^d$. Here we take the dimension $d = 2$. The important point is ξ_m has to be temperature independent at $T < T_{\mathrm{scr}}$ to be consistent with numerical results. The coherent part S_{coh} may have strong temperature dependence even below T_{scr}. In fact, S_{coh} represents the long-ranged power-law decay of correlation and reflects the Fermi surface effect at low temperatures. Therefore S_{coh} grows in general as incommensurate and rather sharp non-Lorentzian peaks below the coherence temperature $T_{\mathrm{coh}} \equiv T_F$. Because K and D_s are temperature independent below T_{scr},

$$\mathrm{Im}\chi(q,\omega) = (1 - e^{-\beta\omega})\mathrm{Re}S(q,\omega) \tag{10}$$

has temperature dependence mainly coming from the Bose factor $(1 - e^{-\beta\omega})$. This leads to the result that $\int dq \mathrm{Im}\chi(q,\omega)$ is universally scaled as

$$\int dq \mathrm{Im}\chi(q,\omega) \propto (1 - e^{-\beta\omega}) \tag{11}$$

at $\omega < \xi_m^{-z}$ when the incoherent contribution is dominant.

3. Scaling Theory of Mott Transition

In general, mean field approximations are justified if the dimension is high enough. However, in the Mott transition of fermionic models, the upper critical dimension beyond which the mean field approximation is justified is not easily estimated as in simple magnetic transitions of the Ising model. The infinite dimensional approach provides an example where the mean field approximation becomes exact at $d = \infty$ by taking account of dynamical fluctuations correctly.

The nimerical results on the charge compressibility and spin correlations in 2D have inspired extensive studies to clarify the nature of the Mott transition in low-dimensional systems. In particular, the singular behavior of the charge compressibility observed in the numerical results was an unexpected and surprising property of the MIT in 2D. To understand the numerical results in a consistent way, it has been proposed [16], [17] that the scaling assumption is satisfied. Below basic points of this proposal are reviewed.

3.1 Hyperscaling Scenario

Because the MIT is controlled by quantum fluctuations, we may take the control parameters such as the electron chemical potential μ or the bandwidth t. Using this control parameter, the distance from the critical point is measured by Δ. The control parameter can either be the chemical potential to control the filling or the bandwidth (or the interaction). The scaling theory assumes the existence of single characteristic length scale ξ which diverges as $|\Delta| \to 0$ and a single characteristic frequency scale Ω which vanishes as $|\Delta| \to 0$. The correlation length ξ is assumed to follow

$$\xi \sim |\Delta|^{-\nu} \tag{12}$$

as $\Delta \to 0$, which defines the correlation length exponent ν. The frequency scale Ω is determined from the quantum dynamics of the system independently of the length scale. The dynamical exponent z determines how Ω vanishes as a power of Δ

$$\Omega \sim \xi^{-z} \sim |\Delta|^{z\nu}. \tag{13}$$

In critical phenomena, the hyperscaling is believed to hold between the upper critical dimension d_c and the lower critical dimension d_l. Above d_c, the mean field description is justified and the hyperscaling description breaks down whereas below d_l the transition itself disappears. The hyperscaling asserts the homogeneity in the singular part of the free energy density f_s in terms of arbitrary length-scale transformation parameter b:

$$f_s(\Delta) \sim b^{-(d+z)} f_s(b^{1/\nu}\Delta) \sim \Delta^{\nu(d+z)}. \tag{14}$$

When the inverse temperature β and the linear dimension of the system size L are both large but finite, finite-size scaling function $\mathcal{F}$ is expected to hold as

$$f_s(\Delta) \sim \Delta^{\nu(d+z)} \mathcal{F}(\xi/L, \xi^z/\beta) \tag{15}$$

in the combination of non-dimensional arguments.

3.2 Drude Weight and Charge Compressibility

The Drude weight defined from the coefficient of the δ-function at $T = 0$ as

$$\sigma(\omega) = D\delta(\omega) + \sigma_{\mathrm{reg}}(\omega) \tag{16}$$

and the charge compressibility in Eq.(3) are the two important and relevant quantities to describe the Mott transition. The Mott insulator has two basic properties, one, insulating property and the other, incompressibility. In particular, the incompressibility is a property in distinction from Anderson-localized insulators. The Drude weight and the charge compressibility are both zero in the Mott insulating phase whereas they are both nonzero in metallic phases. Therefore, these two quantities must be totally nonanalytic at the Mott transition point. The existence of nonanalyticities in these two quantities at the transition point is also common in the transition to the band insulator.

To understand the fundamental importance of the Drude weight and the charge compressibility, it is helpful to represent them as stiffness constants to the twist of the phase in boundary conditions in the path integral formalism. In fact, it can be shown that the spatial and temporal rigidities of the phase of wavefunctions are associated with nonzero Drude weight and nonzero charge compressibility, respectively [14], [17], [18].

An infinitesimally small twist of the phase, $\tilde{\phi}_L$, imposed between two boundaries $r_x = 0$ and $r_x = L$ may increase the ground state energy. It has been shown [19], [20], [21] that this type of phase twist measures the stiffness which is proportional to the Drude weight. The change in the ground state energy up to $\tilde{\phi}_L^2$ is obtained from the second order perturbation as

$$\delta E = \frac{1}{2\pi e^2} \frac{\tilde{\phi}_L^2}{L^2} D + O\left(\left(\frac{\tilde{\phi}_L}{L}\right)^4\right). \tag{17}$$

Eq.(17) shows that the stiffness constant for the spatial phase twist is proportional to the Drude weight. An important point to keep in mind is that the Drude weight is obtained by taking the limit $\tilde{\phi}_L \to 0$ first and $L \to \infty$ afterwards. If the opposite limit $\lim_{\phi_L \to 0} \lim_{L \to \infty}$ is taken, one has to take account possible level crossings at finite $\tilde{\phi}_L$ in the limit $L \to \infty$ which yields in principle different quantity, namely, the superfluid density [22], [23].

An infinitesimally small twist of the phase, $\tilde{\phi}_\beta$ imposed between two boundaries $\tau = 0$ and $\tau = \beta$ in the imaginary-time direction may also change the energy. The free energy as a function of the imposed small twist should have the form

$$f(\dot{\phi}) = f(\dot{\phi} = 0) + \frac{i\tilde{\phi}_\beta}{\beta}\rho + \frac{\tilde{\phi}_\beta^2}{2\beta^2}\kappa + O(\tilde{\phi}_\beta^3) \tag{18}$$

where the electron density $\rho = -\partial f/\partial \mu$ and the charge compressibility $\kappa = \partial \rho/\partial \mu$ are obtained as expansion coefficients. Equation(18) shows that κ indeed measures the stiffness constant of the phase in the temporal direction.

3.3. Scaling of physical quantities

Combining expressions (17) and (18) for the Drude weight and the compressibility, respectively, with the finite-size scaling form derived from the hyperscaling, one obtains useful scaling forms for physical quantities [14], [17], [18], [24]. This is because the finite-size scaling form (15) is also valid for δf, namely, the difference of the free energy between the cases in the presence and the absence of the twists. From the comparison of Eq.(17) and (15) in the order of L^{-2}, the scaling of D is obtained:

$$D \propto \Delta^\zeta \tag{19}$$

with $\zeta = \nu(d + z - 2)$. Comparison of (15) and (18) yields, in the order of β^{-1} and β^{-2},

$$\delta \propto \Delta^{-\alpha+1} \tag{20}$$

and

$$\kappa \propto \Delta^{-\alpha} \tag{21}$$

with $\alpha = \nu(z - d)$, where δ denotes the doping concentration, namely, the density measured from the Mott insulating phase.

When the hyperscaling holds, scaling of other physical quantities are also derived from Eqs.(14) and (15). In the insulating phase, the charge excitation gap E_g and the localization length ξ_l defined in (41) should be determined from the single characteristic energy and length scale Ω and ξ, respectively. Then,

$$E_g \propto |\Delta|^{z\nu} \tag{22}$$

and

$$\xi_l \propto |\Delta|^{\nu} \tag{23}$$

immediately follows, where we have used the fact that ν and z are identical in both sides of the transition point. In the metallic side, the scaling of the specific heat $C = \beta^2 \partial^2 (\ln Z)/\partial \beta^2$ is given from (15):

$$C = \Delta^{\nu(d+z)} T \frac{\partial^2}{\partial T^2} \mathcal{F}(\xi/L = 0, \xi^z T). \tag{24}$$

At the critical point, the specific heat is given from the Δ independent term as

$$C \propto T^{d/z}. \tag{25}$$

If $C = \gamma T$ is satisfied at low temperatures in the metallic phase, the coefficient γ follows

$$\gamma \propto \Delta^{\nu(d-z)}. \tag{26}$$

Another interesting quantity is the coherence temperature T_F below which the electron motion becomes quantum mechanical and degenerate. It is clear that T_F has to approach zero as $\Delta \to 0$ in the continuous transition. In case of the Fermi liquid, T_F is nothing but the Fermi temperature. The existence of single characteristic energy scale with singularity at $\Delta = 0$ leads to the scaling of T_F in the form

$$T_F \propto \Delta^{\nu z}. \tag{27}$$

When the control parameter Δ is the chemical potential μ measured from the critical point μ_c, it represents the filling control MIT. In this case, it is possible to derive a useful scaling relation. Because the doping concentration is $\delta = -\frac{\partial f_s}{\partial \mu} = -\frac{\partial f_s}{\partial \Delta}$, it scales as

$$\delta \sim \Delta^{\nu(d+z)-1} \tag{28}$$

From the comparison of Eq.(28) and Eq.(20),

$$\nu z = 1 \tag{29}$$

is derived. The characteristic length scale is then

$$\xi \sim \delta^{-1/d}, \tag{30}$$

which is the length scale of the "mean hole distance". This is a natural consequence because the mean hole distance is indeed a characteristic length scale which diverges at $\Delta = 0$.

In two dimensions (2D) or three dimensions (3D), we have no exact solution for this problem. Therefore, we have to rely on combined analyses of several different numerical results to derive critical exponents. The most important question to be answered is whether 2D and 3D are lower than the upper critical dimension d_c or not. If $d_c < 2$, some type of mean field theory would be justified even in 2D, whereas we have to go beyond the mean field level if $d_c > 2$. Below we summarize how the proposal for the hyperscaling description of this MIT has been brought about as a consistent description of the numerical results.

Quantum Monte Carlo calculations first performed in the metallic side have shown $\kappa \propto \delta^{-1}$ [3], [6]. When we assume the above scaling theory, this implies that there exists a new universality class $z = 1/\nu = 4$ [16], [17]). The numerical results indicates that the MIT is accompanied by simultaneous antiferromagnetic order transition in the two dimensional Hubbard model. In this case, the antiferromagnetic correlation follows a particular scaling behavior near the transition point. The origin of the destruction of the magnetic order in the metallic region is ascribed to the charge dynamics. In this case, the scaling of the magnetic correlation length ξ_m follows $\xi_m \sim \xi$ where ξ is the length scale of MIT itself discussed before and hence the mean distance of doped holes. In other words, the magnetic correlation length is controlled by ξ because the long-range order is destroyed by the charge motion. We need a care for the definition of ξ_m in metals because the magnetic excitation is gapless in metals leading to a power-law decay at asymptotically long distance at $T = 0$ as discussed above. In general, at the critical point of the magnetic transition, the spin correlation follows

$$\langle \mathbf{S}(0) \cdot \mathbf{S}(\mathbf{r}) \rangle \sim r^{-(d+z-2+\eta)} e^{i\mathbf{Q}\cdot\mathbf{r}} \tag{31}$$

with the ordering vector $\mathbf{Q}$ and the exponent η. In the present case, the magnetic correlation in the insulating phase has to be kept unchanged until the critical point as a function of the control parameter, namely, the chemical potential. This means that the staggered magnetization defined by

$$M_s = \frac{1}{N} \sum_i S_i^z e^{i\mathbf{Q}\cdot\mathbf{r}_i} \tag{32}$$

is a constant in the insulating phase. Therefore the magnetization exponent β_s defined by

$$M_s \propto |\Delta|^{\beta_s} \tag{33}$$

has to satisfy $\beta_s = 0$ and $\eta = 2 - d - z$. Note that this is a continuous transition because ξ_m diverges continuously although M_s jumps at $\Delta = 0$. In the metallic region, this constant staggered magnetization is cut off at ξ_m and the antiferromagnetic correlation has to decay as a power-law beyond ξ_m. The exponent for the power-law at $r > \xi_m$ is determined from the Fermi surface effect in the coherent region of metals.

With this nontrivial definition of ξ_m in mind, the scaling form for the singular part of the free energy is generalized to include the dependence on the staggered magnetic field h_s with the wavenumber $\mathbf{Q}$ as

$$f_s(\Delta, h_s) \propto |\Delta|^{\nu(d+z)} Y(h_s/|\Delta|^{\varphi}) \tag{34}$$

at $T = 0$ where φ is the exponent for h_s. From this expression, the staggered magnetization is derived as

$$M_s = \frac{\partial f_s}{\partial h_s}\Big|_{h_s \to 0} = |\Delta|^{\beta_s}. \tag{35}$$

with the exponent

$$\beta_s = \nu(d+z) - \varphi_s. \tag{36}$$

Therefore, from $\beta_s = 0$, we obtain $\varphi_s = \nu(d+z)$. Then the scaling form of the staggered susceptibility $\chi(\mathbf{Q}, \omega = 0)$ is given by

$$\chi(\mathbf{Q}, \omega = 0) = \frac{\partial M_s}{\partial h_s}\Big|_{h_s \to 0} \sim \Delta^{-\gamma_s} \tag{37}$$

with $\gamma_s = \nu(d+z)$. $\mathrm{Re}\chi(q,\omega)$ and $\mathrm{Im}\chi(q,\omega)/\omega$ both have the peak around $q = \mathbf{Q}$ at $\omega = 0$ with the same critical behavior of the peak height $\propto \Delta^{-\gamma_s}$ and widths $\sim \xi^{-z}$ in the ω space and $\sim \xi^{-1}$ in the k space. The dynamical structure factor is given from the fluctuation dissipation theorem

$$\mathrm{Im}\chi(q,\omega) = (1 - e^{-\beta\omega})\mathrm{Re}S(q,\omega). \tag{38}$$

At $T = 0$, $\mathrm{Re}S(q,\omega)$ also has the peak height $\propto \Delta^{-\gamma_s}$ with the widths ξ^{-z} and ξ^{-1} in the ω and q space, respectively. The consistency of this scaling with the spin correlation $\langle \mathbf{S}(\mathbf{r})\mathbf{S}(0)\rangle$ discussed above is confirmed from the relation

$$\mathrm{Re}S(\mathbf{Q}, t = 0) = \int_{-\infty}^{\infty} \mathrm{Re}S(\mathbf{Q},\omega)d\omega \propto \Delta^{-\gamma_s}\xi^{-z}$$
$$\sim \Delta^{1-\gamma_s} \sim \xi^d \sim \delta^{-1}. \tag{39}$$

This is indeed the scaling of the numerically observed structure factor. Although this is an indirect consistency of the scaling theory with the numerical results, this has given the first and strong motivation to see the universality class in more detail. In the next section, we see more direct and critical numerical test of the scaling theory.

4. Is the Scaling Theory the Correct Description of 2D Metal-Insulator Transition?

In §2, we have discussed the numerical results of the Mott transition from the metallic side. To examine whether the scaling theory correctly describes the MIT in 2D, a more direct test is desired. Recently, a method to probe the transition from the insulator side has been developed by Assaad and Imada [4], [5]. The transition approached from the insulator side also clarifies the character in the metallic side. A large advantage to observe the transition from the insulator side is that we can avoid the negative sign problem known in the quantum Monte Carlo method.

The insulator undergoes a transition to a metal when the chemical potential μ approaches the critical point μ_c from the region of the charge gap. The ground state wavefunction $|\Phi_g\rangle$ in the insulating phase should not change when μ changes because the number of particles is fixed at N within the gap while the Hamiltonian contains μ only through the term μN irrespective of the eigenvectors in the Hilbert space of N particles. Two-body correlation function is also independent of μ within the charge gap. However, the single-particle Green function defined as

$$\mathcal{G}(r, \tau) = \langle Tc(r, \tau)c^\dagger(0, 0)\rangle \tag{40}$$

depends on μ because it propagates a particle created above the gap in $N + 1$-particle sector for the interval of $\tau > 0$. It yields a factor of $e^{-\tau(E_n - \mu)}$ in (40) for a created particle with the energy E_n. Then μ dependence of $\mathcal{G}(r, \tau)$ is simply $e^{\mu\tau}$. Therefore when $\mathcal{G}(r, \tau)$ at $\mu = 0$ is calculated, $\mathcal{G}(r, \tau)$ at $\mu \neq 0$ is given as $\mathcal{G}(r, \tau, \mu \neq 0) = \mathcal{G}(r, \tau, \mu = 0)e^{\mu\tau}$. This means all the $\mathcal{G}(r, \tau)$ for μ within the charge gap can be obtained by calculating $\mathcal{G}(r, \tau)$ once at $\mu = 0$. Because it could contain numerical instability for large τ due to the factor $e^{\mu\tau}$, a numerical stabilization technique for the matrix calculation is necessary [4].

In any case, the calculated Green function provides the localization length ξ_l defined by

$$\mathcal{G}(r, \omega = \mu) \equiv \int_0^\infty \mathcal{G}(r, \tau, \mu)d\tau \sim e^{-r/\xi_l}. \tag{41}$$

ξ_l may be regarded as the localization length of the wavefunction of virtually created state at the chemical potential. The localization length ξ_l has to diverge at the MIT. The scaling of ξ_l near μ_c calculated with the above method by Assaad and Imada [5] for the two-dimenional Hubbard model shows

$$\xi_l \sim |\mu - \mu_c|^{-\nu}$$
$$\nu = 0.26 \pm 0.05. \tag{42}$$

This exponent $\nu \sim 1/4$ is consistent with the exponent of the compressibility analysis given in (4) in the metallic side under the hyperscaling assumption. This direct estimate of the exponent ν lends a strong support for the validity of the scaling theory and the existence of a new universality class.

All the data of the 2D Hubbard model are consistently explained by the scaling theory with the universality class of $z = 4$ and $\nu = 1/4$. On the contrary, these two results can be explained neither by the Hartree Fock approximation nor by the $d = \infty$ results. Indeed, neither the Hartree Fock approximation nor the $d = \infty$ results do not satisfy the hyperscaling assumption (14). For example, the Hartree-Fock approximation predicts $\nu = 1/2$. Furthermore, the scaling of the localization length ξ_l excludes the possibility that the MIT in one dimension and two dimensions belong to the same universality class, because ξ_l scales following $\nu = 1/2$ in one dimension.

5. Suppression of Coherence

In case of the new universality class especially in two dimensions discussed in the previous sections, physical quantities show many anomalous features in terms of the standard MIT. These unusual properties are summarized as the suppression of coherence. In this section we discuss several unusual aspects derived from the new universality class.

The scaling theory predicts that the Drude weight is $D \propto \delta^{1+\frac{z-2}{d}}$. As is well known, $D \propto \delta$ is satisfied in the usual MIT, which is consistent with the scaling theory when $z = 2$. When the universality class is characterized by $z = 1/\nu > 2$, D is suppressed stronger than δ-linear dependence at small δ. In fact, when $z = 4$ is satisfied in 2D, $D \propto \delta^2$ is derived. This is one of the indications for the unusual suppression of coherence in this universality class. Relative suppression of D as compared to the value D_0 for the band-insulator-metal transition, D/D_0, is scaled as $\delta^{\frac{z-2}{d}}$.

From the sum rule, the ω-integrated conductivity gives the averaged kinetic energy

$$-\langle K \rangle = \int_0^\infty \sigma(\omega)d\omega. \tag{43}$$

In the strong coupling limit as in the t-J model, the total kinetic energy $\langle K \rangle$ is expected to be proportional to δ. From the above sum rule, the Drude part $D \propto \delta^{1+\frac{z-2}{d}}$ becomes negligibly small at small δ in the total weight $-\langle K \rangle \propto \delta$ if $z > 2$. This means most of the total weight $-\langle K \rangle$ is exhausted in the incoherent part $\sigma_{\mathrm{reg}}(\omega)$ in (16). In the scaling theory, the form of $\sigma_{\mathrm{reg}}(\omega)$ is not specified and indeed it may depend on details of systems including the interband transition. However, for the intraband contribution, $\sigma(\omega)$ is expected to follow $\sigma(\omega) \sim (1 - e^{-\beta\omega})/\omega$ at very high temperatures $t < T$ because the current correlation function $C(\omega)$ defined by

$$C(\omega) \equiv \int_{-\infty}^\infty dt \, e^{i\omega t}\langle j(0)j(t)\rangle \simeq \frac{\sigma_0/\tau}{-i\omega + \frac{1}{\tau}} \tag{44}$$

becomes temperature and frequency insensitive for $\omega < t$. In fact, $\sigma(\omega) \sim 1/T$ for small ω is a general feature at high temperatures. When the temperature is lowered, this incoherent part $\sim (1 - e^{-\beta\omega})/\omega$ is in general transferred to the Drude part $\sim \frac{1}{-i\omega+\gamma}$ below the coherence temperature. In the usual band-insulator-metal transition, this transfer can take place for the dominant part of the weight because $-\langle K \rangle$ and D both may be proportional to δ. Therefore, the incoherent part is totally reconstructed to the Drude part with decreasing temperature. However, for $z > 2$, this transfer takes place only in a tiny part of the total weight because the relative weight of the Drude part to the whole weight is $\delta^{(z-2)/d}$ which vanishes as $\delta \to 0$ for $z > 2$. Even at $T = 0$, the major part of the conductivity weight must be exhausted in the incoherent part. Since we have no reason to have a dramatic change of the form for $\sigma_{\mathrm{reg}}(\omega)$ at any temperatures from the scaling point of view, the optical conductivity more or less follows the form

$$\sigma_{\mathrm{reg}}(\omega) \sim C \frac{1 - e^{-\beta\omega}}{\omega} \tag{45}$$

even for $t > \omega$ and $t > T$ where we have neglected model dependent feature such as the interband transition. This "intraband" incoherent weight may be proportional to $C \propto \delta$. Such dominance of the incoherent part is consistent with the numerical results of 2D systems [25]. The scaling of $\sigma(\omega)$ by (45) is another indication for the support of $z > 2$ in 2D. On the contrary, the numerical results in 1D by Stephan and Horsch [26] support that the incoherent part is small at low temperatures. This may be due to $z = 2$ in 1D because the majority of the weight seems to be exhausted in the Drude weight.

Another suppression of coherence is seen in the coherence temperature T_{coh}. The scaling of T_{coh} is given by $T_{coh} \propto \delta^{z/d}$. Because the standard MIT is characterized by $T_{coh0} \propto \delta^{2/d}$, the relative suppression T_{coh}/T_{coh0} is again proportional to $\delta^{\frac{z-2}{d}}$. When $z = 4$ as suggested by numerical results in 2D, we obtain $T_{coh} \propto \delta^2$ in 2D in contrast with $T_{coh0} \propto \delta$ for $z = 2$.

In the Mott insulating phase, entropy coming from the spin degrees of freedom is released essentially to zero when the antiferromagnetic order exists. The entropy is also released when the spin excitation has a gap as in the spin gap phase. Even without such clear phase change, the growth of short-ranged correlation progressively releases the entropy with decreasing temperature. When carriers are doped, an additional entropy due to the charge degrees of freedom is introduced in proportion to δ. This additional entropy is assigned not solely to the charge degrees of freedom but also to the spin entropy because they are coupled. This additional entropy proportional to δ is the origin of the destruction of the antiferromagnetic long-range order. This additional entropy $\propto \delta$ has to be released below the coherence temperature T_{coh}. Therefore, a natural consequence is that if T-linear specific heat characterizes the degenerate (coherent) temperature region, the coefficient γ should be given by $\gamma \sim \delta/T_{coh} \sim \delta^{1-\frac{z}{d}}$. This is indeed the scaling law we obtained in (26). From this heuristic argument, it turns out that the metallic phase near the Mott insulator is characterized by large residual entropy at low temperatures which is also related with the suppression of the antiferromagnetic correlation as compared to the insulating phase. It may also be said that the anomalous suppression of the coherence at small δ is caused by short-ranged antiferromagnetic correlations which scatters carriers incoherently.

In the Mott insulator, because the single-particle process is suppressed due to the charge gap, we have to consider the two-particle process expressed by the superexchange interaction. Similarly to this insulating case, even in the metallic region, such suppressions of coherence discussed above for the single-particle process lead to the necessity to consider the two-particle process explicitly. In contrast with the insulating phase, the two-particle processes in metals contain an effective pair hopping term and a term described by

$$\mathcal{H}_W = -t_W \sum_i [\sum_{\delta\sigma}(c_{i\sigma}^\dagger c_{i+\delta\sigma} + c_{i+\delta\sigma}^\dagger c_{i\sigma})]^2 \tag{46}$$

becomes relevnt. The coherence is not suppressed in case of two-particle process because the universality class of the two-particle transfer is given by $z = 1/\nu = 2$. Then it becomes more relevant than the single-particle transfer for small δ. In fact, this term does not suffer from the mass enhancement effect. This new term may lead to the appearance of the superconducting phase [27]. The transition from superconductor to the Mott insulator is indeed characterized by $z = 1/\nu = 2$. This shows an instability of metals near the Mott insulator to the superconducting pairing or in more general an instability of $z = 4$ universality phase to a symmetry-broken state.

Acknowledgements

The author thanks F.F. Assaad, N. Furukawa and D.J. Scalapino for collaborations and fruitful discussions.

REFERENCES

[1] N. Furukawa and M. Imada, J. Phys. Soc. Jpn. **60**, 3669 (1991).

[2] A. Moreo, E. Dagotto and D. J. Scalapino, Phys. Rev. B **43** 11442 (1991)

[3] N. Furukawa and M. Imada, J. Phys. Soc. Jpn. **61** 3331(1992).

[4] F.F. Assaad and M. Imada, J. Phys. Soc. Jpn., **65**, 189 (1996).

[5] F.F. Assaad and M. Imada, Phys. Rev. Lett., **76**, 3176 (1996).

[6] N. Furukawa and M. Imada, J. Phys. Soc. Jpn. **62** 2557 (1993).

[7] J. Jaklič and P. Prelovšek, Cond-mat/9603081.

[8] M. Kohno, Phys. Rev., **55**, 1435 (1997).

[9] S. White, D. J. Scalapino, R. Sugar, E. Loh, J. Gubernatis and R. Scalettar, Phys. Rev. B**40**, 506 (1989).

[10] M. Imada, N. Furukawa and T. M. Rice, J. Phys. Soc. Jpn. **61** 3861 (1992).

[11] Y. Iino and M. Imada, J. Phys. Soc. Jpn., **64**, 4392 (1995).

[12] S. Chakraverty, B. I. Halperin and D. R. Nelson, Phys. Rev. B **39**, 2344 (1989).

[13] M. S. Makivić and H. -Q. Ding, Phys. Rev. B**43**, 3562 (1991).

[14] M. Imada, J. Phys. Soc. Jpn. **63**, 851 (1994).

[15] J. Jaklič and P. Prelovšek, Phys. Rev. Lett. **74**, 3411 (1995).

[16] M. Imada, J. Phys. Soc. Jpn. **63**, 4294 (1994).

[17] M. Imada, J. Phys. Soc. Jpn. **64**, 2954 (1995).

[18] M. Imada, J. Low Temp.Phys. **99**, 437 (1995).

[19] W. Kohn, Phys. Rev. A**133** 171 (1964).

[20] D. J. Thouless, Phys. Rep. **13C** 94 (1974).

[21] B. S. Shastry and B. Sutherland, Phys. Rev. Lett., **65**, 243 (1990).

[22] N. Byers and C.N. Yang, Phys. Rev. Lett. **7**, 46 (1961).

[23] D. J. Scalapino, S. R. White and S. Zhang, Phys. Rev. B**47** 7995 (1993).

[24] M. A.. Continentino, Phys. Rep. **239** 179 (1994).

[25] J. Jaklič and P. Prelovšek, Phys. Rev. B **52**, 6903 (1995).

[26] W. Stephan and P. Horsch, Phys. Rev. B **42** 8736 (1990).

[27] F. F. Assaad, M. Imada and D. J. Scalapino, Phys. Rev. Lett. **76** 4592 (1996) .

Quantum Zero-Point Critical Fluctuations in Arrays of Ultrasmall Josephson Junctions

J.V. José

Physics Department and Center for the Interdisciplinary Research
on Complex Systems, Northeastern University, Boston, MA 02115, USA

We briefly review results from extensive quantum Monte Carlo
and semiclassical analytic studies of models of ultrasmall Josephson
junction arrays (JJA). Specifically, we mention results: (i) close to
the renormalized semi-classical critical temperature and its successful comparison to experiment. (ii) The existing renormalization-
group, self-consistent harmonic approximation and QMC evidence
for a low temperature quantum induced phase transition in a mutual capacitance dominated JJA in zero magnetic field. (iii) duality transformations and phase structure of two capacitively coupled
JJA that leads to a structure reminiscent of a fractional quantum
Hall effect type system.

I. INTRODUCTION

Josephson junction arrays (JJA) have been the source of many theoretical
and experimental studies in the last few years [1]. This interest has been in
part, because JJA represent experimental realizations of the two-dimensional
XY model, one of the most studied theoretical spin models. Furthermore, due
to the recent advances in submicrometer technology it has been possible to
fabricate relatively large arrays of ultrasmall SIS (superconductor to insulator
to superconductor) Josephson junctions [2–5]. These arrays can have junctions
with areas that can vary from a few microns to submicron sizes, with effective capacitances that can be smaller than a few femtoFarads (fF=10×10^{-15}
Farads). Under these circumstances the electric field between the areas that
form the junctions has to be quantized [11]. The nature of the possible order
present in these devices depends on the competition between the Josephson
energy E_J, that tries to establish long range superconducting phase coherence,
and the capacitive charging energy E_{C_m} that disrupts this order. The relevant
quantity to study this competition is the quantum parameter

$$\alpha_m = \frac{E_{C_m}}{E_J}.$$

(1)

In this paper I briefly present the results of work done in collaboration with C.
Rojas (more details are given in [7], in his Ph.D. thesis [8] and in [9]). We have
studied the phase coherence of models of JJA that quantitatively represent the

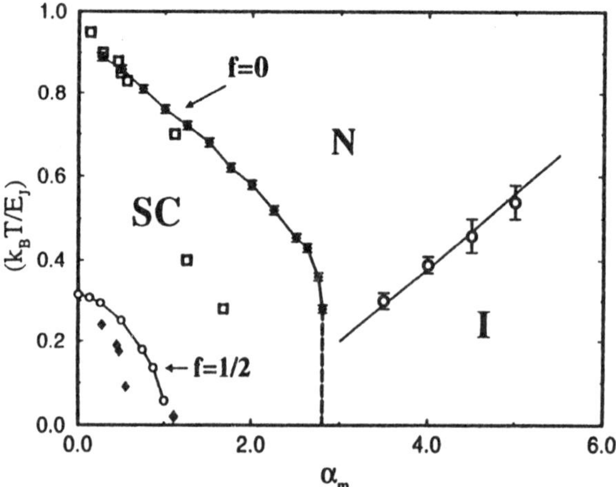

FIG. 1. Phase diagram T vs. quantum parameter α_m, for $f = 0$ and for full frustration $f = 1/2$. SC stands for superconducting, N=normal and I=insulating phases, respectively. The experimental results are denoted by squares. The quantum Monte Carlo results are denoted by crosses (f=0) and circles (f=1/2) joined by continuous lines as guides to the eye. The QMC results include the statistical error bars in the calculations.

experimentally fabricated devices, including the specific experimental values for α_m [3]. For the most part the experimental systems have been two-dimensional, but quasi-three-dimensional samples have also been fabricated. Here quasi- means two layers of JJA capacitively coupled at each lattice site [10]. From the type of junctions fabricated there are two main contributions to the charging energy. The charging energy associated with the cost of adding a single charge to a superconducting island given by $E_{C_s} = \frac{e^2}{2C_s}$, and the energy necessary to transfer a charge from an island to a nearby one given by $E_{C_m} = \frac{e^2}{2C_m}$, with e the electronic charge. For the Delft experiments, the self capacitances were typically on the order $C_s \sim 3 \times 10^{-17}$F, while the mutual capacitances were $C_m \sim 1 \times 10^{-15}$F. This means that C_m can be two orders of magnitude larger than C_s.

The resulting temperature vs α_m phase diagram, for zero, $f = 0$, and full, $f = 1/2$, frustration [1] are shown in Fig. (1). For small α_m there is a superconducting phase in which the Cooper pairs are delocalized and the vortex excitations are localized. On the right hand side the system has delocalized

---

[1] Here $f \equiv \frac{\Phi}{\Phi_0}$, where Φ is the magnetic flux through a plaquette and $\Phi_0 = \frac{h}{2e}$ is the quantum of flux

vortices and thus it is an insulator. In the following sections we discuss experimental aspects of this phase diagram, their comparison to our quantum Monte Carlo simulation results as well as a combination of approximate analytic techniques that also allowed us to check the reliability of the QMC results.

II. TWO-DIMENSIONAL CASE

A. Model

The appropriate model Hamiltonian representing the JJA ($f = 0$)is

$$\hat{\mathcal{H}} = \frac{q^2}{2} \sum_{\vec{r}_1,\vec{r}_2} \hat{n}(\vec{r}_1)\tilde{\mathbf{C}}(\vec{r}_1,\vec{r}_2)\hat{n}(\vec{r}_2) + E_J \sum_{<\vec{r}_1,\vec{r}_2>} [1 - \cos(\hat{\phi}(\vec{r}_1) - \hat{\phi}(\vec{r}_2))],$$

(2)

where $q = 2e$; $\hat{\phi}(\vec{r})$ is the quantum phase operator, $\hat{n}(\vec{r})$ is its canonically conjugate number operator which measures the excess number of Cooper pairs in the $\vec{r}$ island. These operators satisfy the commutation relations [11] $[\hat{n}(\vec{r}_1), \hat{\phi}(\vec{r}_2)] = -i\delta_{\vec{r}_1,\vec{r}_2}$. The matrix $\tilde{\mathbf{C}}(\vec{r}_1,\vec{r}_2)$ is the electric field Green function and its inverse, $\mathbf{C}(\vec{r}_1,\vec{r}_2)$, is the geometric capacitance matrix, which must in principle be obtained from solving the Poisson equation subject to the appropriate boundary conditions. This is not easy to do in general and typically this matrix is approximated, both theoretically and in the analysis of the experiments, by diagonal plus nearest neighbor contributions: $\mathbf{C}(\vec{r}_1,\vec{r}_2) = (C_s + zC_m)\delta_{\vec{r}_1,\vec{r}_2} - C_m \sum_{\vec{d}}\delta_{\vec{r}_1,\vec{r}_2+\vec{d}}$, where the $\vec{d}$ vector runs over nearest neighboring islands, and z is the coordination number.

Here we are interested in calculating the thermodynamic properties of the model defined by $\hat{\mathcal{H}}$. The quantity of interest is the partition function $Z \equiv \mathrm{Tr}\left\{e^{-\beta\hat{\mathcal{H}}}\right\}$. The trace is taken either over the phase variables, $\hat{\phi}$, or over the charge number operators, $\hat{n}$. To evaluate the partition function we used its path integral representation [12]. This means that we add one more dimension to the problem, the imaginary time dimension τ, with a range $\tau\epsilon[0,\beta\hbar]$, where $\hbar = h/2\pi$, with h Planck's constant and $\beta = 1/k_BT$, where T is the absolute temperature and k_B Boltzmann's constant. At zero temperature we have an anisotropic three-dimensional system, while at finite temperatures it is quasi-three dimensional. To evaluate the partition function in this representation we discretize the imaginary time axis in L_τ slices with spacing $\epsilon = \frac{\beta\hbar}{L_\tau}$.

To write a convenient expression for the partition function in the imaginary time representation we use the Poisson summation formula [18] and obtain the explicit form [7]

$$Z = \prod_{\tau=0}^{L_\tau-1} \sqrt{\det[C]} \prod_{\vec{r}} \int_0^{2\pi} \sqrt{\frac{L_\tau}{2\pi\beta q^2}}\, d\phi(\vec{r},\tau) \sum_{m(\vec{r},\tau)=-\infty}^{\infty} \exp\left[-\frac{1}{\hbar} S[\{\phi\},\{m\}] \right].$$

$$\tag{3}$$

Here we defined the action

$$\frac{1}{\hbar} S[\{\phi\},\{m\}] = \sum_{\tau=0}^{L_\tau-1} \left[\frac{\beta}{L_\tau} H_J(\{\phi(\tau,\vec{r})\}) + \frac{L_\tau}{2\beta q^2} \sum_{\vec{r}_1,\vec{r}_2} [\phi(\tau+1,\vec{r}_1) - \phi(\tau,\vec{r}_1) \right.$$

$$\left. +2\pi m(\tau,\vec{r}_1)] \times \mathbf{C}(\vec{r}_1,\vec{r}_2)[\phi(\tau+1,\vec{r}_2) - \phi(\tau,\vec{r}_2) + 2\pi m(\tau,\vec{r}_2)] \right] +$$

$$+O(1/L_\tau^2), \tag{4}$$

where the important quantization condition $\phi(L_\tau,\vec{r}) = \phi(0,\vec{r})$ is implicit. To recover the quantum solutions at low temperatures, where we expect novel things to occur, we need to take the continuum limit in the imaginary time direction, which can be done by taking fixing ϵ and taking L_τ very large. This is what makes the QMC computer simulations of this problem difficult and that's why we need to have alternative analytic ways to check the numerical results wherever possible. We discuss this in more detail in the following sections.

III. THE SIMULATION APPROACH

To carry out our QMC calculations, we tried different algorithms but ended up settling with the standard Metropolis method since we needed to update not only the phases but also the integers given in the partition function Z (Eqs. (3), (4)) together with the quantum boundary conditions. The advantage of this approach is that it is general enough to be used over all the α_m ranges covered in the phase diagram. We then have a set of angles $\phi(\tau,\vec{r}) \in [0, 2\pi)$, defined at the nodes of a three-dimensional cubic lattice, with two space dimensions, L_x and L_y, and one imaginary time dimension, L_τ. The quantum imaginary time periodic boundary condition appears from the trace condition in Z, and we also took periodic boundary conditions along the spatial directions. The link variables $m(\tau,\vec{r})$ are defined in the bonds between the lattice sites along the τ-direction and they can take any integer value.

As the phases are updated we restrict their values to the interval $[0, 2\pi)$, so that if after an update a $\phi(\tau,\vec{r})$ is outside this interval, we carry out the transformations,

if $\phi(\tau,\vec{r}) < 0$, then $\qquad\qquad$ if $\phi(\tau,\vec{r}) > 2\pi$, then

$$
\begin{array}{llll}
\phi(\tau,\vec{r}) & \to & \phi(\tau,\vec{r}) + 2\pi, & \qquad \phi(\tau,\vec{r}) \to \phi(\tau,\vec{r}) - 2\pi, \\
m(\tau,\vec{r}) & \to & m(\tau,\vec{r}) + 1, & \qquad m(\tau,\vec{r}) \to m(\tau,\vec{r}) - 1, \\
m(\tau+1,\vec{r}) & \to & m(\tau+1,\vec{r}) - 1. & \qquad m(\tau+1,\vec{r}) \to m(\tau+1,\vec{r}) + 1.
\end{array}
\tag{5}
$$

From Eq.(4), we see that the action is invariant under these transformations. Moreover, the shifts in the column and individual phase moves are adjusted to keep the acceptance rates in the range $[0.2, 0.3]$.

When α_m is small, the fluctuations of the phases along the imaginary time axis as well as the fluctuations in the m's are suppressed by the second term in Eq. (4). Attempts to change a phase variable have then a very small success rate. Therefore we implemented two types of Monte Carlo moves in the phase degrees of freedom. In one sweep of the array we update $L_x \times L_y$ imaginary time columns, by shifting all the phases along a given column by the same angle. This move does not change the second term in Eq. (4), and thus it probes only the Josephson energy [13]. To account for phase fluctuations along the imaginary time axis, which become more likely as (α_m/T) increases, we also make local updates of the phases along the columns. We did check the reliability of this procedure by comparing with our analytic RG-WKB and self-consistent harmonic approximation results [7]. Another important aspect of the implementation of the MC algorithm is the order in which we visit the array. This is relevant for the optimization of the computer code for different computer architectures. In a scalar machine we used an algorithm that updates column by column in the array. For a vector machine we used the fact that for local updates, like the ones we used, the lattice can be separated into four sublattices in a checkerboard-like pattern. This partition is done in such a way that each of the sublattices can be updated using a long vector loop without problems of data dependency. Using this last visiting scheme, the cpu time grows sublinearly with the size of the array. One of the problems that this type of visiting scheme has in a vector machine, like the Cray C90, is that the array's dimensions have to be even, and this produces memory conflicts. We have not made further attempts to optimize this part of the code.

We also replaced the $U(1)$ symmetry of the phases by the subgroup $Z(N)$, with $N = 5000$ [13]. This allows the use of integer arithmetic for the values of the phase variables, and to store lookup tables for the Josephson cosine part of the Boltzmann factors. This can not be done for the charging energy contribution to the Boltzmann factors, except in the $C_{\mathrm{m}} = 0$ case where the m's can be summed up in a virtually exact form. In the latter case we can also store lookup tables using the following definition of an effective potential V_{eff},

$$\exp\left[-\left(\frac{L_\tau C_{\mathrm{s}}}{q^2\beta}\right) V_{\mathrm{eff}}(\phi)\right] = \sum_{m=-\infty}^{\infty} \exp\left[-\frac{1}{2}\left(\frac{L_\tau C_{\mathrm{s}}}{q^2\beta}\right)(\phi + 2\pi m)^2\right]. \qquad (6)$$

These sums can be evaluated numerically to any desired accuracy.

We calculated the thermodynamic averages after doing N phase updates and M, m updates across the whole lattice. Typically, for α_{m} small we used $N = 4$ and $M = 1$. In the opposite limit we used $N = 1$ and $M = 8, 10, \ldots$ depending on the α_m and T values. We needed to do this since, due to the long range interaction among the charges, our local m updating algorithm had serious decorrelation time problem. We often found that in order to get reasonably small statistical errors, we needed to perform about at least $N_{\mathrm{meas}} = 2^{12} = 4096$ measurements of the thermodynamical quantities, while other times we made up to $N_{\mathrm{meas}} = 2^{13} = 8192$ measurements.

A. High temperature results

In this section we present some of our high temperature quantum Monte Carlo results. In this limit the results are quite reliable since essentially $\beta\hbar \ll 1$ and thus the quantum fluctuations act mostly to renormalize the classical results, as originally found many years ago [15]. A way to measure the long range phase coherence in the model is by calculating the helicity modulus, which is defined by $\Upsilon = \left.\frac{\partial^2 F}{\partial A^2_{\vec{r},\vec{r}+\hat{x}}}\right|_{A=0}$. Here $e^{-\beta F} = Z$, defines the free energy F, and A denotes a twist of the phases along the $\hat{x}$ direction, so that Υ gives the response of the system to this twist, and it is proportional to the superfluid density. We calculated Υ in the small α_{m} region and the inverse dielectric constant ϵ^{-1} in the large α_{m} regime, and both quantities in the intermediate region.

Most of the calculations we performed used the capacitance values from experiment. These were the **only** fitting parameters used in the calculations. In particular, the ratio between the self and mutual capacitance was kept fixed around the values $C_{\mathrm{s}}/C_{\mathrm{m}} \approx 0.01$ and 0.03, with the bulk of the calculations carried out for 0.01. We found that in the helicity modulus case both values give essentially the same results. Almost all of the calculations were done by lowering the temperature, in order to reduce the chances for the system to be trapped in metastable states [14]. Our results are given in Fig. (1). There we see that the experimental and numerical results are indeed quantitatively close in the $f = 0$ case and have the correct qualitative trend in the $f = 1/2$ case.

We have also carried out renormalization group WKB (RG-WKB) calculations valid for $\beta\hbar \ll 1$ in the mutual capacitance dominated regime, and they agree quite well with experiment and QMC calculations in the small α_m regime of Fig. (1). The RG-WKB results are in principle only valid for $\beta\hbar \ll 1$. However a persistent property of the results is the suggestion of a low temperature instability. We discuss this possibility next.

B. Low temperature results

As we mentioned above, motivated by the RG-WKB results, there has been an extensive search for what we called a **QUIT** (QUantum Induced Transition). Originally this prediction was made for the case where the capacitive matrix only has a self-capacitive term. This means that there is **only** a superconducting phase in the T-vs-α phase diagram. For the specific type of fabricated

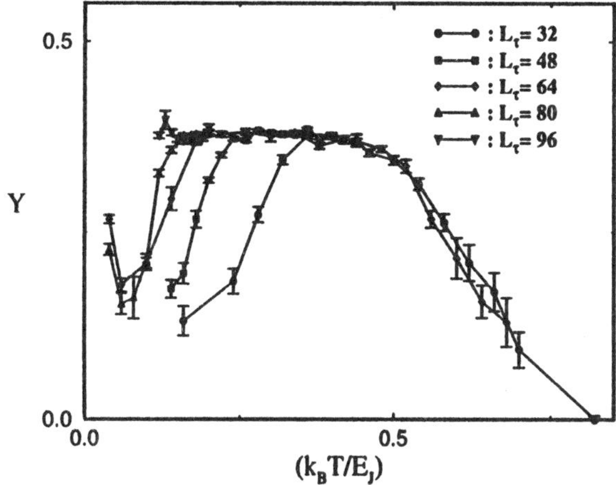

FIG. 2. Υ vs temperature for $\alpha_m = 2.25$, $L_x = L_y = 20$, and $L_\tau = 32$, 48, 64, 80, and 96. At this value of α_m there is a drop in the helicity modulus at low temperatures. This abrupt drop is probably due to having a finite L_τ.

samples it was then necessary to explicitly consider the case where both C_s and C_m where included and where C_m is two orders of magnitude larger than C_s. We carried out the corresponding RG-WKB calculations and again they led to the possibility of having a low temperature QUIT [7]. An important aspect of the analytic results is that to leading order in α_m the $T_{QUIT} \sim \alpha_m$, independent of an applied magnetic field. This means that T_{QUIT} for zero and for $f = 1/2$ fields, to leading order in α, must be the same. This result is important since to test the existence of a QUIT in a field, as was tried in Ref. [14] much lower temperatures need to be simulated. Here we concentrate in the $f = 0$ case to keep things clearer, since the $f = 1/2$ case has its own special excitations that may confuse the T_{QUIT} issue. Our approach was then to calculate Υ for relatively large values of α_m so as to enhance the possibility of seeing the transition but this implied that much larger lattices and runs had to be carried out. In Fig. (2) we show typical results for Υ for fixed values of α but varying L_τ. There we see that there is a reentrant type behavior for Υ, but it shows that as L_τ increases the reentrant temperature decreases. The important question is what happens in the $L_\tau \to \infty$ limit. In Fig. (3) we show the $L_\tau \to \infty$ for $\alpha_m = 2.0$. There we see that the extrapolated $\Upsilon(L_\tau \to \infty)$ gives evidence for $T_{QUIT}(\alpha_m = 2.0) \neq 0$ the same is true for a calculation with $\alpha_m = 2.25$, however for $\alpha_m = 2.5$ the extrapolated value for T_{QUIT} becomes negative indicating that it must be zero. The conclusion we draw from these calculations is that there is a critical value for α_m below which there is a finite $QUIT$ and above which it is zero. We also carried out an analytic self consisted harmonic approximation (SCHA) analysis to carefully study the important dependence on L_τ. We did find quantitative agreement between our

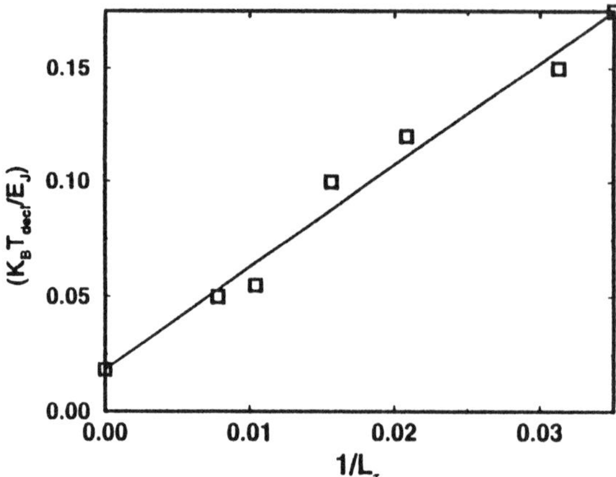

FIG. 3. Estimate of the $QUIT$ transition vs $1/L_\tau$ extracted from results similar to those of Fig. 2 for $\alpha_m = 2.0$, $L_x = L_y = 20$ and $L_\tau = 48, 64, 80, 96, 128$. The line from a least squares fit suggests a non zero T_{QUIT} in the limit $1/L_\tau = 0$.

analytic $SCHA$ results and those obtained from our QMC analysis thus giving further validity to the QMC results obtained [7]. Therefore we conclude that our evidence points to the existence of a $QUIT$ for the system where the mutual capacitance dominates. There is also some experimental suggestion for a low temperature instability but it is not yet conclusive [3]. More experimental and theoretical work is needed to ascertain the real existence of this $QUIT$.

IV. QUASI-THREE-DIMENSIONAL ARRAYS

In the previous section we presented our results of a two-dimensional array of ultrasmall Josephson junctions. Here we are interested in the problem of a quasi-three-dimensional array composed of **two** JJA layers coupled at each site by a very small capacitor. Our studies were motivated by the fabrication of this type of sample at Delft [10]. The fabricated samples had for each layer $L_x = 230$, $L_y = 60$, $C_m \approx 2.3 fF$, and $C_{int} \approx 0.6 fF$, where C_{int} is the local interaction capacitance between the the two JJA layers.

The motivation for studying this system is potentially quite interesting since, as mentioned above, by independently varying each of the layers parameters each array can be either in the Semiclassical (SC) or Quantum (Q) regimes. The two-array system can then be in one of three distinct configurations: To wit, SC_1-SC_2, SC_1-Q_2, or Q_1-Q_2. Each one of the JJA layers is either charge or vortex dominated, thus the question arises as to how will the interactions affect the behavior of the full two-layered system. Here we are assuming that one can measure the properties of each array independently.

We find that a general effect of the electrostatic coupling between the two JJA, if they are in the SC$_1$-SC$_2$ case, is to increase the critical temperature of the superconducting state, as one would have expected. We obtained this result by approximate analytic analysis as well as QMC simulations. This reasoning is briefly mentioned below. The more interesting SC$_1$-Q$_2$ case is more complicated but within the Villain approximation, and techniques used by Fazio and Shön [6], we show that there are interesting duality transformation in this system as well as interesting fractional quantum Hall effect like scenarios for this JJA configuration. These results have been found independently by two groups. [16,8,17,9].

A. The model

The system consists of two planar arrays on top of each other. The intra-array interaction between the superconducting islands in each array is of the same form as discussed in the previous sections, including capacitive and Josephson couplings. The two planes are only capacitively coupled. The model Hamiltonian reads,

$$\hat{H} = \frac{q^2}{2} \sum_{\vec{r}_1,\vec{r}_2} \sum_{\mu,\nu} \hat{n}_\mu(\vec{r}_1)\tilde{\mathbf{C}}_{\mu,\nu}(\vec{r}_1,\vec{r}_2)\hat{n}_\nu(\vec{r}_2) + F_1(\{\hat{\phi}_1\}) + F_2(\{\hat{\phi}_2\}). \tag{7}$$

where $q = 2e$ is the charge of the Cooper pairs, and the index $\mu = 1,2$ labels the two arrays. The dynamical variables for the array μ are $\hat{\phi}_\mu$, $\hat{n}_\mu$. The functions F_μ are the Josephson interaction terms,

$$F_\mu(\{\hat{\phi}\}) = E_J^{(\mu)} \sum_{<\vec{r}_1,\vec{r}_2>} (1 - \cos(\hat{\phi}_\mu(\vec{r}_1) - \hat{\phi}_\mu(\vec{r}_2))). \tag{8}$$

Here $E_J^{(\mu)}$ is the Josephson coupling for the junctions in array μ. The inverse capacitance matrix can be found from the capacitance supermatrix $\mathbf{C}_{\mu,\nu}$ which is made of four blocks labeled by $\mu,\nu = 1,2$,

$$\mathbf{C}_{\mu,\nu}(\vec{r}_1,\vec{r}_2) = \begin{cases} \left(C_{\rm s}^{(\mu)} + zC_{\rm m}^{(\mu)} + C_{\rm int}\right) & \text{if } \mu = \nu \text{ and } \vec{r}_1 = \vec{r}_2, \\ -C_{\rm m}^{(\mu)}, & \text{if } \mu = \nu \text{ and } \vec{r}_1 = \vec{r}_2 \pm \vec{d}, \\ -C_{\rm int}, & \text{if } \mu \neq \nu \text{ and } \vec{r}_1 = \vec{r}_2, \\ 0, & \text{otherwise.} \end{cases} \tag{9}$$

The diagonal blocks of this matrix are the intra-array capacitance matrices. It is then easy to find the inverse $\tilde{\mathbf{C}}$ of Eq. (9), defined by

$$\sum_\nu \sum_{\vec{r}_2} \tilde{\mathbf{C}}_{\mu,\nu}(\vec{r}_1,\vec{r}_2)\mathbf{C}_{\nu,\rho}(\vec{r}_2,\vec{r}_3) = \delta_{\mu,\rho}\, \delta_{\vec{r}_1,\vec{r}_3}. \tag{10}$$

The diagonal blocks $\mathbf{C}_{\mu,\mu}$ represent the intra-array capacitances and we denote them by $\mathbf{C}_\mu = \mathbf{C}_{\mu,\mu}$. The Hamiltonian given in Eq.(7) is then the starting point in our study.

B. High temperature semiclassical limit

We began by studying the change in each array's T_c as a function of the coupling capacitance C_{int}. As in the one-array problem, we started by using a semiclassical expansion to study the system for small ratios of α_μ [15]. If $C_{\text{int}} = 0$, the arrays are independent. For the experimental values of the Delft samples the ratio (C_{int}/C_s) is also small. Therefore, in addition to the $\alpha_\mu \ll 1$ semiclassical approximation we also considered the change in T_c as a function of $C_{\text{int}} < 1$. We found that in order to have a consistent semiclassical expansion in C_{int}, we needed to carry out the expansion up to second order in $\tilde{\mathbf{C}}$, which implies that we needed to perform the expansion up to second order in q^2. The end result of our calculation is a $Z = \int d\bar{\phi} \, \exp\{-S_{\text{eff}}[\bar{\phi}]\}$, with the effective classical action given by [8],

$$
S_{\text{eff}}[\phi] = \beta \sum_\mu \left\{ 1 - \frac{\beta q^2}{12} \left[\tilde{\mathbf{C}}_{\mu,\mu}(|\vec{0}|) - \tilde{\mathbf{C}}_{\mu,\mu}(|\vec{d}|) \right] + \right.
$$

$$
\left. + \frac{(q^2\beta)^2}{1152} \left[\tilde{\mathbf{C}}_{\mu,\mu}(|\vec{0}|) - \tilde{\mathbf{C}}_{\mu,\mu}(|\vec{d}|) \right]^2 \right\} F_\mu(\phi_\mu)
$$

$$
+ \frac{(q^2\beta)^2\beta^2}{720} \text{Tr} \left\{ \sum_{\mu,\nu} \tilde{\mathbf{C}}_{\mu,\nu} \frac{\partial^2 F_\nu}{\partial \phi_\nu^2} \tilde{\mathbf{C}}_{\nu,\mu} \frac{\partial^2 F_\mu}{\partial \phi_\mu^2} \right\}. \tag{11}
$$

Note that if we keep terms up to first order in q^2, we recover the single layer results, and that the contribution to second-order in q^2 has the opposite sign compared with the sign of the first-order. Of importance then is the fact that since higher orders in q^2 coincide with higher orders in β then, up to second order, the superconducting state would be stable at low temperatures. This conclusion does not, by itself, rule out the possibility of having a $QUIT$ for the two-array problem but it does show that this situation may be very different than for the one array case, although higher order terms in the semiclassical expansion may reverse the low-temperature limit.

Unlike the first-order expansion in q^2, Eq. (11) does not have a simple form since the third term introduces nonlocal interactions, making a direct calculation of T_c rather difficult. In order to ascertain the role of C_{int} on T_c, we performed a variational calculation with two possible "trial" functions for the semiclassical partition function given in Eq. (11). In particular for the self-capacitance only limit, and with identical C_s for the two arrays, the calculation of the effective or renormalized quantum parameter α is

$$\alpha_{\text{eff}} = \left(\frac{1 + C_{\text{int}}/C_{\text{s}}}{1 + 2C_{\text{int}}/C_{\text{s}}} \right) \alpha. \tag{12}$$

The first conclusion drawn from this analysis is that, in the semiclassical limit, the inter-plane capacitance makes the problem less quantum mechanical by increasing the critical temperature. Another interesting conclusion from these equations is that as the ratio $(C_{\text{int}}/C_{\text{s}})$ grows, T_c increases, reaching asymptotically a plateau. We have also performed quantum Monte Carlo calculations in this limit using the semiclassical action given in Eq. (11) confirming the analytic variational results [8].

Here we have only presented the case where both arrays are equal and the mutual capacitance is zero, because the analytic variational calculation was easier in that case. However we have also considered the more general semiclassical limit using quantum Monte Carlo simulations and the trends in the more general case are qualitatively the same.

The explicit change of T_c up to second order in the effective quantum parameter α_{eff} can be inferred from Eq. (12) and we get,

$$\left(\frac{k_B T_c}{E_J} \right) = \left(\frac{k_B T_c^{(0)}}{E_J} \right) - \left(\frac{2}{3} \right) \alpha_{\text{eff}} + \left(\frac{189}{180} \right) \alpha_{\text{eff}}^2 + O(\alpha_{\text{eff}}^3), \tag{13}$$

where the known variational result for the classical 2-D XY model is $(k_B T_c^{(0)}/E_J) = 2$. As usual, the variational approximation overestimates the value of the critical temperature, but the qualitative trend is the correct one.

C. The Quantum Limit

In the previous section we considered the semiclassical limit of two arrays with small α's, that is both were in the semiclassical regime. In this section we will consider the case C_1-Q_2, that is when the dynamics of one of the arrays is dominated by the motion of charges, the superconducting phase, and the other by the motion of the vortices, the insulating phase. Charges and vortices are related via the uncertainty relation between the phase of the order parameter and the Cooper pairs number operator. These two kinds of excitations dominate the spectrum in different parts of the T-vs-α parameter space. The charge excitations dominate in the semiclassical regime, that is to say, when the Josephson energy is much larger than the charging energy, while the vortices dominate in the opposite limit, the quantum mechanical regime, see Fig. (1). Within the same array then, vortices and Cooper pairs are for the most part mutually exclusive. This means that in the region of the parameter space where vortices are well defined, the Cooper pairs are not and vice-versa. In this case to probe the charge-vortex system we could add an external magnetic field to the system.

By contrast in the capacitively coupled JJA problem we find that the coupling between the charge and vortex excitations in different layers strongly depends on C_{int}. Furthermore, the effective vortex-charge interaction is delocalized and of finite range. This is, however, only true if the arrays are coupled capacitively. We briefly describe the results of this analysis and more details will be reported elsewhere [16,8,9].

Our analysis uses and extends to the two-array problem the techniques developed for one array by Fazio et. al. [6]. An equivalent analysis and similar results to ours have also been independently derived by Blanter et al. [17]. Below we outline the reasoning of our analysis.

The path integral expression for one array reads,

$$Z = \prod_{\vec{r}} \prod_{\tau} \sum_{n(\tau,\vec{r})} \int_0^{2\pi} \frac{d\phi(\tau,\vec{r})}{2\pi} \exp\left\{ -\int_0^{\beta\hbar} d\tau \times \left[\right.\right.$$

$$\frac{q^2}{2} \sum_{\vec{r}_1,\vec{r}_2} n(\tau,\vec{r}_1)\tilde{\mathbf{C}}(\vec{r}_1,\vec{r}_2)n(\tau,\vec{r}_2) + i \sum_{\vec{r}} n(\tau,\vec{r})\frac{d\phi}{d\tau}(\tau,\vec{r}) \tag{14}$$

$$\left.\left. +E_J \sum_{<\vec{r}_1,\vec{r}_2>} \left(1 - \cos\phi_{\vec{r}_1,\vec{r}_2}(\tau)\right) \right]\right\}, \tag{15}$$

where we have used the notation $\phi_{\vec{r}_1,\vec{r}_2}(\tau) = \phi(\tau,\vec{r}_1) - \phi(\tau,\vec{r}_2)$, and formally taken the limit $L_\tau \to \infty$ to write the imaginary time summation as an integral. A more convenient expression is obtained after integrating the ϕ's. The resulting form for the partition function gives [18],

$$Z \approx \sum_{\{n(\tau,\vec{r})\}} \sum_{\{\vec{m}(\tau,\vec{r})\}} \exp\left\{ -\sum_\tau \left[\frac{\epsilon q^2}{2} \sum_{<\vec{r}_1,\vec{r}_2>} n(\tau,\vec{r}_1)\mathbf{C}^{-1}(\vec{r}_1,\vec{r}_2)n(\tau,\vec{r}_2) \right.\right.$$

$$\left.\left. +\frac{1}{2\epsilon E_J} \sum_{\vec{r}} |\vec{m}(\tau,\vec{r})|^2 \right] \right\}, \tag{16}$$

where n and m are integer variables. These equations have implicit constraints over the sums that have to be implemented during the calculations [18]. Once the constraints are explicitly included we arrive at a convenient representation for the partition function in terms of integer vortex v and charge n degrees of freedom. The resulting partition function reads,

$$Z = \sum_{\{n\}} \sum_{\{v\}} \exp\left[- S_{\text{eff}}(n,v) \right] \quad \text{where,}$$

$$S_{\text{eff}}(n,v) = \sum_{\vec{r}_1,\vec{r}_2,\tau} \left[\frac{q^2\epsilon}{2\pi} n(\tau,\vec{r}_1)\mathbf{C}(\vec{r}_1,\vec{r}_2)n(\tau,\vec{r}_2) + \pi\epsilon E_J v(\tau,\vec{r}_1)\mathbf{G}(\vec{r}_1,\vec{r}_2)v(\tau,\vec{r}_2) \right.$$

$$\left. +i\, n(\tau,\vec{r}_1)\mathbf{\Theta}(\vec{r}_1,\vec{r}_2)\Delta_\tau v(\tau,\vec{r}_2) + \frac{1}{4\pi\epsilon E_J}\Delta_\tau n(\tau,\vec{r}_1)\mathbf{G}(\vec{r}_1,\vec{r}_2)\Delta_\tau n(\tau,\vec{r}_2) \right], \tag{17}$$

where

$$\mathbf{G}(\vec{r}_1, \vec{r}_2) \approx \ln|\vec{r}_1 - \vec{r}_2|, \tag{18}$$

$$\Theta(\vec{r}_1, \vec{r}_2) \approx \arctan\left(\frac{y_1 - y_2}{x_1 - x_2}\right). \tag{19}$$

In Eq. (17) Δ_τ denotes the discrete gradient along the imaginary time axis. In the large E_J limit, the last term in this equation shows that the time derivative of the charges are soft, i.e., the coupling constant of this term is very small resulting in strong quantum fluctuations for the n's. Therefore, in this limit the n's are not well defined and the v's dominate. We will call this a vortex-dominated state. On the other hand, for small E_J, the last term in the effective action places a considerable restriction on the values for the charge time derivatives, making them well defined. In this case, the v's have large fluctuations; we will call this a charges-dominated state. If the system is in one of these two states, say a vortex dominated state, the fact that the charges are not well defined means that they have a soft self coupling and can have large fluctuations. This makes them behave as an effective continuous Gaussian model. Integrating the continuous variables, we find an effective action for the still integer vortex conjugate variables.

One important aspect of the derivation of Eq. (17) is that the charging energy contribution was not touched. Therefore, in the calculation for the two-array problem, where the only interaction between the arrays was electrostatic, we can write the effective action immediately by just repeating the same steps for each array, as described above, and just adding the extra charging term that couples the two arrays. Then, the two array equivalent equation to Eq. (17) is,

$$
\begin{aligned}
S_{\text{eff}}(n^{(1)}, v^{(1)}; n^{(2)}, v^{(2)}) = \sum_{\vec{r}_1, \vec{r}_2, \tau} &\left[\frac{q^2 \epsilon}{2\pi} n^{(1)}(\tau, \vec{r}_1) \tilde{\mathbf{C}}_{1,1}(\vec{r}_1, \vec{r}_2) n^{(1)}(\tau, \vec{r}_2) + \right. \\
&+\pi\epsilon E_J^{(1)} v^{(1)}(\tau, \vec{r}_1) \mathbf{G}(\vec{r}_1, \vec{r}_2) v^{(1)}(\tau, \vec{r}_2) + \\
&+i\, n^{(1)}(\tau, \vec{r}_1) \Theta(\vec{r}_1, \vec{r}_2) \Delta_\tau v^{(1)}(\tau, \vec{r}_2) + \\
&\left. +\frac{1}{4\pi\epsilon E_J^{(1)}} \Delta_\tau n^{(1)}(\tau, \vec{r}_1) \mathbf{G}(\vec{r}_1, \vec{r}_2) \Delta_\tau n^{(1)}(\tau, \vec{r}_2') \right] + \\
+ \sum_{\vec{r}_1, \vec{r}_2, \tau} &\left[\frac{q^2 \epsilon}{2\pi} n^{(2)}(\tau, \vec{r}_1) \tilde{\mathbf{C}}_{1,1}(\vec{r}_1, \vec{r}_2) n^{(2)}(\tau, \vec{r}_2) + \right. \\
&+\pi\epsilon E_J^{(2)} v^{(2)}(\tau, \vec{r}_1) \mathbf{G}(\vec{r}_1, \vec{r}_2) v^{(2)}(\tau, \vec{r}_2) + \\
&+i\, n^{(2)}(\tau, \vec{r}_1) \Theta(\vec{r}_1, \vec{r}_2) \Delta_\tau v^{(2)}(\tau, \vec{r}_2) + \\
&\left. +\frac{1}{4\pi\epsilon E_J^{(2)}} \Delta_\tau n^{(2)}(\tau, \vec{r}_1) \mathbf{G}(\vec{r}_1, \vec{r}_2) \Delta_\tau n^{(2)}(\tau, \vec{r}_2) \right] + \\
+ \sum_{\vec{r}_1, \vec{r}_2, \tau} &\left[\frac{q^2 \epsilon}{\pi} n^{(1)}(\tau, \vec{r}_1) \tilde{\mathbf{C}}_{1,2}(\vec{r}_1, \vec{r}_2) n^{(2)}(\tau, \vec{r}_2) \right].
\end{aligned}
\tag{20}
$$

Here we have used the definition of $\tilde{\mathbf{C}}$ given in Eq. (10).

The result given above is valid for any of the three independent possible cases mentioned before. However, here we will discuss the interesting case where one of the arrays is in the semiclassical regime (a vortex-dominated state) and the other is in the fully quantum regime (a charge-dominated state). This leads to a study of the interaction between vortices in the former and charges in the latter. Furthermore, we assume that the arrays are dominated by the mutual capacitance between nearest neighbor islands. Taking array 1 as being vortex-dominated and array 2 charge-dominated we have the limits, $E_{C_m}^{(1)} \ll E_J^{(1)}$ and $E_{C_m}^{(2)} \gg E_J^{(2)}$. To continue the analysis we started by integrating out the vortices in array 2, and the charges in array 1 by using. the Poisson summation formulai [18]. After the integrations we are left with the following expression for the partition function $Z = \sum_{\{v^{(1)}\}} \sum_{\{n^{(2)}\}} \exp\left[-S_{\text{eff}}(v^{(1)}, n^{(2)}) \right]$, where the effective action for vortices in array 1 and charges in array 2 is given by

$$
\begin{aligned}
S_{\text{eff}}[v^{(1)}, n^{(2)}] =& \sum_{\tau, \vec{r}_1, \vec{r}_2} \left[\pi\epsilon E_J^{(1)} \, v^{(1)}(\tau, \vec{r}_1) \mathbf{G}(\vec{r}_1, \vec{r}_2) \, v^{(2)}(\tau, \vec{r}_2) + \right. \\
& \left. + \frac{1}{2\pi\epsilon E_J^{(2)}} \Delta_\tau n^{(2)}(\tau, \vec{r}_1) \mathbf{G}(\vec{r}_1, \vec{r}_2) \Delta_\tau n^{(2)}(\tau, \vec{r}_2) \right] + \\
& + \sum_{\tau, \tau', \vec{r}_1, \vec{r}_2} \left[\frac{\epsilon}{2} n^{(2)}(\tau, \vec{r}_1) \mathbf{G_n}(\tau, \tau'; \vec{r}_1, \vec{r}_2) n^{(2)}(\tau', \vec{r}_2) + \right. \\
& + i\, n^{(2)}(\tau, \vec{r}_1) \tilde{\mathbf{\Theta}}(\tau, \tau'; \vec{r}_1, \vec{r}_2) \Delta_\tau v^{(1)}(\tau', \vec{r}_2) + \\
& \left. + \frac{\pi}{2q^2\epsilon} \Delta_\tau v^{(1)}(\tau, \vec{r}_1) \mathbf{G_v}(\tau, \tau'; \vec{r}_1, \vec{r}_2) \Delta_\tau v^{(1)}(\tau', \vec{r}_2) \right],
\end{aligned} \tag{21}
$$

where we have defined the following interaction potentials

$$
\mathbf{G_n}(\tau, \tau'; \vec{r}_1, \vec{r}_2) = \frac{q^2}{\pi} \left[\tilde{\mathbf{C}}_{2,2}(\vec{r}_1, \vec{r}_2) \delta_{\tau, \tau'} - \right.
$$
$$
\left. - \frac{q^2\epsilon}{\pi} \sum_{\vec{r}_3, \vec{r}_4} \tilde{\mathbf{C}}_{1,2}(\vec{r}_1, \vec{r}_3) \mathbf{M}^{-1}(\tau, \tau'; \vec{r}_1, \vec{r}_2) \tilde{\mathbf{C}}_{1,2}(\vec{r}_4, \vec{r}_2) \right], \tag{22}
$$
$$
\tilde{\mathbf{\Theta}}(\tau, \tau'; \vec{r}_1, \vec{r}_2) = -\frac{q^2\epsilon}{\pi} \sum_{\vec{r}_3, \vec{r}_4} \mathbf{\Theta}(\vec{r}_2, \vec{r}_3) \mathbf{M}^{-1}(\tau, \tau'; \vec{r}_1, \vec{r}_2) \tilde{\mathbf{C}}(\vec{r}_4, \vec{r}_2), \tag{23}
$$
$$
\mathbf{G_v}(\tau, \tau'; \vec{r}_1, \vec{r}_2) = \sum_{\vec{r}_3, \vec{r}_4} \mathbf{\Theta}(\vec{r}_1, \vec{r}_3) \mathbf{M}^{-1}(\tau, \tau'; \vec{r}_3, \vec{r}_4) \mathbf{\Theta}(\vec{r}_4, \vec{r}_2). \tag{24}
$$

We will give an approximate local expression for the $\mathbf{M}$ matrix below, although when it is written down explicitly it is nonlocal.

128

In Eq. (21) we have an effective interaction between the vortices in array 1 and the charges in array 2. In order to understand this interaction we will follow Ref. [6] and study one particular case. Let's say that a vortex and a charge move following the trajectories $\vec{R}(\tau)$ and $\vec{X}(\tau)$, respectively. Then the space-time distribution of vortices and charges can be described by

$$v^{(1)}(\tau,\vec{r}) = \delta[\vec{r} - \vec{R}(\tau)] \qquad n^{(1)}(\tau,\vec{r}) = \delta[\vec{r} - \vec{X}(\tau)], \qquad (25)$$

from this equation, the imaginary time derivative is

$$\Delta_\tau v^{(1)}(\tau,\vec{r}) = \Delta_\tau \delta[\vec{r}_1 - \vec{R}(\tau)] = -\sum_\mu \Delta_\mu \delta[\vec{r}_1 - \vec{R}(\tau)] \, \Delta_\tau \vec{R}(\tau). \qquad (26)$$

The last term in this equation relates the time derivative to a summation over space derivatives. We can now write the interaction term in the action in the following way

$$S_{\text{int}} = -i \int_0^{\beta\hbar} d\tau \int_0^{\beta\hbar} d\tau' \sum_\mu \Delta_\mu \tilde{\Theta}\left(\tau,\tau';\vec{X}(\tau),\vec{R}(\tau')\right) \frac{dR_\mu(\tau')}{d\tau}. \qquad (27)$$

This has the nature of a gauge coupling, so that we can define the corresponding vector potential

$$\vec{a}(\tau,\vec{r}) = \int_0^{\beta\hbar} d\tau \vec{\Delta}\tilde{\Theta}\left(\tau,\tau';\vec{r},\vec{X}(\tau)\right). \qquad (28)$$

From this equation, the interaction part of the action, given in Eq. (27), reads

$$S_{\text{int}} = -i \int_0^{\beta\hbar} d\tau' \, \vec{a}\left(\tau',\vec{R}(\tau')\right) \cdot \frac{d\vec{R}(\tau')}{d\tau}. \qquad (29)$$

In these equations we have chosen to view the interaction in a representation where a vortex moves under the influence of the gauge field $\vec{a}$ produced by the charge. This is equivalent to the representation where a charge is moving in the field produced by the vortex. Since we have a vector field, we can define the corresponding associated effective magnetic field

$$\vec{b}(\tau,\vec{r}) = \vec{\Delta} \times \vec{a}(\tau,\vec{r}),$$
$$= -\frac{q^2}{\pi} \int_0^{\beta\hbar} d\tau' \sum_{\vec{r}_3,\vec{r}_4} \left(\vec{\Delta} \times \vec{\Delta}\Theta(\vec{X}(\tau'),\vec{r}_3)\right) \mathbf{M}^{-1}(\tau,\tau';\vec{r}_3,\vec{r}_4)\tilde{\mathbf{C}}_{1,2}(\vec{r}_4,\vec{r}).$$
$$(30)$$

Now we can use the following property of the kernel Θ

$$\vec{\Delta} \times \vec{\Delta}\Theta(\vec{r}_1,\vec{r}_2) = 2\pi \, \delta_{\vec{r}_1,\vec{r}_2} \, \hat{k}. \qquad (31)$$

This kernel is the field solution of a point vortex located at the origin. From all these we can finally write the expression for the effective magnetic field as [8,9]

$$\vec{b}(\tau,\vec{r}) = -\frac{q^2\epsilon}{\pi} \int_0^{\beta\hbar} d\tau' \sum_{\vec{r}_1} \mathbf{M}^{-1}(\tau,\tau';\vec{X}(\tau'),\vec{r}_1)\tilde{\mathbf{C}}_{1,2}(\vec{r}_1,\vec{r})\,\hat{k}. \tag{32}$$

Up to this point the effective action in Eq. (21), the gauge effective vector potential, and the effective magnetic field interaction are nonlocal in time. Approximating $\mathbf{M}$ by a local one we can write the effective magnetic field as

$$\vec{b}(\vec{r}) \approx -C_{\text{int}}\,\mathbf{C}_2^{-1}(\vec{r},\vec{X}(\tau))\,\hat{k}. \tag{33}$$

This means that if we have a charge in $\vec{X}(\tau)$ and a vortex in $\vec{R}(\tau)$, then the vortex will feel an effective magnetic field produced by the charge with magnitude $-C_{\text{int}}\mathbf{C}_2^{-1}(\vec{X}(\tau),\vec{R}(\tau))$. Notice that we have arrived at a system of vortices and charges that interact via a gauge field that has a finite range. This type of structure, as we said above, is rather similar to what one finds in electronic two-dimensional quantum Hall effect type systems. A more general and thorough analysis of this connection will be presented elsewhere.

V. CONCLUSIONS

We have briefly presented results from a thorough study of the α_m vs. T phase diagram for an array of ultrasmall Josephson junctions in zero and $f = 1/2$ external magnetic fields. One of our main goals was to perform different calculations for these arrays using experimentally realistic parameters. Our calculations were based in our path integral formulations of the quantum partition function for the JJA. We used a WKB-RG approximation to find the first order correction in α_m to the classical partition function. From this calculation we found an effective 2-D XY classical partition function, where the coupling constant is modified by the quantum fluctuations. The high temperature limit of this calculation, for $f = 0$, compares quite well both with the quantum MC results and with experiment, for the capacitance ratios given by experiment beeing the only adjustable parameters. The results for $f = 1/2$ agreed only qualitatively though. The RG-WKB calculation in the mutual capacitance dominated regime suggests the existence of a low temperature instability. In this limit we provided quantum MC evidence for the existence of a $QUIT$. We stressed, however, that still more experimental and theoretical work is needed to further elucidate the properties and reality of this $QUIT$. One of the important conclusions from the quantum MC calculations is that the general trend in the low temperature results, in particular for the helicity modulus, can be strongly dependent on the degree of discretization along the imaginary time axis.

We also performed some quantum Monte Carlo calculations of the inverse dielectric constant of the charged gas in the $\alpha_m > 1$ regime, in order to study the conducting to insulating crossover. We found that the present Monte Carlo path integral implementation of our model does not allow us to make fully reliable calculations of this transition, very much analogous to what happened experimentally. Our results for this transition are then only qualitative. The reason for this is that in the simulations we have to update simultaneously phases and charges, and the latter ones have long range interactions. Further technical improvements are needed in order to make solid quantitative statements about the insulating phase.

We also discussed results from our studies of two capacitively coupled arrays in different regimes. This is a difficult problem but one that promises to lead to interesting new physics. We studied this problem first using a semiclassical approximation and then carried out QMC simulations of the effective action in this limit. There we found that if the two arrays are in the semiclassical limit the interacting capacitance has the effect of increasing the critical temperature of the arrays.

We have also briefly discussed the interesting limit where one of the arrays is in the semiclassical limit while the other is in the quantum one. Equivalently, one is charge dominated while the other is vortex dominated. We found that the effect of the capacitive interaction is equivalent to having a charge and a vortex interacting via a gauge potential proportional to the interacting capacitance. Here we followed the work of Fazio *et al.* [6] and obtained an effective action for two interacting Coulomb gases; one for the vortices and the other for the Cooper pairs. The key step here was to notice that in the derivation for the one-array case, the charging energy was not modified. Therefore the generalization to the two-array case was direct. We wrote the effective action in terms of four interacting Coulomb gases. Once we had the effective action for vortices and Cooper pairs in both arrays, we considered the case were one the arrays is charge dominated while the other was vortex dominated. These two conditions imply that the vortices in the charge dominated array and the charges in the vortex dominated array can be integrated out. The result after Gaussian integrations is that of an effective action that couples the vortices in one array to the charges in the other. The interaction turned out to be mediated by a nonlocal gauge field. This result was also independently derived by Blanter et al. [17].

Finally, these types of systems hold the promise of leading to a great variety of novel experimentally observable phenomena. In particular the vortex-charge interaction discussed at the end of last section, deals with the interplay of quantum-classical effects, and it may lead to a possible fractional quantum Hall effect either in the gas of charges or that of vortices.

ACKNOWLEDGMENTS

This work has been partially supported by *NSF* grant DMR-9521845.

REFERENCES

[1] For a review of superconducting networks see *Proceedings of the 2nd CTP Workshop on Statistical Physics: KT Transition and Superconducting Arrays*, Edited by D. Kim, et al. (Min Eum Sa, Seoul, Korea, 1993).

[2] H. S. J. van der Zant, *et al.* Phys. Rev. Lett, **69**, 2971 (1992).

[3] H. S. J. van der Zant, Ph. D. Thesis, Delft (1991) and Phys. Rev. **B54**, 10081 (1996). preprint (1996).

[4] T. S. Tighe, *et al.*, Phys. Rev. **B47**, 1145 (1993).

[5] C. D. Chen, *et al.* Physica Scripta **T42**, 182 (1992).

[6] R. Fazio and G. Schön, Phys. Rev. **B43**, 5307 (1991).

[7] C. Rojas, J. V. José, Phys. Rev. **B54**, 12361 (1996)

[8] C. Rojas, Ph.D. thesis, Northeastern University, 1996.

[9] J. V. José and C. Rojas, to be published.

[10] L. L. Sohn, et al., Physica **B194-196**, 125 (1994).

[11] P. W. Anderson, in *Lectures in The Many Body Problem*, edited by E. R. Caianiello, (Academic, New York, 1964), Vol. 2.

[12] L. S. Schulman, *Techniques and applications of path integration* New York: Wiley, (1981).

[13] Jacobs, J. V. José, M. A. Novotny, and A. M. Goldman, Phys. Rev. **B38**, 4562 (1988); L. Jacobs, et al., Phys. Rev. Lett. **53**, 2177 (1984)

[14] J. Mikalopas, et al, Phys. Rev. B **50**, 1321 (1994).

[15] J. V. José, Phys. Rev. **B29**, 2836 (1984)

[16] C. Rojas and J. V. José, and A. M. Tikofsky Bull. Am. Phys. Soc. **40**, 68, B11-7 (1995).

[17] Ya. M. Blanter and F. Schön, Phys. Rev. **B53**, 14534 (1996).

[18] J. V. José, L. P. Kadanoff, S. Kirkpatrick, and D. R. Nelson, Phys. Rev. **B16**, 1217 (1977).

The Effect of Randomness on Long-Range Order
in the Two-Dimensional Half-Filled Hubbard Model

C. Huscroft[1], R.T. Scalettar[1], and M. Ulmke[2]

[1]Physics Department, University of California, Davis, CA 95616, USA
[2]Theoretische Physik III, Institut für Physik, Universität Augsburg,
 D-86135 Augsburg, Germany

Abstract. Quantum Monte Carlo simulations are used to determine the effects of disorder on magnetic, charge density wave, and superconducting correlations in the two–dimensional half–filled Hubbard model. In the repulsive case, random site energies and random intersite hybridizations destroy long range antiferromagnetic order by interfering with moment formation and by the creation of independent singlet bonds, respectively. In the attractive model, random site energies break the symmetry between charge density and pairing fluctuations in favor of the superconducting phase. Finally, we suggest a bimodal distribution of repulsive on–site interactions might allow the separation of magnetic and Mott transitions.

1. Introduction

The repulsive Hubbard Hamiltonian is a widely studied model of the magnetic and metal–insulator phase transitions that can occur in itinerant electron systems. Recently, a large amount of work has focused on the two–dimensional square lattice, a possible model of the magnetic properties of the CuO_2 sheets in superconducting oxides and of two dimensional magnetic systems generally.[1] It is now known that the ground state has long range antiferromagnetic correlations at half–filling.[2] The existence of other ordered phases, including exotic superconducting ones, such as those predicted within mean field theories and other approximate analytic approaches is still uncertain.[3]

Similarly, the attractive Hubbard Hamiltonian has been used to model qualitative features of superconducting and charge density wave (cdw) phase transitions in correlated electron systems.[4,5] Here the attractive on–site interaction is imagined to originate at a microscopic level through the mediation of some other, unspecified, degrees of freedom, for example the interaction of the electrons with phonons. The phase diagram in two dimensions is believed to consist of simultaneous charge density wave (cdw) and superconducting order at half-filling and zero temperature, with a finite temperature Kosterlitz–Thouless phase transition to a purely superconducting phase off half-filling.[6]

In this paper we discuss the effect of randomness on the Hubbard model. Consider the Hamiltonian,

$$H = - \sum_{\langle i,j \rangle \sigma} t_{ij} c_{i\sigma}^{\dagger} c_{j\sigma} + \sum_{i} \left[U_i (n_{i\uparrow} - \frac{1}{2})(n_{i\downarrow} - \frac{1}{2}) + (\epsilon_i - \mu)(n_{i\uparrow} + n_{i\downarrow}) \right]. \quad (1)$$

Springer Proceedings in Physics, Volume 83
Computer Simulation Studies in Condensed-Matter Physics X
Eds.: D. P. Landau, K.K. Mon, H. -B. Schüttler
© Springer-Verlag Berlin Heidelberg 1998

Here $c_{i\sigma}, c_{i\sigma}^\dagger$ are the destruction and creation operators for electrons of spin σ on a two dimensional square lattice. t_{ij} is a hybridization between nearest neighbor sites on that lattice, and U_i is an on-site interaction. ϵ_i is an on-site energy and μ is the chemical potential.

In real materials the perfect crystal structure is always more or less distorted, leading to randomness in the parameters t_{ij}, U_i, and ϵ_i. What happens to magnetic, cdw, and superconducting correlations as the distributions of t_{ij}, U_i, and ϵ_i become increasingly broad? We will use quantum Monte Carlo (qmc) to study four specific cases:

- $t_{ij} = t$ and $U_i = U > 0$ (repulsive), with site energies chosen to be random with uniform distribution $-\Delta/2 < \epsilon_i < +\Delta/2$.

- $\epsilon_i = 0$ and $U_i = U > 0$ (repulsive), with the hybridizations chosen to be random with uniform distribution $t - \Delta/2 < t_{ij} < t + \Delta/2$.

- $\epsilon_i = 0$ and $t_{ij} = t$, with a random (repulsive) U_i chosen from a bimodal distribution $P(U_i) = (1 - f)\delta(U_i - U) + f\delta(U_i)$.

- $t_{ij} = t$ and $U_i = U < 0$ (attractive), with site energies chosen to be random with uniform distribution $-\Delta/2 < \epsilon_i < +\Delta/2$.

In all cases we chose $t = 1$ to set our scale of energies.

We first review some of the physics of the clean Hubbard model, as well as the relationship between the attractive and repulsive models.

Each site i of the lattice can appear in one of four configurations: empty, doubly occupied, or singly occupied with an $\uparrow$ or $\downarrow$ spin electron. In the first two cases, the magnetic moment on site i vanishes, $\langle(n_{i\uparrow} - n_{i\downarrow})^2\rangle = 0$, while in the latter two configurations there is a well defined moment, $\langle(n_{i\uparrow} - n_{i\downarrow})^2\rangle = 1$. At half–filling, where the average density per site $\langle n\rangle = \langle n_{i\uparrow} + n_{i\downarrow}\rangle = 1$, and in the absence of any interactions ($U=0$), each of the four configurations is equally likely, and half the sites exhibit a local moment. As an on–site repulsion U is turned on, the probabilities for doubly occupied sites and (due to particle number conservation) for empty sites are suppressed, and, at large U and half–filling, each site is singly occupied and hence possesses a magnetic moment. The situation is reversed in the attractive case. The empty and doubly occupied configurations dominate as the interaction is turned on.

Will the moments that are formed in the repulsive case at half–filling arrange themselves with neighboring spins parallel or antiparallel? Second order perturbation theory in the hopping t indicates that the spins prefer to align antiferromagnetically. Anti–aligned spins allow the system to lower its energy by $J = 4t^2/U$ on each bond via the virtual process in which a doubly occupied site is created in the intermediate state. This hopping, and associated lowering of the energy, is forbidden by the Pauli principle if the neighboring spins are parallel. The identical argument in the attractive case shows the system favors a cdw phase where sites are alternately doubly occupied and empty.

Will this tendency to "local" antiferromagnetism on neighboring bonds ever result in long range antiferromagnetic order in two dimensions? A more formal consideration of the above strong coupling discussion shows that the repulsive Hubbard model at half–filling and large U maps onto the spin–1/2 antiferromagnetic Heisenberg model. The coupling is isotropic, and the resulting symmetry of the order parameter is such that order, if it occurs, can only take place in the ground state.[7] Numerical studies of both the 2D Heisenberg and Hubbard models have shown that at $T = 0$ long range order does occur.[2,8]

A weak coupling analysis within the random phase approximation (RPA) complements this large U treatment. The magnetic susceptibility in the presence of interactions is expressed approximately as $\chi(\mathbf{q}) = \chi_0(\mathbf{q})/[1 - U\chi_0(\mathbf{q})]$, where the susceptibility of the noninteracting model $\chi_0(\mathbf{q}) = -\sum_{\mathbf{p}}[f(\epsilon_{\mathbf{p+q}}) - f(\epsilon_{\mathbf{p}})]/[\epsilon_{\mathbf{p+q}} - \epsilon_{\mathbf{p}}]$. At $\langle n \rangle = 1$, the noninteracting susceptibility is peaked at $\mathbf{q} = (\pi, \pi)$ so that the RPA predicts the Hubbard model exhibits antiferromagnetic order at half–filling. The RPA phase diagram also includes a ferromagnetic phase for U sufficiently large and $\langle n \rangle \neq 1$. Note, however, RPA erroneously predicts transitions at finite temperature, in contradiction to the Mermin–Wagner theorem.[8,9]

Some of the physics of the attractive model is best understood by introducing a formal mapping to the repulsive case. Consider a particle–hole transformation: $c_{\mathbf{i}\downarrow} \leftrightarrow c^{\dagger}_{\mathbf{i}\downarrow}(-1)^{|\mathbf{i}|}$ on the down spin fermions with $|\mathbf{i}| \equiv i_x + i_y$. The phase factor $(-1)^{|\mathbf{i}|}$ changes sign as one goes between the two sublattices of our (bipartite) square lattice. The kinetic energy is invariant under this transformation, and the interaction term changes sign. (Note that these statements are true even for disordered $U_{\mathbf{i}}$ and $t_{\mathbf{ij}}$.) Thus as long as $\mu = \epsilon_{\mathbf{i}}$, the attractive and repulsive Hubbard Hamiltonians exactly map onto each other.

Under this same transformation, the z component of spin, $m_{\mathbf{i}z} = n_{\mathbf{i}\uparrow} - n_{\mathbf{i}\downarrow}$, in the repulsive model maps onto the charge, $n_{\mathbf{i}} = n_{\mathbf{i}\uparrow} + n_{\mathbf{i}\downarrow}$, in the attractive case, and vice versa. The transverse spin operators $m_{\mathbf{i}+} = c^{\dagger}_{\mathbf{i}\uparrow}c_{\mathbf{i}\downarrow}$ and $m_{\mathbf{i}-} = c^{\dagger}_{\mathbf{i}\downarrow}c_{\mathbf{i}\uparrow}$, map onto the pair creation and destruction operators, $\Delta^{\dagger}_{\mathbf{i}} = (-1)^{|\mathbf{i}|}c^{\dagger}_{\mathbf{i}\uparrow}c^{\dagger}_{\mathbf{i}\downarrow}$ and $\Delta_{\mathbf{i}} = (-1)^{|\mathbf{i}|}c_{\mathbf{i}\downarrow}c_{\mathbf{i}\uparrow}$. The repulsive model is isotropic in the spin. Therefore, by analogy, cdw and superconducting correlations should coexist in the attractive model, and long range order is possible only at $T = 0$.

Away from half–filling, the particle–hole transformation maps a chemical potential μ in the attractive model into a field h_z in the z direction in the repulsive model. Such a field breaks the rotational symmetry of the corresponding antiferromagnetic Heisenberg model and selects out order in the xy plane, since neighboring anti–aligned spins can tilt out of the plane and take advantage of the field energy. The reduction of the symmetry from the Heisenberg model to the XY model suggests that in two dimensions there could now be a finite temperature Kosterlitz–Thouless phase transition into a magnetic state.[6] In the language of the attractive model, then, doping will select out superconducting fluctuations over cdw order, and $T_{c,\,\mathrm{pair}} > 0$.

What are some of the expected effects of disorder? In the repulsive model, random site energies will interfere with moment formation, since singly occupied sites with large positive $\epsilon_{\mathbf{i}}$ will have a tendency to transfer their electrons and doubly occupy sites with large negative ϵ_i. As the moments themselves are eliminated, long range magnetic order disappears.

On the other hand, when the randomness is in the hopping t_{ij}, there is little tendency for the moments to be reduced. Indeed, at strong coupling the model becomes a quantum spin–1/2 Heisenberg Hamiltonian with random exchange constants J_{ij}. It is known that this model has an order–disorder transition which occurs when singlets form on the strong bonds and effectively decouple from each other.[9] (A similar phenomenon occurs in bilayer Hubbard and Heisenberg models.[10])

As pointed out above, the clean attractive Hubbard model has a special symmetry between cdw and superconducting order at half–filling. How is this affected by random site energies? One might argue that since the randomness is coupling directly to the charge density on a site, it is more likely to disrupt cdw order. Furthermore, Anderson's theorem suggests that random site energies might not interfere with superconductivity since pairing can occur between time–reversed states of the disordered single particle Hamiltonian instead of the $(\mathbf{k}, -\mathbf{k})$ pairs of the clean system.[11]

These arguments give a qualitative picture of what site and bond randomness might do to long range magnetic, pairing, and cdw correlations. Qmc simulations can be used to check these predictions, and also to locate quantitatively the positions of the phase transition.

2. Determinant Quantum Monte Carlo

All finite temperature qmc simulations proceed by a conversion of the trace over the quantum mechanical degrees of freedom to a sum over classical variables which can be sampled stochastically by the usual methods developed for classical Monte Carlo.

Consider a calculation of

$$\langle \hat{A} \rangle = \frac{1}{Z} \mathrm{Tr}[\hat{A} \, \exp(-\beta \hat{H})]. \tag{2}$$

with the partition function $Z = \mathrm{Tr}[\exp(-\beta \hat{H})]$. One might know separately the eigenstates of the kinetic, $\hat{K}$, and potential, $\hat{P}$, pieces in $\hat{H}$, but the non–commutativity of the operators prevents us from a division of the exponential to isolate $\hat{K}$ and $\hat{P}$. However, at high temperatures, (small β), it is a good approximation[12] to express

$$\exp(-\beta \hat{H}) \approx \exp(-\beta \hat{K})\exp(-\beta \hat{P}). \tag{3}$$

This suggests we introduce a small parameter $\tau = \beta/L$ and write the full exponential as a product of L pieces,

$$\exp[-\beta \hat{H}] = \left[\exp[-\tau \hat{H}]\right]^L \approx \left[\exp[-\tau \hat{K}]\exp[-\tau \hat{P}]\right]^L. \tag{4}$$

τ is the "imaginary time" increment. At this point there are various choices of how to proceed which result in the different qmc algorithms. The determinant approach makes use of the fact that the trace over a product of quadratic forms in fermion operators, can be performed analytically.[13]

$$\mathrm{Tr}\left[\exp[\sum_{nm} c_n^\dagger A_{nm} c_m]\exp[\sum_{nm} c_n^\dagger B_{nm} c_m]\ldots\right] = \det[I + e^A e^B \ldots] \qquad (5)$$

Note that the original "Tr" is a quantum mechanical sum over a complete set of states of dimension 4^N in the Hilbert space, where N is the number of spatial sites in the lattice, whereas "det" is a determinant of the real valued matrices $A, B, \ldots$ which have dimension N.

Our Hamiltonian, unfortunately, contains interaction terms which are quartic in the fermion creation and destruction operators. However, such terms can be made quadratic by introducing a Gaussian integration over an auxiliary variable,

$$\exp[-U\tau(n_\uparrow - \frac{1}{2})(n_\downarrow - \frac{1}{2})] = \frac{e^{-U\tau/4}}{\sqrt{\pi}} \int_{-\infty}^{+\infty} dx \exp[-x^2 - 2\lambda x(n_\uparrow - n_\downarrow)]. \qquad (6)$$

Here $\lambda^2 = U\tau/2$. A similar identity holds for $U < 0$ where $n_\uparrow - n_\downarrow$ is replaced by $n_\uparrow + n_\downarrow - 1$. If such an integral is introduced for each of the N spatial sites of each of the L imaginary time slices, all the quartic terms are replaced by quadratic ones.

What is left is an expression for Z which consists of an integral over a classical field x whose components are indexed by space and imaginary time.

$$Z = \int \prod_{i\tau} dx_{i\tau} \det M_\uparrow[x] \det M_\downarrow[x]. \qquad (7)$$

The integrand, the "Boltzmann weight" for our simulation, is a highly nonlocal product of determinants. Evaluating how it changes when a single field component is modified is very expensive computationally. A single sweep through the lattice to update all the NL classical degrees of freedom scales as $N^3 L$.

The particle–hole transformation described in the introduction has important implications for the qmc simulation. In the case $\mu = \epsilon_i$, the up and down spin determinants have the same sign and their product is positive definite. However if the difference $\mu - \epsilon_i$ is not zero on every site, there is no guarantee that the integrand is positive, and hence it cannot in general serve as a probability for the field configuration. This "sign problem" has not been resolved in any satisfactory manner and is the central algorithmic bottleneck in the field.[2] For a more detailed description of the sign problem and of the complete determinant algorithm, see [2,13].

We are interested in measuring expectation values $\langle \hat{A} \rangle$ where $\hat{A}$ is some string of fermion creation and annihilation operators. It can be shown that $\langle c_{i\sigma} c_{j\sigma}^\dagger \rangle$ is obtained by averaging the matrix elements $(M_\sigma^{-1})_{ij}$ in the configurations generated using the determinants of M_σ as the Boltzmann weight. Observables $\hat{A}$ containing larger numbers of operators can also be expressed in terms of such matrix elements, using an analogue of Wick's theorem.[14] The observables we will evaluate in the simulation are the spin, charge, and pair correlations as a function of separation on the lattice,

$$s(\mathbf{j} - \mathbf{l}) = \langle\, (n_{\uparrow\mathbf{l}} - n_{\downarrow\mathbf{l}})(n_{\uparrow\mathbf{j}} - n_{\downarrow\mathbf{j}}) \,\rangle,$$
$$c(\mathbf{j} - \mathbf{l}) = \langle\, (n_{\uparrow\mathbf{l}} + n_{\downarrow\mathbf{l}} - 1)(n_{\uparrow\mathbf{j}} + n_{\downarrow\mathbf{j}} - 1) \,\rangle,$$
$$p_s(\mathbf{j} - \mathbf{l}) = \langle\, \Delta_{\mathbf{l}} \Delta_{\mathbf{j}}^{\dagger} \,\rangle, \tag{8}$$

and their associated structure factors,

$$S_{\mathrm{af}} = \frac{1}{N} \sum_{\mathbf{j},\mathbf{l}} s(\mathbf{j} - \mathbf{l})(-1)^{|\mathbf{j}-\mathbf{l}|},$$
$$S_{\mathrm{cdw}} = \frac{1}{N} \sum_{\mathbf{j},\mathbf{l}} c(\mathbf{j} - \mathbf{l})(-1)^{|\mathbf{j}-\mathbf{l}|},$$
$$S_{\mathrm{pair}} = \frac{1}{N} \sum_{\mathbf{j},\mathbf{l}} p_s(\mathbf{j} - \mathbf{l}). \tag{9}$$

Along with the possibility of magnetism, the Hubbard model also exhibits a Mott metal–insulator transition. At strong coupling and half filling, each site is singly occupied, and for an electron to move, it must doubly occupy a site, at an energy cost of U. If U is much greater that the kinetic energy gained by delocalization, then the system will be insulating. One simple way to examine this is to look at the average kinetic energy $\langle K \rangle$ which should have a sharp minimum at $\langle n \rangle = 1$.

To evaluate this gap to charge excitations, one can also measure the density of electrons $\langle n \rangle$ as a function of chemical potential μ. For less than one particle per site, $\langle n \rangle < 1$, the density smoothly increases as μ increases. However, right at $\langle n \rangle = 1$, the density no longer increases in response to μ, since at this point adding particles suddenly costs the large interaction energy U. Thus a plot of $\langle n \rangle$ versus μ exhibits a plateau at half–filling. This phenomenon can be equivalently phrased in terms of the compressibility $\kappa = \partial\langle n \rangle / \partial\mu$ which will vanish in the gapped insulating phase at half–filling and be nonzero in the doped metallic phase.

3. Repulsive Model – Site Disorder

We first present resuts for the case of $U > 0$ and random $\epsilon_{\mathbf{i}}$. Fig. 1 shows the local moment $c(0,0)$ on a 4x4 lattice as a function of disorder. As expected, increasing Δ destroys the moments, and $c(0,0)$ equals the noninteracting value $\frac{1}{2}$ when $2\Delta \approx U$. As the moments disappear, magnetic order is destroyed, and the spin-spin correlation function at largest separation $c(2,2)$ and the antiferromagnetic structure factor $S(\pi,\pi)$ vanish.

We can also examine the effect of disorder on the Mott metal–insulator transition. In Fig. 2 the compressibility κ is shown for a noninteracting system $U = 0$ as a function of disorder, and also for $U = 4$. The behavior at zero

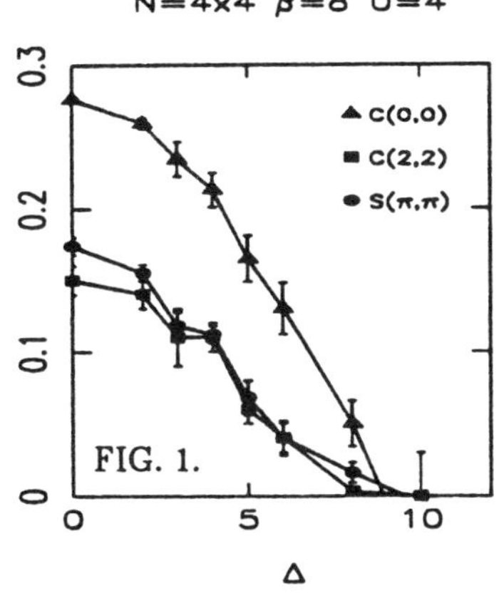

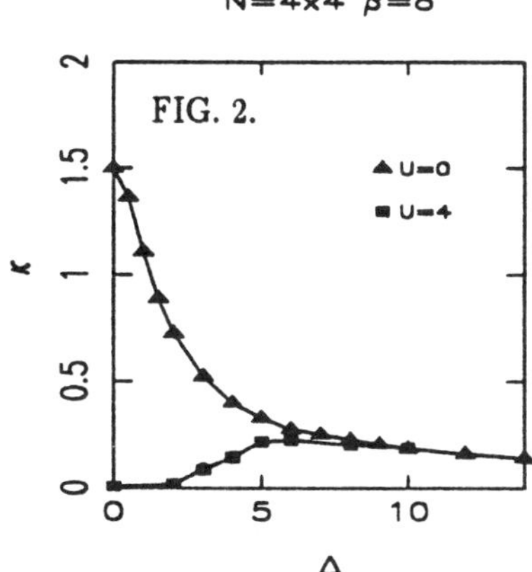

FIG. 1. Local moment, longest range spin–spin correlations function, and antiferromagnetic structure factor as a function of site disorder. The value given for $c(0,0)$ is actually its difference from the noninteracting result $c(0,0) = \frac{1}{2}$.

FIG. 2. The compressibility κ as a function of disorder strength.

disorder is very different in the two cases. The interacting system is a Mott insulator with $\kappa = 0$, while the non–interacting system has a large compressibility. As expected, when the disorder becomes the dominant energy scale, κ becomes the same. In the interacting case site disorder leads to a strong *enhancement* of κ.

In principle what one would like to do is an appropriate finite size scaling analysis on both the magnetic correlations and the compressibility to determine whether long range order exists for any nonzero Δ. However, the sign problem prevents us from studying larger lattices at the low temperatures necessary to do this.[15]

4. Repulsive Model – Bond Disorder

We turn instead to the case where the Hamiltonian is particle–hole symmetric and there is no sign problem. Fig. 3 shows the local moment $c(0,0)$ on a 10x10 lattice as a function of disorder. In contrast to the case of site disorder (Fig. 1), the moment is fairly insensitive to increasing Δ. Nevertheless, magnetic order is still destroyed, as seen by the vanishing of the spin-spin correlation function at largest separation $c(5,5)$ and the antiferromagnetic structure factor $S(\pi, \pi)$.

Fig. 4 shows the spin–spin correlations as a function of separation on a 10x10 lattice for different values of Δ. One sees a well defined oscillatory pattern which is characteristic of antiferromagnetic order, and for the smaller values of Δ these correlations do not appear to be falling off too rapidly with system size, suggesting there might be long range order.

In order to make this statement more precise, we recall that spin–wave theory predicts a particular lattice size dependence of the spin correlations and

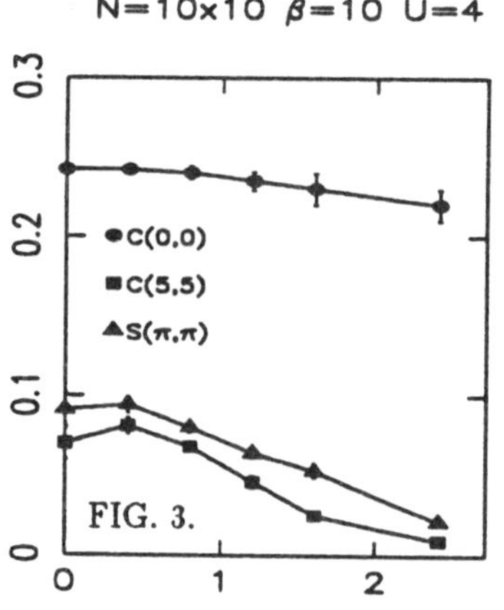

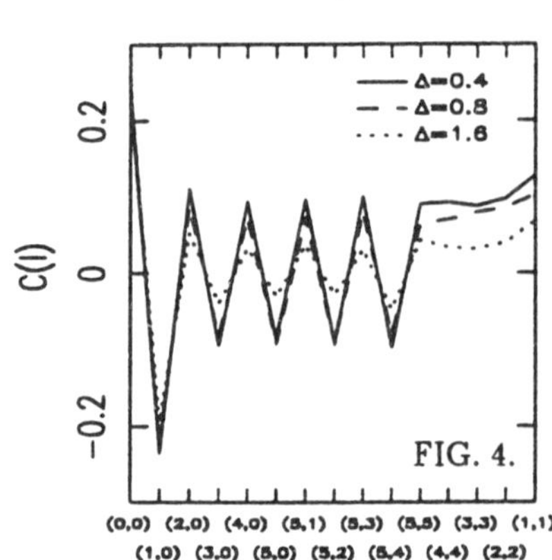

FIG. 3. Local moment, longest range spin–spin correlations function, and antiferromagnetic structure factor as a function of bond disorder. The value give for $c(0,0)$ is actually its difference from the noninteracting result $c(0,0) = \frac{1}{2}$.

FIG. 4. Spin–spin correlations as a function of lattice separation.

structure factor,

$$\frac{1}{N}S_{\text{cdw}} = m^2/3 + a/L$$
$$c(L/2, L/2) = m^2/3 + b/L, \tag{10}$$

where m^2 is the order parameter.[16] The results of an analysis using Eq. 10 are shown in Figs. 5,6, and 7. At weak disorder, the magnetic correlations extrapolate to a non–zero value for the order parameter in the thermodynamic limit. But for stronger disorder, $\langle m^2 \rangle$ vanishes. The results of this finite size scaling analysis for different Δ results in a value of the critical amount of hopping disorder $\Delta_c \approx 1.5$ to destroy antiferromagnetism in the Hubbard Hamiltonian. This value for Δ_c is in good agreement with that obtained in the random bond Heisenberg model.[9,15]

5. Repulsive Model – Interaction Disorder

Half–filling is the density where antiferromagnetic order is most clearly present (and perhaps the only place it is present in two dimensions) in the Hubbard model. It is also where the Mott metal–insulator transition occurs. Can these two transitions occur separately?

The Hubbard Hamiltonian with a bimodal distribution of interaction strengths is a candidate model for such a situation. Consider first the effect of the presence of a fraction f of sites with $U_i = 0$ on the position of the Mott

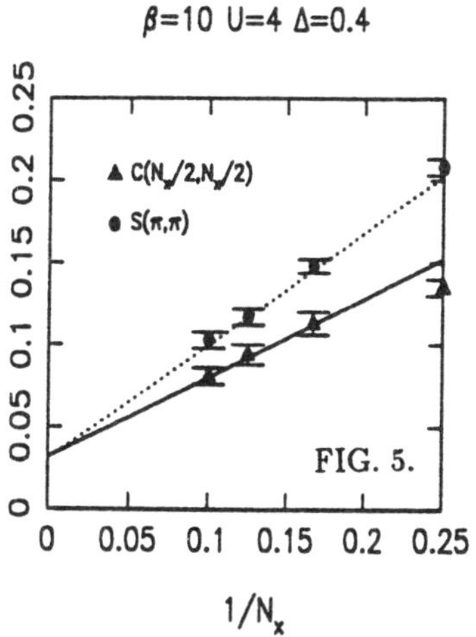

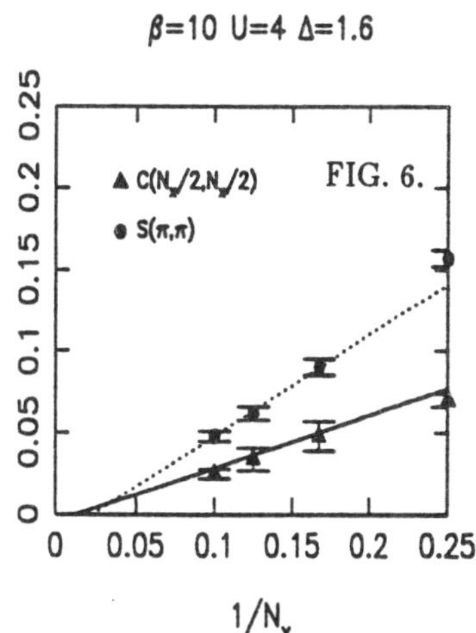

FIG. 5. Scaling plot for the longest range spin–spin correlations and the structure factor at weak disorder. The extrapolations to infinite systems should give the same value, and are a check on the numerics.

FIG. 6. Same as Fig. 5 but for strong disorder.

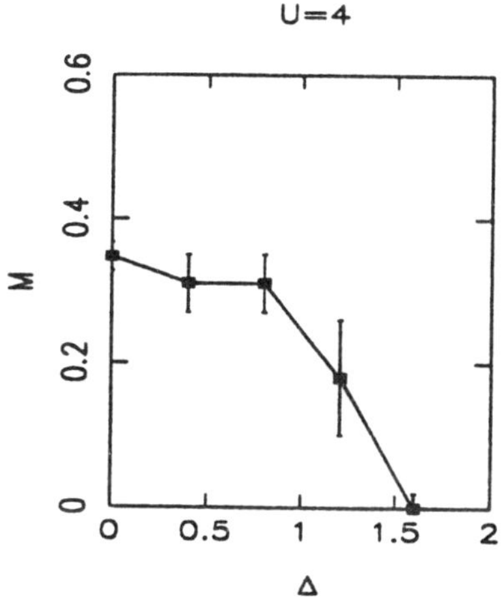

FIG. 7. Order parameter as a function of bond randomness.

transition. The jump in μ will now occur at a density $\langle n \rangle = 1 + f$, since an additional fraction f of the sites can be doubly occupied without invoking U. This is shown in Fig. 8. The reduction of the mobility of the electrons is illustrated in Fig. 9 where the kinetic energy is shown to have a sharp minimum at at a density $\langle n \rangle = 1 + f$ also shifted away from half–filling.

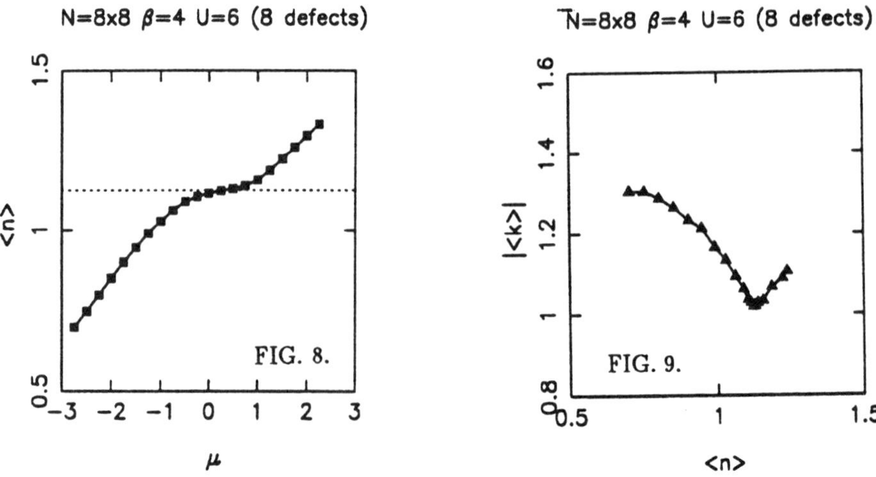

FIG. 8. The density versus chemical potential shows a Mott plateau at a filling $\langle n \rangle = 1 + f$ where $f = 0.125$ is the fraction of sites with $U_i = 0$.

FIG. 9. The kinetic energy is small at the filling where the insulating gap opens.

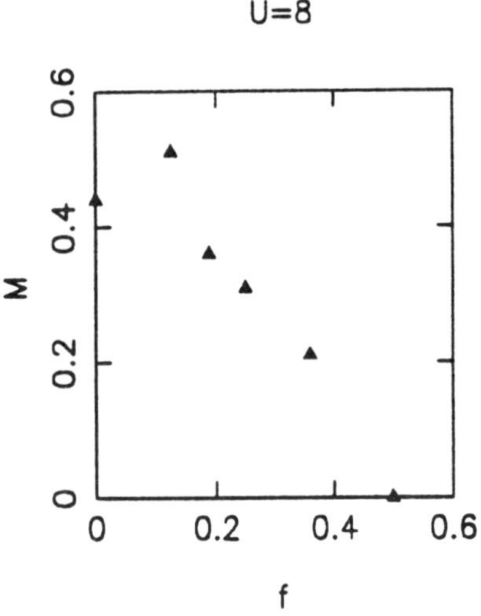

FIG. 10. Order parameter as a function of randomness in the interaction.

Does antiferromagnetic order remain at half–filling? A finite size scaling similar to that of section 4 shows that it does. Indeed, the order parameter remains non–zero out to $f_c \approx 0.4$, as shown in Fig. 10. Thus we appear to have a situation where the magnetic and Mott transitions may occur at different densities.

Actually, the argument is complicated a bit by the particle–hole symmetric form chosen for the interaction in Eq. 1. In this form the plateau at $\langle n \rangle = 1$ in the clean system where $f = 0$ splits into two symmetric gaps at $\langle n \rangle = 1 \pm f$ when f is non–zero. This is a consequence of the fact that one has chosen, in effect, a special situation in which the randomness in interactions is correlated with a randomness in the site energy terms which exists when the particle–hole symmetric form of the interaction is introduced. Nevertheless, this somewhat artificial situation has computational advantages since at half–filling there is no sign problem and the existence of long range antiferromagnetic order can be convincingly addressed. The particle–hole symmetric form was used in the analysis leading to Fig. 10, while Figs. 8,9 actually show data for the asymmetric case. However, the Mott plateau also exists (and in fact is a bit more robust) in the symmetric case as well.

6. Attractive Model – Site Disorder

Finally we turn to the attractive model and the issue of the breaking of the cdw and pairing correlations by site disorder. In Fig. 11 we have plotted the pair–pair correlations as a function of separation at a fixed low temperature and different degrees of randomness. We see that disorder initially leaves the pair correlations almost unchanged, but sufficiently large randomness eventually suppresses the pair correlations. The pair correlations are much more robust to disorder than the cdw correlations, which in a similar plot to Fig. 11 (not shown) are significantly suppressed with disorder as low as $\Delta = 1.00$.

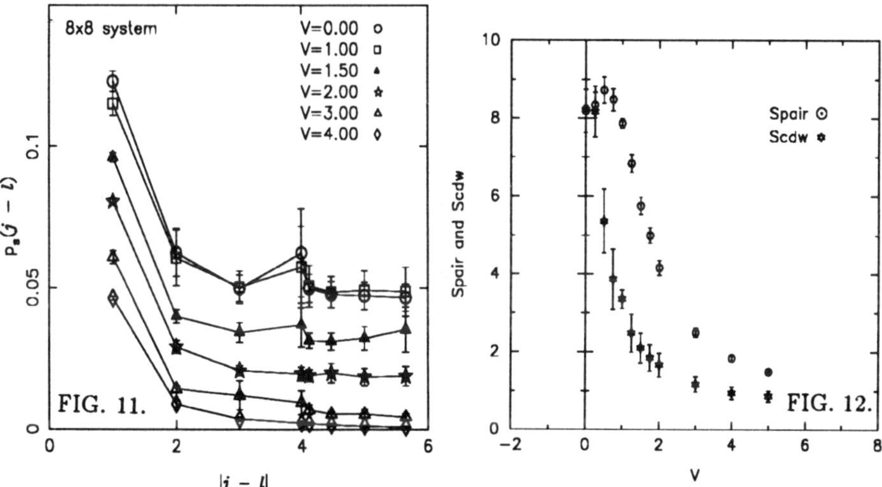

FIG. 11. Pair correlations as a function of separation at $\beta = 10$, $U = -4$ for different values of disorder strength, here denoted by V.

FIG. 12. Pair field and cdw structure factors on the 8×8 lattice as a function of disorder V at $\beta = 10$, $U = -4$.

We again see this difference in the effect of disorder on the cdw and pair–pair correlations in Fig. 12, wherein the structure factors are plotted as a function of disorder. S_{cdw} decreases much more rapidly than S_{pair}.

We may understand the difference in the effect of site disorder on cdw and pair-pair correlations by considering the clean versus the disordered systems. In the clean, $\Delta = 0$, system, pairs of electrons form, due to the attractive interaction energy U, and these pairs favor occupying alternating sites due to the kinetic energy. Site disorder does not break up the pairs (in fact it helps them), but it does destroy the alternation pattern. This has a drastic effect on the cdw structure factor, which sums the density correlation with a very specific phase (π, π). However, it does not decrease the pair structure factor which has no such phases and, in effect, counts only the number of doubly occupied sites.

A finite size analysis similar to that in the previous sections confirms this picture and leads to the conclusion the long range cdw order is immediately destroyed, i.e. $V_c = 0$. However pair order survives out to $V_c \approx 1.5$.[17]

Acknowledgements

We gratefully acknowledge the support of National Science Foundation, grant NSF-DMR-9520776 (RTS), and the Department of Energy Accelerated Strategic Computing Initiative (CH), as well as computer time at the San Diego Supercomputing Center.

REFERENCES

1. For a recent review, see *The Hubbard Model,* Arianna Montorsi (ed), World Scientific, 1992.

2. J.E. Hirsch and S. Tang, Phys. Rev. Lett. **62**, 591 (1989); S.R. White, D.J. Scalapino, R.L. Sugar, E.Y. Loh, J.E. Gubernatis, and R.T. Scalettar, Phys. Rev. **B40**, 506 (1989).

3. D.J. Scalapino, Physics Reports **250**, 330 (1995).

4. Two reviews include R. Micnas, J. Ranninger, and S. Robaszkiewicz, Rev. Mod. Phys. **62**, 113 (1990); M. Randeria in *Bose Einstein Condensation,* A. Giffin *et al.* (eds), Cambridge University Press (1994).

5. R.M. Fye, M.J. Martins, D.J. Scalapino, J. Wagner, and W. Hanke, Phys. Rev. **B45**, 7311 (1992); M. Imada, Physica C **185-189**, 1447 (1991). Raimundo R. dos Santos, Phys. Rev. **B50**, 635 (1994).

6. R.T. Scalettar, E.Y. Loh, Jr., J.E. Gubernatis, A. Moreo, S.R. White, D.J. Scalapino, R.L. Sugar, and E. Dagotto, Phys. Rev. Lett. **62**, 1407 (1989); A. Moreo and D.J. Scalapino, Phys. Rev. Lett. **66**, 946 (1991).

7. N.D. Mermin and H. Wagner, Phys. Rev. Lett. **17**, 1133 (1966).

8. J.D. Reger and A.P. Young, Phys. Rev. **B37**, 5978 (1988).

9. A. Sandvik and M. Vekić, Phys. Rev. Lett. **74** 1226 (1995).

10.   A.W. Sandvik and D.J. Scalapino, Phys. Rev. Lett. **72**, 2777 (1994). R.T. Scalettar, J. W. Cannon, D. J. Scalapino, and R.L. Sugar, Phys. Rev. **B50**, 13419 (1994).

11. P.W. Anderson, J. Phys. Chem. Solids **11**, 26 (1959).

12.   H.F. Trotter, Proc. Am. Math. Soc. **10**, 545 (1959); M. Suzuki, Comm. Math. Phys. **51**, 183 (1976); M. Suzuki, Phys. Lett. **113A**, 299 (1985).

13. J. W. Negele and H. Orland, *Quantum Many Particle Systems*, Addison–Wesley, 1988.

14. R. Blankenbecler, D. J. Scalapino, and R.L. Sugar, Phys. Rev. **D24**, 2278 (1981); G. Sugiyama and S.E. Koonin, Ann. Phys. **168**, 1 (1986); S. Sorella, S. Baroni, R. Car, and M. Parrinello, Europhys. Lett. **8**, 663 (1989).

15. M. Ulmke and R.T. Scalettar, Phys. Rev. **B55**, 4149 (1997).

16. D.A. Huse, Phys. Rev. **B37**, 2380 (1988).

17. C. Huscroft and R.T. Scalettar, Phys. Rev. **B55**, 1185 (1997).

Quantum Critical Exponents of a Planar Antiferromagnet

M. Troyer and M. Imada

Institute for Solid State Physics, University of Tokyo, Roppongi 7-22-1,
Tokyo 106, Japan

We present high precision estimates of the exponents of a quantum phase transition in a planar antiferromagnet. This has been made possible by the recent development of cluster algorithms for quantum spin systems, the loop algorithms. Our results support the conjecture that the quantum Heisenberg antiferromagnet is in the same universality class as the O(3) nonlinear sigma model. The Berry phase in the Heisenbrg antiferromagnet do not seem to be relevant for the critical behavior.

I. INTRODUCTION

Instead of classical transitions controlled by temperature T a quantum phase transition between a symmetry broken phase with long-range Nèel order and a quantum disordered state with a finite spin excitation gap may be realized at $T = 0$ by controlling a parameter g to increase quantum fluctuations. Criticalities around such quantum phase transitions at $g = g_c$ may reflect inherent quantum dynamics of the system and yield unusual universality classes with rich physical phenomena.

The most prominent example are the high temperature superconductors. There the quantum spin fluctuations are thought to lead to d-wave superconductivity as soon as antiferromagnetism is suppressed by hole doping. This close connection between antiferromagnetism, quantum fluctuations and high temperature superconductivity has triggered many theoretical investigations.

Most of these investigations are based on a mapping to an effective field theory, the 2D O(3) quantum nonlinear sigma model. This sigma model is in the same universality class as the 3D O(3) classical sigma model or the 3D classical Heisenberg model. A large number of detailed predictions about quantum critical behavior has been made for the sigma model [1,2]. However the spin-1/2 quantum antiferromagnet generally contains Berry phase terms [3] that are not present in the sigma model. The relevancy of these terms is not clear.

In order to shed light onto this question we have simulated a two dimensional quantum antiferromagnet (2D QAFM) that exhibits a quantum phase transitions. We calculate the critical exponents to determine the universality class and to check predictions made based on the nonlinear sigma model. First results have been published in Ref. [4].

Springer Proceedings in Physics, Volume 83
Computer Simulation Studies in Condensed-Matter Physics X
Eds.: D. P. Landau, K.K. Mon, H. -B. Schüttler
© Springer-Verlag Berlin Heidelberg 1998

A. Quantum critical exponents

The critical exponents of a quantum phase transition at $T = 0$ can be defined similar to a classical finite temperature phase transition. The quantum mechanical control parameter g plays the role of the temperature in the classical system. Approaching the quantum critical point from the disordered side $(g > g_c)$ the correlation length diverges as

$$\xi \propto (g - g_c)^{-\nu}. \tag{1}$$

By the Trotter-Suzuki mapping the d-dimensional quantum system can be mapped onto a $d + 1$-dimensional classical system. At zero temperature the system is infinite also in the additional imaginary time direction. The space and time dimensions are however not necessarily equivalent, and the correlation length in the time direction diverges in general with a different exponent

$$\xi_\tau \propto (g - g_c)^{-z\nu}, \tag{2}$$

where z is the dynamical exponent. In a Lorentz invariant system space and time directions are equivalent and $z = 1$. Related to the divergence of the correlation length is a vanishing of the spin excitation gap

$$\Delta \propto (g - g_c)^{z\nu}. \tag{3}$$

When passing through the critical point long range order is established. The order parameter in the case of a Néel ordered antiferromagnet is the staggered magnetization

$$m_s = \left| \frac{1}{N} \sum_{\mathbf{r}} e^{i\mathbf{Q}\mathbf{r}} S_{\mathbf{i}}^z \right|^2, \tag{4}$$

where N is the number of spins in the lattice, $S_{\mathbf{r}}^z$ the z-component of the spin at site $\mathbf{r}$ and $\mathbf{Q} = (\pi, \pi)$. Close to the critical point the staggered magnetization behaves as

$$m_s \propto (g_c - g)^\beta. \tag{5}$$

At the critical point itself the real space correlation show a power-law falloff

$$\langle S_{\mathbf{0}}^z S_{\mathbf{r}}^z \rangle \propto e^{i\mathbf{Q}\mathbf{r}} r^{-(d+z-2+\eta)}, \tag{6}$$

where η is the correlation exponent. These three exponents are related by the usual scaling law

$$2\beta = (d + z - 2 + \eta)\nu, \tag{7}$$

where the effective dimension is $d + z$ in a quantum system.

TABLE I. Critical exponents β, ν, and η. Listed are both the estimates without making any assumption for z, and the best estimate if Lorentz invariance ($z = 1$) is assumed. For comparison the exponents of the 3D classical Heisenberg (O(3)) model, the 3D Ising model and the 2D quantum mean field exponents are listed.

model	ν	β	η
2D QAFM	0.685 ± 0.035	0.345 ± 0.025	0.015 ± 0.020
Lorentz invariant 2D QAFM	0.695 ± 0.030	0.345 ± 0.025	0.033 ± 0.005
3D O(3) [5]	0.7048 ± 0.0030	0.3639 ± 0.0035	0.034 ± 0.005
3D Ising [6]	0.6294 ± 0.0002	0.326 ± 0.004	0.0327 ± 0.003
mean field	1	$1/2$	0

TABLE II. Universal prefactor $\Omega_1(\infty)$ in the linear temperature dependence of the uniform susceptibility at criticality. Listed are the results for the quantum nonlinear sigma model in a $1/N$ expansion, the results by classical Monte Carlo simulation on a 3D classical rotor model and the result of the present study.

method	Ref.	$\Omega_1(\infty)$
1/N expansion	[2]	0.2718
classical Monte Carlo	[2]	0.25 ± 0.04
quantum Monte Carlo	this study	0.26 ± 0.01

B. Predictions from the nonlinear sigma model

As mentioned above most analytic calculations of quantum critical behavior are based on the 2D O(3) quantum nonlinear sigma model (QNLσM). Here we want to review the critical properties of the sigma model relevant for the current study.

The critical exponents of the QNLσM can be determined from simple symmetry, universality and scaling arguments [1,2]. Lorentz invariance implies that $z = 1$. Furthermore the 2D QNLσM is equivalent to the 3D classical sigma model. This in turn is in the universality class of the 3D classical O(3) model, or the classical 3D Heisenberg ferromagnet. The exponents β, ν and η should thus be the same as the well known classical exponents of these models (see Tab. I).

Chakravarty, Halperin and Nelson have discussed the phase diagram of a planar Heisenberg antiferromagnet in the framework of the QNLσM. They concentrate on the ordered phase and describe it as a classical 2D antiferromagnet with renormalized parameters.

Chubukov, Sachdev and Ye have investigated the quantum critical regime of the QNLσM in close detail [2]. They make some further predictions based on scaling arguments. On the ordered side the spin stiffness ρ_s vanishes as

$$\rho_s \propto (g_c - g)^{(d+z-2)\nu} = (g_c - g)^\nu, \qquad (8)$$

where the second equivalence comes from the prediction that $z = 1$. Additionally it follows from general scaling arguments that the uniform susceptibility at the critical point is universal:

$$\chi_u = \Omega_1(\infty) \left(\frac{g\mu_B}{\hbar c}\right)^2 T. \tag{9}$$

Here c is the spin wave velocity and $\Omega_1(\infty)$ a universal constant. Estimates for $\Omega_1(\infty)$ are listed in Tab. II.

The spin wave velocity c scales as

$$c \propto (g_c - g)^{\nu(z-1)} \tag{10}$$

and is thus regular at the critical point if $z = 1$.

C. What about Berry phases?

The equivalence of the 2D QAFM to the 2D QNLσM however is still an open question because of the existence of Berry phase terms in the QAFM that are not present in the QNLσM [3]. It has been argued that these terms cancel in special cases, such as in the bilayer model [7,8]. Then it is plausible that the quantum phase transition is in the same universality class as the QNLσM. This was confirmed by quantum Monte Carlo calculations of Sandvik and coworkers [7,9,10]. They have investigated the finite size scaling of the ground state structure factor and susceptibilities on lattices with up to $10 \times 10 \times 2$ spins. Although these lattices are quite small they still found good agreement of the exponents z and η with the QNLσM predictions [7,9].

In another study Sandvik *et al.* [10] have investigated finite temperature properties of the bilayer QAFM on larger lattices and also found good agreement with the QNLσM predictions. In the absence of Berry phase terms the equivalence of the QAFM and the QNLσM is quite well established by these simulations.

But in general these Berry phase terms exist. Chakravarty *et al.* argue that they can change the critical behavior and lead to different exponents [1,11]. Chubukov *et al.* on the other hand argue that the Berry phase terms are dangerously irrelevant [2] and do not influence the critical behavior.

Previous numerical simulations on dimerized square lattices [9,12] are indeed not consistent with the QNLσM predictions. Sandvik and Vekić [9] find a dynamical exponent $z \neq 1$, but their largest system was only 10×10 spins. The deviation could be a problem with scaling arising from inequivalent spatial directions.

Katoh and Imada however found $z \approx 1$, compatible with Lorentz invariance. Additionally they calculated the correlation length ξ and from it the exponent ν. The validity of their result $\nu \approx 1$ is however again questionable because

because of the restriction to very small lattices of 12 × 12. On the other hand
the discrepancy could be an effect of the Berry phase terms that are present in
the dimerized square lattice but probably not in the bilayer.

The main purpose of the simulations reported is to she light onto this question
and to clarify the role of the berry phase terms. Our results support the ideas
of Chubukov *et. al.* [2] that the Berry phase terms are dangerously irrelevant.

II. ALGORITHM AND PARALLELIZATION

Using the new quantum cluster algorithms, the loop algorithms [13,14] it
is possible to simulate much larger lattices at lower temperatures, just as the
corresponding classical cluster algorithms have allowed the simulation of critical
classical spin systems. With these algorithms it has for the first time become
possible to study quantum critical spin systems in detail.

A disadvantage of the cluster methods however is that they cannot be vectorized as easily as the local update algorithms. Using powerful vector machines
is therefore not an option. Fortunately however most of the modern super-
computers are parallel machines, and Monte Carlo methods are nearly ideally
suited for that architecture.

One of the authors has developed an object oriented Monte Carlo library
in C++ [15]. Using this library it is very simple to parallelize a Monte Carlo
program and to port it to new parallel computers. The library automatically
parallelizes any Monte Carlo simulation at the two "embarrassingly parallel"
levels. The first level of trivial parallelism is the parameter parallelism. Simulation with different parameters, such as system size, coupling or temperature
can be performed independently in parallel. At this level there is practically
no overhead due to the parallelization. We get perfect speedup and the library
takes care of load balancing.

A single simulation can similarly be parallelized by running it in parallel
with different initial states and random seeds on each of the processors. The
simulations run nearly independent. Communication is required only at the
start and the end of the simulation. This level of parallelization incurs some
overhead however. The overhead is the time used to thermalize a simulation.
We loose efficiency if this thermalization time becomes comparable to the time
actually needed for the simulation.

The third and deepest level of parallelization cannot be automatically done
by the library since it depends on the algorithm used for Monte Carlo. The
lattice used for one simulation can be spread over many processors. This parallelization has to be done by the programmer of the algorithm, but it is supported by various functions of the library. It is worthwhile to invest time in
this parallelization only in two cases. The first is when, as mentioned above,
thermalization is slow. Often the main reason is however different one. Large
lattices simply might not fit into the memory of one processor.

In our simulations reported here we have used the 1024-node, 300 GFlop Hitachi SR2201 massively parallel computer of the university of Tokyo. At the time of its introduction this machine was the fastest general purpose computer in the world. Each processor has 256 MByte of local memory, enough to simulate quantum spin systems with 20000 spins at temperatures as low as $T = 0.01$. This was large enough for the present study and we did not spend time on the third level of parallelization but used only the first two levels provided by the library.

The algorithm used was the continuous time loop algorithm [14]. The loop algorithms, first developed by Evertz et al. [13] are quantum version of the classical cluster algorithms. The continuous time version is preferable over the earlier discrete time versions since it eliminates the need to extrapolate in the finite Trotter time step $\Delta\tau$. In our experience we found that this leads to a four-fold speed increase. Additionally the continuous time algorithm uses only 10% of the memory compared to the discrete time algorithm, allowing the simulation of larger lattices.

III. THE CAV$_4$O$_9$ LATTICE

As the universality class of a phase transition does not depend on the microscopic details of the lattice structure we are free to choose the best lattice for our purposes. We have chosen the CaV$_4$O$_9$ lattice, a 1/5-th depleted square lattice depicted in Fig. 1 for our calculations. There are three reasons for this choice. Firstly the Berry phase terms are present on this lattice [16]. Next both space directions are equivalent, in contrast to the dimerized square lattice [9,12]. This makes the scaling analysis easier. Finally at the quantum critical point all the couplings are nearly equal in magnitude, which is also optimal

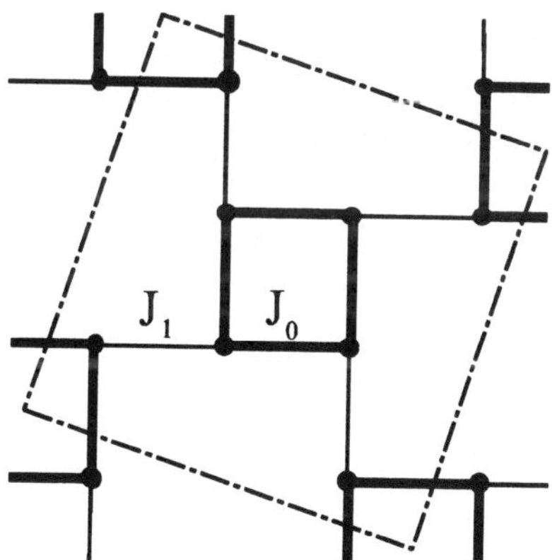

FIG. 1. Lattice structure of the 1/5-th depleted square lattice of CaV$_4$O$_9$. The dashed square indicates the eight spin unit cell used in our calculations.

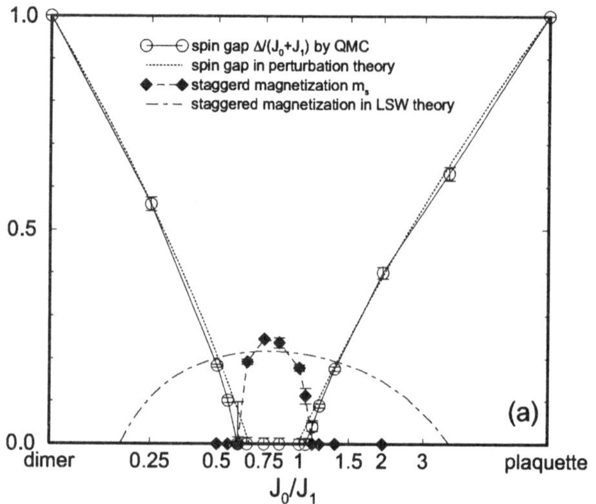

FIG. 2. Phase diagram of the CaV_4O_9 spin lattice as a function of the ratio J_0/J_1, reprinted from Ref. [17]. The leftmost point corresponds to the dimer limit $J_0 = 0$ and the rightmost point to the plaquette limit $J_1 = 0$. Circles indicate quantum Monte Carlo results for the spin gap, normalized by $J_0 + J_1$. Diamonds show the staggered magnetization. As reference the perturbation theory estimates for the gap [18] and the linear spin wave theory (LSW) estimates for the staggered moment have been included.

from a numerical point of view. We have performed our simulations on square lattices with $N = 8n^2$ spins, where n is an integer. Our largest lattices contained 20 000 spins. For the following discussion it is useful to introduce the linear system size L in units of the bond lengths a of the original square lattice: $L \equiv \sqrt{5N/4}a$.

The phase diagram of this lattice has been discussed in detail in Ref. [17] and is shown in Fig. 2. By removing every fifth spin we obtain a lattice consisting of four-spin plaquettes linked by dimer bonds. We label the couplings in a plaquette J_0 and the inter-plaquette couplings J_1. By controlling the ratio of these couplings J_1/J_0 we can tune from Néel order at $J_1 = J_0$ to a quantum disordered "plaquette RVB" ground state with a spin gap $\Delta = J_0$ at $J_1 = 0$.

IV. RESULTS

A. The critical point

The first step in the determination of the critical behavior is a high precision estimate of the critical coupling ratio J_1/J_0. We have calculated the second moment correlation length ξ_L on systems of various sizes L. This can be de-

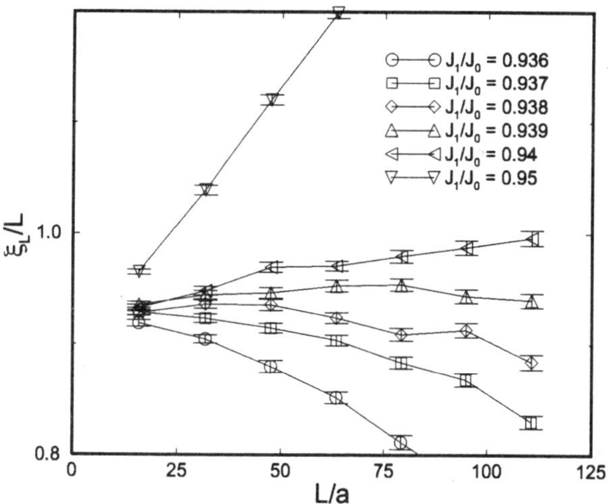

FIG. 3. Plot of the ratio of correlation length divided by system size ξ_L/L. At the critical point the correlation length calculated in a finite system is proportional to the system size. This is the case for $(J_1/J_0)_c = 0.939 \pm 0.001$.

termined in the usual way from the magnetic structure factor $S(\mathbf{q})$ close to the Néel peak at $\mathbf{Q}$:

$$S(\mathbf{Q} + \delta\mathbf{q}) = \frac{S(\mathbf{Q})}{1 + (\|\delta\mathbf{q}\|\xi)^2} + O(\delta\mathbf{q}^4). \tag{11}$$

The temperature was chosen to be $T = J_0/L$, keeping the finite $2+1$ dimensional system in the cubic regime. From standard finite size scaling arguments it follows that this correlation length ξ_L scales proportional to the system size L at criticality. We have calculated the ratio ξ_L/L (shown in Fig. 3) for a variety of couplings and system sizes up to $N = 9600$. Independence of the system size was seen at the critical coupling ratio $(J_1/J_0)_c = 0.939 \pm 0.001$.

B. The exponents

Next we have calculated the finite size scaling of both the staggered structure factor $S(\mathbf{Q}) = L^2 m_s$ and of the corresponding staggered susceptibility. At criticality they scale like

$$S(\mathbf{Q}) \propto L^{2-z-\eta} \tag{12}$$

$$\chi_s \propto L^{2-\eta} \tag{13}$$

The temperature was chosen to be $T = J_0/(6L)$, low enough to see ground

153

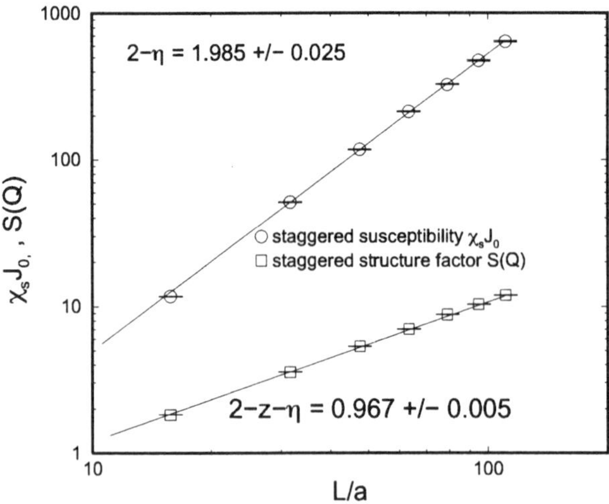

FIG. 4. Finite size scaling of the staggered structure factor and susceptibility at the critical point.

state properties within our accuracy. By fitting the results shown in Fig. 4 we obtain the estimates $z = 1.018 \pm 0.02$ and $\eta = 0.015 \pm 0.020$. This is perfectly consistent with the Lorentz invariance ($z = 1$) expected from a mapping to the QNLσM. We will discuss η below together with the other exponents. From these fits it is also obvious that at least $N = 800$ spins are necessary to obtain good scaling.

The remaining exponents β and ν are best calculated from the magnetization m_s and the spin stiffness ρ_s on the ordered side. Good estimates for m_s and ρ_s can be obtained by the Hasenfratz-Niedermayer equations [19]. These authors have calculated the *exact* finite-size and finite-temperature values of the low-temperature uniform and staggered susceptibilities χ_u and χ_s for the ordered phase of a 2D QAFM on a lattice with the symmetries of a square lattice. Their equations for the staggered susceptibility

$$\chi_s(T, L) = \frac{\mathcal{M}_s^2 L^2}{3T} \left\{ 1 + 2 \frac{\hbar c}{\rho_s L l} \beta_1(l) + \left(\frac{\hbar c}{\rho_s L l} \right)^2 [\beta_1(l)^2 + 3\beta_2(l)] + ... \right\} \quad (14)$$

and for the uniform susceptibility

$$\chi_u(T, L) = \frac{2\rho_s}{3(\hbar c)^2} \left\{ + \frac{1}{3} \frac{\hbar c}{\rho_s L l} \tilde{\beta}_1(l) + \right. \quad (15)$$

$$\left. \frac{1}{3} \left(\frac{\hbar c}{\rho_s L l} \right)^2 \left[\tilde{\beta}_2(l) - \frac{1}{3} \tilde{\beta}_1(l)^2 - 6\psi(l) \right] + ... \right\}.$$

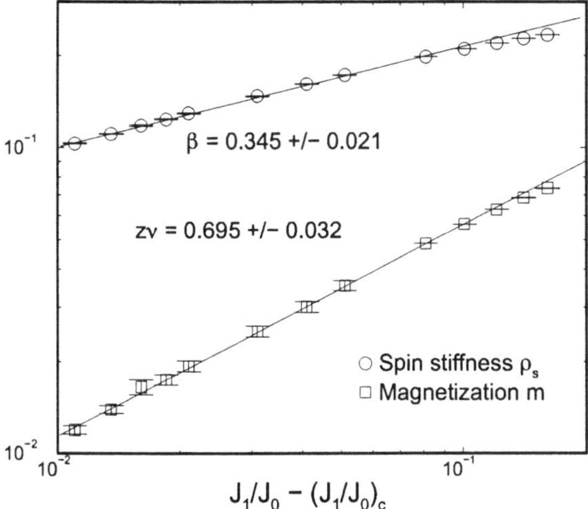

FIG. 5. Staggered magnetization m_s and spin stiffness ρ_s calculated by a fit of the low temperature susceptibilities on finite lattices to the Hasenfratz-Niedermayer equations.

are correct for the low temperature regime $k_B T \ll 2\pi\rho_s$ with cubic geometry $l^3 = \frac{\hbar c}{TL} \approx 1$. Up to second order in T (or $1/L$ respectively) the susceptibilities are universal, determined by only three parameters: the staggered magnetization m_s, the spin stiffness ρ_s and the spin wave velocity c. The shape functions β_1, β_2, $\tilde{\beta}_1$, $\tilde{\beta}_2$ and ψ are known exactly for square lattice geometries. Two high precision quantum Monte Carlo studies have confirmed the validity of these equations for the square lattice QAFM [14,20].

We have calculated the susceptibilities for a wide range of couplings $0.95 < J_1/J_0 < 1.1$, lattice sizes $800 < N < 16200$ and temperatures $0.006 < T/J_0 < 0.1$. The fits to the Hasenfratz-Niedermayer equations are all excellent, with $\chi^2/\text{d.o.f.} \approx 1.5$. This is another confirmation of the universality of the Hasenfratz-Niedermayer equations. From the fits we obtain the staggered magnetization m_s, the spin stiffness ρ_s and the spin wave velocity c. The exponents β and ν can then be obtained in a straightforward way (see Fig. 5) and are listed in Tab. I.

C. Discussion

Let us now discuss the results. First we observe that the exponents satisfy the scaling relation Eq. (7), indicating the validity of the scaling ansatz for this quantum phase transition. The exponents β, ν and η are in excellent agreement with the exponents of the 3D classical O(3) or Heisenberg model. They are however incompatible with the mean field exponents calculated by Katoh and Imada on small lattices.

Assuming Lorentz invariance $z = 1$ we can improve our estimates for the other three exponents. The agreement of the improved estimates with the 3D O(3) exponents becomes even better. We can also rule out the 3D Ising universality class whose exponents are also listed in Table I for a comparison.

This excellent agreement is a strong indication that the Berry phase terms in the 2D QAFM are indeed dangerously irrelevant as suggested by Chubukov, Sachdev and Ye. To further confirm their predictions we have calculated the uniform susceptibility close to criticality down to $T = 0.02$, more than an order of magnitude lower than Ref. [7]. We have extrapolated the finite size results on lattices with up to $N = 20000$ spins to the thermodynamic limit. Looking for the coupling at which a linear behavior occurs gives an independent estimate of the critical point: $(J_1/J_0)_c = 0.939 \pm 0.002$, in excellent agreement with the above estimate. The linear slope is $\Omega_1(\infty)(aJ_0/\hbar c)^2 = 0.238 \pm 0.003$. By extrapolating the spin wave velocity determined in the ordered phase by the Hasenfratz-Niedermayer fit to the critical point we get $\hbar c/aJ_0 = 1.04 \pm 0.02$ and thus $\Omega_1(\infty) = 0.26 \pm 0.01$, again in excellent agreement with Chubukov et al. (see Tab. II).

V. SUMMARY AND OUTLOOK

To summarize, we have performed a large scale quantum Monte Carlo simulation of a quantum phase transition in a planar antiferromagnet. The new quantum cluster algorithms, in particular the continuous time loop algorithm allow high precision simulation of critical quantum systems.

The critical exponents that we have calculated (listed in Table I) agree within our errors with the exponents of the classical 3D O(3) or Heisenberg model. The dynamical exponent $z = 1.018 \pm 0.02$, consistent with Lorentz invariance. This is compelling numerical evidence for the conjecture that the quantum Heisenberg antiferromagnet is in the same universality class as the 2D quantum nonlinear sigma model and the 3D Heisenberg ferromagnet. The Berry phase terms that are present in the Heisenberg antiferromagnet for non-integer spin do not seem to be influence the critical behavior. This supports the conjecture by Chubukov, Sachdev and Ye [2] that they are dangerously irrelevant.

While the accuracy achieved in the present simulation is remarkable for a simulation of a quantum system it is not very good compared to the best classical results that are an order of magnitude more accurate. If one wishes for higher accuracy the best approach could be to generalize the histogram methods to quantum systems and to do a finite size scaling study of cumulants, similar to the ones done in the classical case [5]. Even then we might not be able to reach the same accuracy as in the classical simulations for two reasons. Firstly we cannot measure the order parameter m_s in a quantum Monte Carlo simulation, but only its square m_s^2. Therefore we can only calculate the cumulants of m_s^2, which will be less favorable numerically. Another difference between a classical and a quantum system is that in the 3D classical system all three space directions are equivalent. In the $(2+1)$D quantum system on the other hand the time direction is not equivalent to the space direction, which makes a scaling analysis more complex.

ACKNOWLEDGMENTS

We want to thank the Computer Center of the university of Tokyo for giving us the ability to use their 1024-node massively parallel Hitachi SR 2201 supercomputer. Being able to use this fast computer has enabled us to perform the simulations reported here. We also want to thank K. Ueda, J.-K. Kim, D.P. Landau, S. Sachdev A.W. Sandvik and U.-J. Wiese for interesting discussions. M.T. was supported by the Japan Society for the Promotion of Science JSPS.

REFERENCES

[1] S. Chakravarty, B. I. Halperin, and D. R. Nelson, Phys. Rev. Lett. **60**, 1057 (1988); Phys. Rev. B **39**, 2344 (1989).
[2] A. V. Chubukov and S. Sachdev, Phys. Rev. Lett. **71**, 169 (1993); A. V. Chubukov, S. Sachdev, and J. Ye, Phys. Rev. B **49**, 11 919 (1994).
[3] F.D.M. Haldane, Phys. Rev. Lett. **61**, 1029 (1988).
[4] M. Troyer, M. Imada and K. Ueda, Report cond-mat/9702077.
[5] K. Chen, A.M. Ferrenberg and D.P. Landau, Phys. Rev. B **48**, 3249 (1993).
[6] A.M. Ferrenberg and D.P. Landau, Phys. Rev. B **44**, 5081 (1991).
[7] A.W. Sandvik and D.J. Scalapino, Phys. Rev. Lett. **72**, 2777 (1994).
[8] C.N.A. van Duin and J. Zaanen, Report No. cond-mat/9701035
[9] A. W. Sandvik and M. Vekić, J. Low. Temp. Phys. **99**, 367 (1995).
[10] A. W. Sandvik, A. V. Chubukov and S. Sachdev, Phys. Rev. B **51**, 16483 (1995).
[11] S. Chakravarty in *Random magnetism and high temperature superconductivity*, ed. by W. P. Beyermann, N. L. Huang-Liu and D. E. MacLaughlin, World Scientific (Singapore 1993).
[12] N. Katoh and M. Imada, J. Phys. Soc. Jpn. **63**, 4529 (1994).
[13] H. G. Evertz, G. Lana and M. Marcu, Phys. Rev. Lett. **70**, 875 (1993).
[14] B. B. Beard and U.-J. Wiese, Phys. Rev. Lett. **77**, 5130 (1996).
[15] M. Troyer and B. Ammon, unpublished. The library is available at the URL http://www.scsc.ethz.ch/~ troyer/alea.html .
[16] S. Sachdev and N. Read, Phys. Rev. Lett. **77**, 4800 (1996).
[17] M. Troyer, H. Kontani and K. Ueda, Phys. Rev. Lett. **76**, 3822 (1996).
[18] K. Ueda, H. Kontani, M. Sigrist and P. A. Lee, Phys. Rev. Lett. **76**, 1932 (1996).
[19] P. Hasenfratz and F. Niedermayer, Z. Phys. B **92**, 91 (1993).
[20] U.-J. Wiese and H. P. Ying, Z. Phys B **93**, 147 (1994).

Part III
Contributed Papers

Fragmentation Scaling of the Percolation Cluster

M. Cheon and I. Chang

Department of Physics, Pusan National University, Pusan 609-735, Korea

Abstract : The scaling behavior for a binary fragmention of percolation clusters near a percolation threshold is investigated by a large cell Monte Carlo real space renormalization group method on two and three dimensions. We obtain the accurate values of critical exponents λ and ϕ describing the scaling of fragmentaion rate and the distribution of cluster masses produced by a binary fragmentation. Our results for λ and ϕ in two dimensions agree with the previous works, and we show in three dimensions that the fragmentation rate is also proportional to the cluster size. We provide the numerical evidence that the conjectured scaling relation $\sigma = 1 + \lambda - \phi$ is satisfied both in two and three dimensions where σ is the crossover exponent of the average cluster number in percolation theory.

I. Introduction

Recently there has been a large interest to understand the fragmentation property of spatially random objects [1-4]. The fragmentation process generates fragments successively as the time elapses. Such a process is governed by a rate equation which is generally an integro -differential equation. Quantities of interest in the fragmentation phenomena are the fragmention rate and the mass distribution of fragments after breakup which are, in fact, the kernels to solve a rate equation. To investigate the scaling behaviors of these quantities without any assumption the percolation model was taken into account [3,5,6], which can describe spatially random objects. This model has an advantage that as a bond occupation probability p changes various clusters - ramified (compact) cluster which is (not) easily broken - can be generated.

On the bond percolation cluster a fragmenting bond is defined as such a bond removal on a cluster results in two disconnected clusters , that is a binary fragmentation. Here the fragmentation rate is proportional to the number a_s of fragmenting bonds of a mother cluster of size s, and the mass distribution is a probability distribution $b_{s's}$, which is a probability to find a daughter cluster of size s' after (binary) fragmentating a mother cluster of size s. On one dimension [4] and Bethe lattice [3] the exact results for these quantities show a scaling behavior for large s without p dependence. For dimensions d

Springer Proceedings in Physics, Volume 83
Computer Simulation Studies in Condensed-Matter Physics X
Eds.: D. P. Landau, K.K. Mon, H. -B. Schüttler
© Springer-Verlag Berlin Heidelberg 1998

> 1 , a_s and $b_{s's}$ don't have scaling form at any p due to the p dependence of cluster structure except at the percolation threshold where there exists a self-similarity, therefore one can expect a scaling behavior at p_c. From the scaling behaviors for $d = 1$ and Bethe lattice the generalized scaling forms for $d > 1$ were suggested, namely $a_s(p = p_c) \sim s^\lambda$ and $b_{s's}(p = p_c) = s^{-\phi}g(s'/s)$ as a cluster size increases [5,6] ($\lambda = 1.0$, $\phi = 1.0$ for $d = 1$ and $\lambda = 1.0$, $\phi = 1.5$ for Bethe lattice). Gyure et.al. performed an intensive Monte Carlo simulation for a binary fragmentation of two-dimensional percolation clusters on the square lattice, and identified these scaling behavior to provide the critical exponents $\lambda = 1.001 \pm 0.006$, $\phi = 1.601 \pm 0.008$ [5,6]. Their results also confirmed the scaling relation $\sigma = 1 + \lambda - \phi$ in two dimensions which was derived using the analogy between red and fragmenting bonds, where σ is the crossover exponent of the cluster number in the standard percolation theory. Henceforth, they conjectured that these scaling behavior exist at p_c with $\lambda = 1.0$ and the scaling relation $\sigma = 1 + \lambda - \phi$ must valid *for all dimensions*.

Our main purpose in this paper is to evaluate the critical exponents λ and ϕ and to verify the conjectured scaling relation in two and three dimensions for a binary fragmentation of bond-percolation clusters. To find the accurate values of critical exponets, we use the large cell-to-cell Monte Carlo renormalization group (RG) method which was developed recently [9].

This paper is organized as follows. In section II we explain the model and RG schems . In section III numerical method and our main results are presented and we conclude in section IV.

II. Real Space Renomalization Group for The Fragmentation

To apply RG for a binary fragmentation of the bond percolation cluster [7,8], let us take a 2×2 cell with eight bonds as the basic block cell, which is a cell of minimal size to have one fragmenting bond at least. We denote such a cell as G which is a d-dimensional cell with E bonds and a linear size L in general. The subgraphs denoted by G' is produced by a dilution of bonds, that is each bond in a cell G is occupied (unoccupied) with a probability p $(1 - p)$, where $0 \leq p \leq 1$, so the number of all possible subgraphs is 2^E. The spanning probability $E(G, p)$ is the sum of all configurational probability $p^b(1 - p)^{E-b}$ over subgraphs G'_p which percolates the cell G, where b is the number of bonds on a subgraph G'_p [9].

$$E(G, p) = \sum_{G'_p \subseteq G} p^{b(G'_p)}(1 - p)^{E - b(G'_p)} \tag{1}$$

$E(G, p)$ under a recursive iteration gives the percolation threshold p^* which is shifted due to the finite size of the block cell.

The average number of fragmenting bonds $F(G,p)$ is the weighted sum of number of all fragmenting bonds over subgraphs G'_p

$$F(G,p) = \sum_{G'_p \subseteq G} p^{b(G'_p)}(1-p)^{E-b(G'_p)} N_f(G'_p) \tag{2}$$

, where $b(G'_p)$ is the number of bonds and $N_f(G'_p)$ is the number of the fragmentating bonds of the percolating cluster on a subgraph G'_p. Thus, $a_s(p_c)$ equals $F(G,p_c)$ at p_c for a large cell. The sum of configurational probability is an increasing and bounded function of p and $N_f(G'_p)$ has the maximum value for ramified cluster ,thus $F(G,p)$ is the function of p having a peak at a certain p. Next, we consider the first moment μ_s of $a_s b_{s's}$ which is the sum of fragments' mass s' after fragmentating a mother cluster of size s. Assuming $a_s(p_c) \sim s^\lambda$ and $b_{s's}(p_c) \sim s^{-\phi}$, $\mu_s(p)$ is expected to have a power law scaling with a critical exponent $2 + \lambda - \phi$ at $p = p_c$.

$$\mu_s(p_c) = \sum_{s'=1}^{(s-1)/2} s' a_s(p_c) b_{s's}(p_c) \sim s^{2+\lambda-\phi} \tag{3}$$

Thus, the average first moment $M(G,p)$ becomes

$$M(G,p) = \sum_{G'_p \subseteq G} p^{b(G'_p)}(1-p)^{E-b(G'_p)} N_\mu(G'_p) \tag{4}$$

in the percolating cluster on the cell G, where $N_\mu(G'_p)$ is the sum of mass of all possible smaller fragments, so $\mu_s(p_c)$ equals $M(G,p_c)$.

In the large cell-to-cell real space RG method, we renormalize a larger block cell G_1 of linear dimension L_1 to a smaller block cell G_2 of linear dimension L_2. So the RG recursion equation for a spanning probability is given by

$$E(G_1,p) = E(G_2,p') \tag{5}$$

,where p' is the renormalized bond occupation probability. This gives a unstable fixed point p^* which approaches the critical point p_c as the block cell size increases. Let's introduce a dimensional mass m_f for a fragmenting bond and m_μ for a bond on a daughter cluster [10]. Then, these total masses of the percolating cluster is conserved under the RG transformation, namely

$$m_f(G_1)F(G_1,p) = m_f(G_2)F(G_2,p') \tag{6}$$

$$m_\mu(G_1)M(G_1,p) = m_\mu(G_2)M(G_2,p') \tag{7}$$

Since $a_s(p_c) \sim s^\lambda$ and $\mu_s(p_c) \sim s^{2+\lambda-\phi}$, m_f and m_μ scales as $s^{-\lambda} \sim L^{-d_f\lambda}$ and $s^{-(2+\lambda-\phi)} \sim L^{-d_f(2+\lambda-\phi)}$ respectively, where d_f is the fractal dimension of the percolating cluster at p_c. Having idenfied p^* from Eq.(5), the critical exponents λ and ϕ are given by

$$d_f\lambda = \frac{log[F(G_1,p^*)/F(G_2,p^*)]}{log[L(G_1)/L(G_2)]} \tag{8}$$

$$d_f(2+\lambda-\phi) = \frac{log[M(G_1,p^*)/M(G_2,p^*)]}{log[L(G_1)/L(G_2)]}. \tag{9}$$

III. Numerical Calculation and Results

We enumerate exactly all possible subgraphs, which percolates in one (horizontal) direction, for 2×2 cell and 3×3 cell in the square lattice to give the exact polynomials for $F(G,p)$ and $M(G,p)$. The polynomials of the 2×2 and 3×3 cells are given by $F_{2\times2}(G,p) = 5p^3q^5 + 31p^4q^4 + 59p^5q^3 + 43p^6q^2 + 10p^7q$, $M_{2\times2}(G,p) = 5p^3q^5 + 31p^4q^4 + 73p^5q^3 + 61p^6q^2 + 16p^7q$, $F_{3\times3}(G,p) = 3p^3q^{15}+66p^4q^{14}+664p^5q^{13}+3946p^6q^{12}+15319p^7q^{11}+40640p^8q^{10}+75260p^9q^9+97725p^{10}q^8 + 88807p^{11}q^7 + 55948p^{12}q^6 + 24142p^{13}q^5 + 6960p^{14}q^4 + 1274p^{15}q^3 + 134p^{16}q^2 + 6p^{17}q$, and $M_{3\times3}(G,p) = 3p^3q^{15} + 66p^4q^{14} + 743p^5q^{13} + 4920p^6q^{12} + 21448p^7q^{11}+63443p^8q^{10}+131064p^9q^9+187425p^{10}q^8+185804p^{11}q^7+124377p^{12}q^6+55668p^{13}q^5+15851p^{14}q^4+2688p^{15}q^3+240p^{16}q^2+8p^{17}q$, where $q = 1-p$. Solving RG recursion equations from 3×3 cell to 2×2 cell gives $p_c = 0.500$, $\lambda = 1.296$,$2 + \lambda - \phi = 1.857$ using $d_f = 91/48$ for 2d bond percolation model [12], A small cell-to-cell RG gives a little difference from Gyure's result, because of the finite size effect of a basic block cell, that $\lambda = 1.001$ and $1 + \lambda - \phi = 1.400$ for 2d. Therefore, we should employ larger cells for RG calculations.

However, if the cell is large ,say $L = 4$ for 2d, the number of all subgraphs to be enumersted is $2^{2\times4\times4}$, hence this work is very time-comsuming and practically impossible for larger L. So we sample the subgraphs G' by Monte Carlo method. In order to introduce Monte Carlo sampling we rearrange the Eq.(2) and Eq.(4) with the summation index for the number of the occupied bonds b as follows.

$$F(G,p) = \sum_{b=1}^{E} p^b(1-p)^{E-b} N_f(b) \tag{10}$$

$$M(G,p) = \sum_{b=1}^{E} p^b(1-p)^{E-b} N_\mu(b) \tag{11}$$

$N_f(b)$ and $N_\mu(b)$ is the sum of $N_f(G'_p)$ and $N_\mu(G'_p)$ over all subgraph with b occupied bonds respectively. First, we sample $M'(b)$ configurations with b

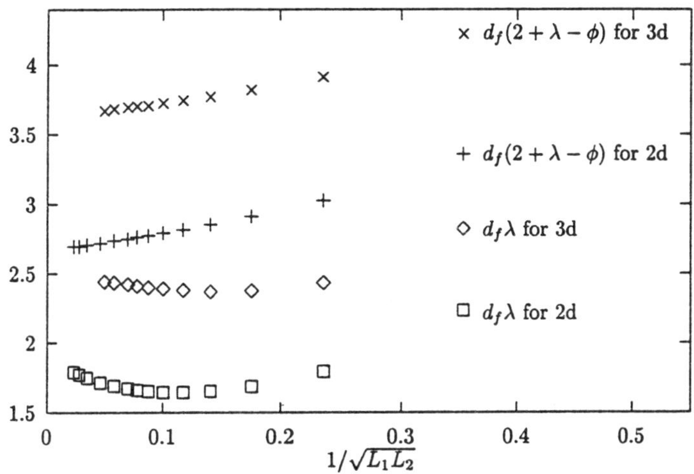

Fig. 1 The critical exponents $d_f\lambda, d_f(2 + \lambda - \phi)$ in 2d and 3d with an abscissa $1/\sqrt{L_1 L_2}$.

bonds by Monte Carlo Method, and compute the number of the fragmenting bond whose sum over the percolating clusters out of $M'(b)$ configurations is denoted by $N'_f(b)$. Hoshen and Kopelman algorithm [11] is used to pick up the percolating cluster. Once one bond is removed, this algorithm is used again to identify fragments and their sizes. If the size of fragments is $s - 1$ this removed bond is not a fragmenting one, otherwise it is a fragmenting one. In this way we compute the $N'_f(b)$ and $N'_\mu(b)$ to approximate the sum $N_f(b)$ of the number of the fragmenting bonds over all subgraphs with b bonds as follows $N_f(b) =_E C_b N'_f(b)/M'(b)$, similarly $N_\mu(b) =_E C_b N'_\mu(b)/M'(b)$. To find $F(G,p)$ and $M(G,p)$ for general p we should have generated subgraphs G' at the sequence of p using histogram method [9]. But, it is sufficient for our purpose to generate subgraphs G' at p_c or several points around p_c.

Various sizes of block cells are used from 2×2 to 60×60 for 2d and from $2 \times 2 \times 2$ to $28 \times 28 \times 28$ for 3d. N different configurations over which we compute $N_f(G'_p)$ and $N_\mu(G'_p)$ are generated at p_c known previously [12]. We need the fixed point p^* from Eq.(5) to obtain the critical exponents λ and ϕ from Eq.(8) $\sim$ (9). We generate N_p different configurations around p_c to find the more accurate value of p^*. So we obtain p_c from $N + N_p$ configurations and $d_f\lambda$,$d_f(2+\lambda-\phi)$ from N configurations. This whole procedure is repeated N_c times to estimate errorbars. These numbers for N, N_p, N_c are $2 \times 10^6, 2.7 \times 10^6, 20$ for 4×4 cell and $2 \times 10^4, 3.4 \times 10^4, 5$ for 60×60 cell in 2d, and $5 \times 10^5, 6 \times 10^5, 20$ for $3 \times 3 \times 3$ cell and $1.5 \times 10^4, 1.2 \times 10^5, 3$ for $28 \times 28 \times 28$ cell in 3d. The rescaling factor is kept by 2 under our RG transformation. Our results show that as the cell size becomes larger, p_c approaches to the known values accurately and the critical exponents converge to the expected values. The results for $2d$ and $3d$ are plotted in Fig. 1 with the inverse of the geometrical mean for sizes of two cells, $1/\sqrt{L_1 L_2}$, and we extrapolate the value in the zero limit of abscissa.

From Fig. 1 , we obtain $\lambda = 0.993 \pm 0.010$ and $2 + \lambda - \phi = 1.405 \pm 0.005$ for 2d, which agree with the Gyure's Monte Carlo result. For the first time in $3d$ we also evaluate $\lambda = 0.996 \pm 0.005$ and $2 + \lambda - \phi = 1.443 \pm 0.008$. Our results support the fact that the number of the fragmenting bonds is proportional to the size of mother cluster and the scaling relation $1 + \lambda - \phi = \sigma$ conjectured to be valid for all dimensions is satisfied here at 3d as well as 2d, where $\sigma = 36/91$ for 2d and 0.45 for 3d [12].

Above 3d we also evaluate the exponents λ and ϕ by the exact series expansion method in general dimensions. The preliminary results above $3d$ also support that $\lambda = 1$ and the conjectured scaling relation is valid for higher dimensions, which we will publish elsewhere.

IV. Conclusion

We have investigated the scaling behavior for a binary fragmention of percolation clusters near a percolation threshold by a large cell Monte Carlo real space renormalization group method on two and three dimensions. We obtain the accurate values of critical exponents λ and ϕ describing the scaling of fragmentaion rate and the distribution of cluster masses produced by a binary fragmentation. Our results for λ and ϕ in two dimensions agree with the previous works. And in three dimensions we show that the fragmentation rate is also proportional to the cluster size, and that the scaling relation $\sigma = 1 + \lambda - \phi$ conjectured to be valid for all dimensions is satisfied.

Acknowledgements : We acknowledge the supports in part from the Korea Science and Engineering Foundation through RCDAMP, the Basic Science Research Institute Program (BSRI-96-2412) and its matching fund program (RIBS-PNU-96-202), and the University research fund at Pusan National University.

References

[1] R. M. Ziff and E. D. McGrady, J. Phys. A **18**, 3027 (1985); E. D. McGrady and R. M. Ziff, Phys. Rev. Lett. **58**, 892 (1987)

[2] Z. Cheng and S. Redner, Phys. Rev. Lett **60**, 2450 (1988) ; J. Phys. A **23**, 1233 (1990).

[3] A. R. Kerstein , J. Phys. A **22**, 3371 (1989).

[4] B. F. Edwards, M. Cai and H. Han , Phys. Rev. A **41**, 5755 (1990) ; M. Cai, B. F. Edwards and H. Han , Phys. Rev. A **43**, 656 (1991).

[5] M. F. Gyure and B. F. Edwards , Phys. Rev. Lett **68**, 2692 (1992).

[6] B. F. Edwards , M. F. Gyure and M. Ferer , Phys. Rev. A **46**, 6252 (1992).

[7] P. J. Reynolds, H. E. Stanley, and W. Klein, Phys. Rev. B **21**, 1223 (1980); J. Phys. C **10**, L167 (1977)

[8] D. C. Hong and H. E. Stanley , J. Phys. A **16**, L525 (1983)

[9] C. K. Hu, Phys. Rev. B **46**, 6592 (1992) ; Phys. Rev. Lett. **69**, 2739 (1992) ; Phys. Rev. B **51**, 3922 (1995)

[10] C. Tsallis, A. Coniglio, and G. Schwachheim , Phys. Rev. B **32**, 3322 (1985)

[11] J. Hoshen and R. Kopelman Phys. Rev. B **14**, 3438 (1976)

[12] D. Stauffer and A. Aharony, *Introduction to Percolation Theory* , 2nd ed. (Taylor & Francis, London, 1992)

Magnons in Heisenberg Chains with Random ±J Nearest-Neighbor Interactions

I. Avgin[1], A. Boukahil[2], and D.L. Huber[2]

[1]Department of Physics, Erciyes University, Kayseri, Turkey
[2]Department of Physics, University of Wisconsin-Madison,
  Madison, WI 53706, USA

Abstract. Recent studies of the distribution and localization of linear magnon modes in Heisenberg spin chains with random ±J nearest-neighbor interactions are summarized. A combination of numerical and analytical techniques are utilized to obtain the limiting behavior at low energies. The magnon contribution to the low temperature specific heat is calculated and compared with the specific heat of a spin-1/2 chain with similar interactions that was calculated by Westerberg *et al.* using real space renormalization group techniques.

1. Introduction

The dynamical behavior of spin chains with random Heisenberg interactions has proven to be a challenging problem with interesting connections to other problems in the theory of disordered systems. Our work [1,2,3] and that of Pimentel and Stinchcombe [4,5,6] have explored the effect of the disorder on the distribution and localization of the low-lying spin excitations found by linearizing the equations of motion of the spins about the classical ground state. This approach, which is expected to apply to quantum spin systems in the limit $S \gg 1$ (i.e. the classical limit), generates harmonic excitations which are characterized by an excitation energy and localization length, the latter being a measure of the distance over which the mode has appreciable amplitude. The goal of the studies has been to understand the energy dependence of the distribution of modes (or density of states) and the localization length for a Heisenberg Hamiltonian with nearest-neighbor interactions and no correlation between different bonds. Calculations were carried out both in zero field [1,3,4,5] and in finite fields [2,3 6]. In the zero-field work, emphasis was placed on the behavior near zero energy where both the localization length and the density of states displayed power law behavior. The studies involved numerical techniques complemented by exact and approximate analytical treatments.

In this note, we focus on a particularly simple realization of the random chain where the nearest-neighbor exchange interaction has the bipolar distribution $P(J) = (1 - c)\delta(J - 1) + c\delta(J + 1)$. The overall sign of the Heisenberg Hamiltonian is such that $c = 0$ corresponds to a ferromagnetic array and $c = 1$ to an antiferromagnetic array. Exact results for the density of states obtained previously for $c = \frac{1}{2}$ [1] and corresponding numerical results for $c \neq \frac{1}{2}$ [2] are used to predict the temperature dependence of the specific heat of the chain in the limit $T \to 0$. Our results are

Springer Proceedings in Physics, Volume 83
Computer Simulation Studies in Condensed-Matter Physics X
Eds.: D. P. Landau, K.K. Mon, H. -B. Schüttler
© Springer-Verlag Berlin Heidelberg 1998

compared with the real space renormalization group analysis for the low-temperature specific heat of the random spin -1/2 chain [7]. We conclude with a brief discussion of the localization of the low energy modes.

2. Calculation

The starting point in the analysis is the Heisenberg equation of motion for the spin operator S_{+j}, $idS_{+j}/dt = [S_{+j}, H]$, where the Hamiltonian H has the form $-\Sigma J_{j,j+1} S_j S_{j+1}$. After linearizing the equation by replacing S_{zj} by its ground-state expectation value denoted by $<S_{zj}>_o$ $(= \pm 1)$ and assuming a harmonic time dependence which is of the form $S_{+j}(t) = U_j \exp(-i\omega t)$, one obtains a set of coupled algebraic equations which can be further simplified by making the Mattis transformation $V_j = <S_{zj}>_o U_j$. Denoting $<S_{zj}>_o$ by ζ_j, one obtains the result

$$(2 - \varsigma_j \omega)V_j = V_{j-1} + V_{j+1}. \tag{1}$$

Although the explicit dependence on the exchange integral has disappeared, it appears implicitly through the parameter ζ_j, which takes the form

$$\varsigma_j = \prod_{k=2}^{j} J_{k-1,k}, \tag{2}$$

for the $\pm J$ model with $|J|=1$.

The dynamical properties associated with Eq. (1) were first studied by Stinchcombe and Pimentel [4] in the symmetric limit, $c = 1/2$. Using a new transfer matrix scaling technique, they established that the system has unconventional hydrodynamics. In particular, they showed that the low energy modes had the dispersion relation $\omega \propto k^{3/2}$, k denoting the wave vector, whereas conventional hydrodynamic theory predicts $\omega \propto k$.. Additional studies in the c=1/2-limit were also carried out by Boukahil and Huber [1]. In their work, they pointed out the connection between the spin glass equation and the discretized one-dimensional Schrödinger equation with random potential λv_j which takes the form

$$(E - \lambda v_j)\psi_j = \psi_{j+1} + \psi_{j-1}. \tag{3}$$

In particular, Eq.(1) is seen to be a special case of (3) corresponding to $E = 2$ and $\zeta_j \omega = \lambda v_j$. By making this connection, they were able to make use of exact results derived by Derrida and Gardner [8] to obtain a low-energy density of states of the form

$$\rho_{1/2}(\omega) = \frac{2\omega^{-1/3}}{\sqrt{2\pi} 6^{1/6} \Gamma(1/6)}. \tag{4}$$

where $\Gamma(x)$ denotes the gamma function.

Subsequently, Avgin and Huber [2] used mode-counting techniques to determine the limiting behavior of the density of states for c in the range $0.1 < c < 0.9$. In their work, they made use of the result that the integrated density of states (counting positive and negative eigenvalues as distinct modes [9]) is given by the number of negative signs in the sequence $V_2/V_1,\ \dots,V_N/V_{N-1}$ divided by N, the total number of spins. They found that the densities of states at low energies ($\omega \leq 0.1$) for chains of 10^7 spins all had the same frequency dependence and varied approximately as

$$\rho_c(\omega) = \exp[-3c/2 + 3/4]\rho_{1/2}(\omega).\qquad(5)$$

3. Specific Heat

The results for the low-energy density of states given in the preceding section can be utilized in a calculation of the low-temperature specific heat of the random spin chain in the independent boson approximation by using the standard procedure of evaluating the mean thermal energy using the Bose weighting factor $[\exp(\omega/T) -1]^{-1}$ and then taking the derivative with respect to temperature [10]. Denoting the specific heat by C_c, the resulting expression takes the form

$$C_c = \frac{10\Gamma(5/3)\varsigma(5/3)T^{2/3}}{3\sqrt{2\pi}\,6^{1/6}\Gamma(1/6)}\exp[-3c/2 + 3/4] \approx 0.719 e^{-3c/2}T^{2/3}.\qquad(6)$$

Equation (6) shows that the specific heat varies as $T^{2/3}$ not only for $c = \frac{1}{2}$ but over a range of c. The exponent 2/3 is larger than the value of 0.44 obtained for the specific heat of the random spin-1/2 chain in Ref. 7 and lies between the value $\frac{1}{2}$, characterizing the specific heat of the ideal ferromagnetic chain (c=0) in the boson approximation, and 1, the specific heat exponent for the ideal antiferromagnetic chain (c=1) in the same approximation. Although our results were obtained for the $\pm J$ model, we conjecture that they hold for a wider class of random interactions, as is the case for the spin-1/2 calculations reported in Ref. 7. We also conjecture that the specific heat exponents for random quantum mechanical chains with S = 1, 3/2, 2, … will increase toward 2/3 with increasing S.

4. Localization

The power law behavior in the density of states is mirrored in the behavior of the localization length. As pointed out by Thouless [11], the inverse localizaton length, or ILL, is given by the sum

$$ILL(\omega) = N^{-1}\sum_{n=2}^{N} \ln|V_n / V_{n-1}|.\qquad(7)$$

In the case of the $\pm J$ model with $c = 1/2$, Avgin and Huber [2] utilized the connection with the discretized Schrödinger [8] to show that this sum reduced to the expression

$$ILL_{1/2}(\omega) = \frac{6^{1/3}}{2\Gamma(1/6)}\sqrt{\pi}\,\omega^{2/3}, \tag{8}$$

in the limit as $\omega \to 0$. By directly evaluating Eq. (7) with chains of 10^7 spins for c in the range $0.1 \leq c \leq 0.9$, they showed that ILL_c was characterized by the same exponent as $ILL_{1/2}$ and was well approximated by the expression:

$$ILL_c(\omega) = \exp[-3c/2 + 3/4]ILL_{1/2}(\omega), \tag{9}$$

analogous to Eq. (5). Taken together, these results show that the low-energy modes become increasingly delocalized as $\omega \to 0$.

5. Discussion

The studies described above involved a combination of exact analytical calculations and numerical modeling. In the special case of $c = \frac{1}{2}$, it is possible to develop an approximate theory based on a mapping between the dynamical equations in zero field and the corresponding equations in the high-field limit, where the ground state is fully aligned [2,12]. In the high-field limit, one can utilize a *coherent exchange approximation* (CEA) in calculating a spin greens function from which one can extract the density of states and inverse localization length. The approximate calculation reproduces the correct exponents and very nearly the correct amplitudes. One has $\rho_{1/2}(\text{CEA})/\rho_{1/2}(\text{exact}) = ILL_{1/2}(\text{CEA})/ILL_{1/2}(\text{exact}) \approx 1.029$.

Acknowledgment Work supported in part by the Scientific and Technical Research Council of Turkey (TUBITAK).

References

1. A. Boukahil and D. L. Huber, Phys. Rev. B **40**, 4638 (1989).
2. I. Avgin and D. L. Huber, Phys. Rev. B **48**, 13 625 (1993).
3. A. Boukahil and D. L. Huber, Phys. Rev. B **50**, 2978 (1994).
4. R. B. Stinchcombe and I. R. Pimentel, Phys. Rev. B **38**, 4980 (1988).
5. I. R. Pimentel and R. B. Stinchcombe, Europhys. Lett. **6**, 719 (1988).
6. I. R. Pimentel and R. B. Stinchcombe, Phys. Rev. B **40**, 4947 (1989).
7. E. Westerberg, A. Furusaki, M. Sigrist and P. A. Lee., Phys. Rev. Lett. **75**, 4302 (1995).

8. B. Derrida and E. Gardner, J. Phys. (Paris) **45**, 1283 (1984).

9. D. L. Huber, Phys. Rev. B **8**, 2124 (1973).

10. I. Avgin and D. L. Huber (unpublished).

11. D. J. Thouless, J. Phys. C: Solid State Phys. **5**, 77 (1972); Physics Reports **13**, 93 (1974).

12. I. Avgin Ph. D. thesis, University of Wisconsin-Madison, 1993.

Eight-State Potts Model on the Quasiperiodic Octagonal Tiling: Free Boundary Effects

D. Ledue[1], D.P. Landau[2], and J. Teillet[1]

[1]LMA UMR 6634 CNRS-Université de Rouen,
 76821 Mont-Saint-Aignan Cédex, France
[2]Center for Simulational Physics, The University of Georgia,
 Athens, GA 30602, USA

Abstract. The effects of free boundary at a temperature-driven transition of the eight-state Potts model on the two-dimensional (2D) quasiperiodic octagonal tiling are investigated using the importance-sampling Monte Carlo method and the single histogram technique. The analysis of the probability distributions of the internal energy evidences that such numerical data suffer from drastic free boundary effects. However, the size dependence of the free energy barrier indicates that the system undergoes a first order transition as in 2D periodic lattices.

1. Introduction

The q-state Potts model [1] in two dimensions has been widely investigated for twenty years [2-4]. The nature of the phase transition in the two-dimensional (2D) q-state Potts model on periodic lattices has been known to be first order for $q > 4$ and second order for $q \leq 4$ [5,6]. Moreover, the transition temperature is known for all q as are the critical exponents for $q \leq 4$. Recently, a rather intriguing feature concerning a change in the nature of the transition for the 2D eight-state Potts model due to the addition of quenched bond randomness has been reported [7,8]. This study has evidenced that the phase transition changes from first order to second order with 2D Ising exponents. This feature has motivated us to investigate the nature of the transition for the eight-state Potts model on 2D quasiperiodic tilings which exhibit unusual long-range order.

Unlike periodic lattices, very few studies about the Potts model on 2D quasiperiodic tilings have been carried out. Actually, mainly the static critical behaviour of the ferromagnetic Ising model ($q = 2$) on the Penrose tiling (5-fold symmetry) and on the octagonal tiling (8-fold symmetry) (Fig. 1) has been investigated. The critical exponents have been found to be the same as for 2D periodic lattices [9-11]. It should be noted that the estimated critical temperatures, $kT_c/J = 2.392\pm0.004$ and $kT_c/J = 2.401\pm0.005$ for the Penrose tiling [9,10], and $kT_c/J = 2.39\pm0.01$ for the octagonal tiling [11] are slightly higher than the critical temperature in the square lattice ($kT_c/J = 2.269$) while the mean coordination number is equal to 4 for all three lattices. Another study of a weakly frustrated ferromagnetic Ising model on the octagonal tiling has provided the same critical exponents [12].

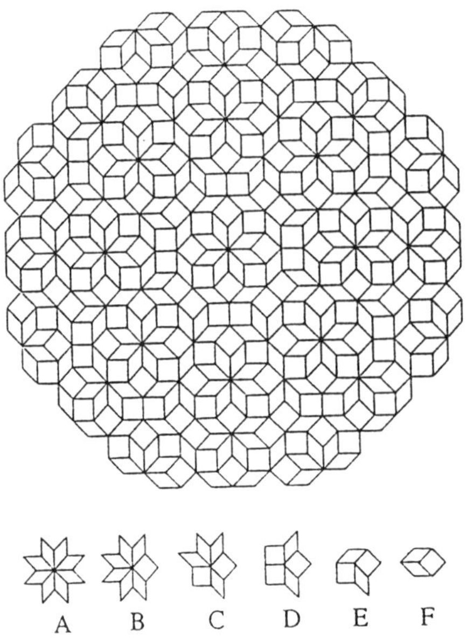

Fig. 1. The octagonal tiling (8-fold orientational symmetry) and its six local environments.

In this paper, we investigate the nature of the transition of the eight-state Potts model on 2D quasiperiodic octagonal tilings [11,13]. The 2D octagonal tilings are constructed by the infinite repetition in space of two distinct " unit cells ": a square and a rhombus (Fig. 1). The set of the vertices of such a tiling is a lattice that exhibits long-range quasiperiodic translational order and long-range orientational order (8-fold). It should be noted that all octagonal tilings are locally isomorphic, that is, they can be made to overlap out to any finite distance by a finite translation [13,14]. This implies that all octagonal tilings should exhibit identical macroscopic physical properties (transition temperature, critical exponents, ground states...). Then, we have considered only one octagonal tiling which is shown on Figure 1.

The model, simulation techniques and a brief review of some methods usable to determine the order of a transition are presented in Section 2. Results are discussed in Section 3 and a summary is given in Section 4.

2. Background

The Hamiltonian of the q-state Potts model [1] is given by :

$$H = -J \sum_{<i,j>} \delta_{S_i S_j} \qquad (J > 0)$$

174

where the spins S_i, which are located at the vertices of the octagonal tiling, take on the values 1, ..., q and δ is the Kronecker delta function. The ferromagnetic interaction J is along the edges of the unit cells of the tiling (the sum goes over all next nearest-neighbour pairs in the tiling), so the mean coordination number is equal to 4. In this study, q = 8.

The procedure we used, is the importance-sampling Monte Carlo (MC) method [15] based on the standard single spin-flip Metropolis algorithm [16]. The analysis of the data has been carried out by the single histogram technique [17]. Our simulations were carried out on finite octagonal tilings of N = 329, 689, 1433, 2481 and 5497 vertices with free boundary conditions. For each run, 5×10^5 Monte Carlo steps (MCS)/spin (N = 329) up to 5×10^6 MCS/spin (N = 5497) were performed (the first 2×10^4 MCS/spin were discarded for equilibration).

$$\text{The specific heat, } C(T,L) = L^{-2}\frac{<E^2>_{T,L} - <E>^2_{T,L}}{kT^2}, \text{ where } L^2 = NS \text{ is}$$

the number of spins in the central part of the tiling which are considered, so called the system size (NS $\leq$ N) and E is the internal energy of the NS spins, can be used, in principle, to determine the order of the transition :

$$C_{max}(L) \sim L^2 A(t\, L^2) \text{ for first order transitions in 2D systems [3,4],}$$

$$C_{max}(L) \sim L^{\alpha/\nu}\, C^0(t\, L^{1/\nu}) \text{ for second order transitions [8],}$$

where $t = |T-T_c|/T_c$ and, α and ν are two infinite system critical exponents [18]. However, it has already been mentioned that a system undergoing a first order transition with the correlation length of the order of the lattice size or larger exhibits a second order type of behaviour [4]. A most powerful method of detecting first order transitions by numerical simulations on finite systems is the Lee-Kosterlitz method [19]. This method suggests that the size dependence of the free energy as a function of the internal energy, $F_L(E) = -\ln(P_L(E))$, where $P_L(E)$ is the probability distribution for a system of linear dimension L, can be used to identify weak first order transitions even when the system size is smaller than the correlation length. If the free energy barrier, $\Delta F_L = F_L(E_2) - F_L(E_1)$, where $F_L(E_2)$ is the maximum between the two wells of equal depth and $F_L(E_1)$ is the minimum (the two wells), grows with increasing system size, the transition will be first order in the thermodynamic limit, otherwise, the transition will be second order. In MC simulations using the single histogram technique, the free energy barrier can be estimated by

$$\Delta F_L = \ln\left(\frac{P_L(E_1)}{P_L(E_2)}\right) \text{ where } P_L(E) \text{ is the reweighted probability distribution}$$

at $T_c(L)$ (which is the location of the maximum in $C(T,L)$), and

$$\Delta F_L = \ln\left(\frac{H(E_1)}{H(E_2)}\right) + \left(\frac{1}{kT_c(L)} - \frac{1}{kT_0}\right)(E_2 - E_1) \quad \text{where } H(E) \text{ is the energy}$$

histogram obtained by the simulation performed at T_0 [8].

3. Results

3.1. Preliminary results (NS = N)

Up to twelve runs (starting from an initial random state or an ordered state) have been performed for the smaller tilings ($N = 329$ up to $N = 1433$) while only two runs were performed for the larger ones ($N = 2481$ and $N = 5497$). For each size, the final thermodynamic quantities have been obtained by averaging over the runs.

In order to investigate the nature of the phase transition, we plotted the size dependence of the maximum value in the specific heat on a log-log scale (continuous line on Fig. 2). The slope of the linear fit is roughly 1.28 which should suggest that the system undergoes a second order transition with $\alpha/\nu \approx 1.28$. In order to check if the observed scaling form of C_{max} is a nonanalytic form, as in Ref. 4, we also paid attention to the shape of the probability distributions of the

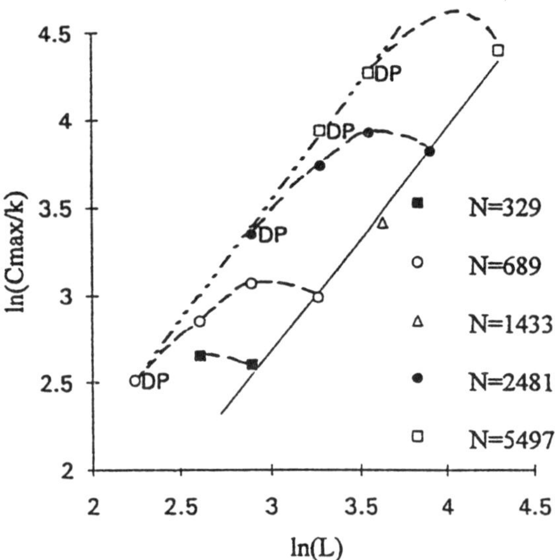

Fig. 2. Log-log plot of the size dependence of the maximum in the specific heat ($L = \sqrt{NS}$) (linear fit for simulations with $NS = N$: continuous line, linear fit for simulations with double-Gaussian probability distributions (DP) : dashed line).

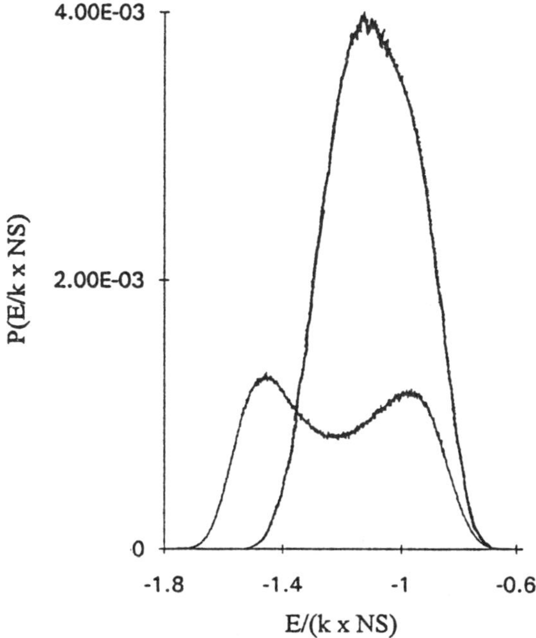

Fig. 3. Probability distributions of the internal energy for NS = N = 689 at
kT/J = 0.8550 (single-Gaussian distribution) and for, NS = 705 and N = 5497 at
kT/J = 0.8730 (double-Gaussian distribution).

internal energy at temperatures very close to the location of the maximum in the
specific heat ($|kT/J-kT_c(L)/J| \le 5\times10^{-4}$). For the smaller systems, the probability
distributions look like single-Gaussian distributions (Fig. 3). For the larger ones,
the probability distributions clearly exhibit a shoulder. However, we never observed
double-Gaussian distributions.

3.2. Complementary results (NS < N)

In order to clarify if our numerical data are strongly disturbed by free boundary
effects, we investigated the specific heat and the probability distributions of the
internal energy for a central part of the tiling (NS < N). For a given NS, we
increased the size of the tiling (N) and we regarded the shape of the probability
distributions. Then, we observed double-Gaussian distributions when N is large
enough relatively to NS, more precisely for (NS = 89,N = 689),
(NS = 329,N = 2481), (NS = 705,N = 5497) (Fig. 3) and (NS = 1217,N = 5497). It
can be seen on Figure 3 that, when NS = N, the states with low energy (ordered
phase) are missing at $T_c(L)$. The free energy barrier, estimated in the four previous
cases, grows with increasing NS which indicates that the transition is first order
[20].

On Figure 2, we also plotted the size dependence of the maximum value in the specific heat on a log-log scale for numerical simulations with NS < N. It should be noted that the slope of the linear fit (≈ 1.36) corresponding to the four simulations which provided double-Gaussian distributions is close to the slope of the linear fit corresponding to NS = N. For a given size of the tiling, the maximum of the specific heat grows with increasing system size NS while it is small relatively to N and decreases when NS tends towards N. The increase of C_{max} can be explained by common size effects, as usually observed in numerical simulation data. The decrease of C_{max} for large values of NS is attributed to free boundary effects which are not negligible when NS tends towards N.

4. Conclusion

Our investigation of the eight-state Potts model on the octagonal tiling has shown that free boundary can have drastic effects on the probability distribution of the internal energy which does not exhibit two peaks as for systems with periodic boundary conditions [4]. These results evidence that such numerical data do not suffer only from size effects but also from strong free boundary effects which prevent the system from being in low energy states at $T_c(L)$ (ordered phase). However, double-Gaussian probability distributions can be observed by analysing a small enough central part of the tiling which is not too much disturbed by the free boundary. Then, the size dependence of the free energy barrier, estimated using the Lee-Kosterlitz method, indicates that the transition is first order as in 2D periodic lattices.

Acknowledgments

The authors wish to thank the Centre de Ressources Informatiques de HAute-Normandie (CRIHAN) for providing computer facilities. This research was supported in part by a NSF/CNRS collaborative research program.

References

1. F. Y. Wu, Rev. Mod. Phys. **54**, 235 (1982)
2. K. Binder, J. Stat. Phys. **24**, 69 (1981)
3. M. S. S. Challa, D. P. Landau and K. Binder, Phys. Rev. B **34**, 1841 (1986)
4. P. Peczak and D. P. Landau, Phys. Rev. B **39**, 11932 (1989)
5. R. J. Baxter, J. Phys. C **6** : L445 (1973)
6. R. J. Baxter, H. N. V. Temperley and S. E. Ashley, Proc. R. Soc. London A **358**, 535 (1978)
7. S. Chen, A. M. Ferrenberg and D. P. Landau, Phys. Rev. Lett. **69**, 1213 (1992)
8. S. Chen, A. M. Ferrenberg and D. P. Landau, Phys. Rev. E **52**, 1377 (1995)
9. Y. Okabe and K. Niizeki, J. Phys. Soc. Jpn **57**, 16 (1988)

10. E. S. Sorensen, M. V. Jaric and M. Ronchetti, Phys. Rev. B **44**, 9271 (1991)

11. D. Ledue, D. P. Landau and J. Teillet, Phys. Rev. B **51**, 12523 (1995)

12. D. Ledue, Phys. Rev. B **53**, 3312 (1996)

13. D. Gratias, in " Du cristal à l'amorphe " (Editions de Physique, Paris, 1988)

14. D. Levine and P. J. Steinhardt, Phys. Rev. B **34**, 596 (1986)

15. D. P. Landau, Phys. Rev. B **13**, 2997 (1976)

16. N. Metropolis, A. E. Rosenbluth, M. N. Rosenbluth, A. H. Teller and E. Teller, J. Chem. Phys. **21**, 1087 (1953)

17. A. M. Ferrenberg and R. H. Swendsen, Phys. Rev. Lett. **61**, 2635 (1988); **63**, 1195 (1989)

18. H. E. Stanley, " Introduction to Phase Transitions and Critical Phenomena " (Clarendon, Oxford, 1971)

19. J. Lee and J. M. Kosterlitz, Phys. Rev. Lett **65**, 137 (1990)

20. Actually, we found that $\Delta F(NS = 1217)$ is lower than $\Delta F(NS = 705)$ but this can be explained by the shape of the probability distribution at $T_c(L)$ for $NS = 1217$ which is not really double-Gaussian indicating that free boundary effects are still disturbing in that case ($NS/N = 22.14\%$)

Generalized Ensemble Simulation of Peptides and Proteins

U.H.E. Hansmann and Y. Okamoto

Department of Theoretical Studies, Institute for Molecular Science (IMS),.
Okazaki, Aichi 444, Japan

Abstract. We discuss the generalized ensemble approach for simulation of peptides and proteins. The simulated molecules are frustrated systems with a complicated energy landscape. The resulting slowing down is alleviated by our ansatz.

1. Introduction

Proteins are linear polymer molecules with the 20 naturally occurring amino acids as monomers. Chains smaller than a few tens of amino acids are called peptides, larger ones proteins. The functions of proteins and peptides are solely determined by their 3D shape. However, while it takes only days to determine the sequence of amino acids, months to years are needed to find out experimentally the corresponding 3D shape. Hence, there is a need for efficient computational methods for predicting the shape of a protein from its sequence of monomers. In the most ambitious approach, one tries to understand the folding of a protein or peptide into its native conformation solely from the underlying physical laws, taking the interactions between all atoms of the protein into account. The intramolecular interactions are modeled by force fields, for instance the ECEPP energy function[1], which was used in our simulations. The complex form of these potential functions containing both repulsive and attractive terms yields to a very rough energy landscape and a huge number of local minima. Hence, at temperatures of experimental interest ($\approx 300K$) traditional methods like Monte Carlo or Molecular Dynamics tend to get trapped in one of the local minima. This is because at temperature T the probability to cross an energy barrier of heights ΔE is proportional to $e^{-\Delta E/T}$ in the canonical ensemble (when local updates are used). Only small parts of the phase space are sampled (in a finite number of simulation steps) and physical quantities cannot be calculated accurately.

In general, one can think of two strategies to alleviate this problem. First, one can look for improved updating scheme like the cluster updates for spin systems [2]. No such global updates are known for proteins. Secondly, one can perform the simulation in a *"generalized ensemble"*, where the above difficulty does not arise. The latter approach was first applied to the protein folding problem in Ref. [3] and will be discussed here.

2. The Generalized Ensemble Approach

Three prominent examples of the generalized ensembles are the multicanonical algorithm [4], $1/k$-sampling [5] and simulated tempering [6]. Common to all three techniques is that a molecular dynamics or Monte Carlo simulation is performed in an artificial ensemble defined in such a way that an uniform distribution of the chosen quantity is obtained. For instance, in the multicanonical algorithm the weights $w(E)$ are chosen so that the

Springer Proceedings in Physics, Volume 83
Computer Simulation Studies in Condensed-Matter Physics X
Eds.: D. P. Landau, K.K. Mon, H. -B. Schüttler
© Springer-Verlag Berlin Heidelberg 1998

Table 1: Lowest energy (in kcal/mol) obtained by 20 annealing runs of multicanonical algorithm (MuCa), $1/k$ sampling, simulated tempering (SiTe) and simulated Annealing (SiAn). For all cases, the total number of Monte Carlo sweeps per run was 50,000. n_{GS} is the number of runs in which a ground-state conformation ($E \leq -11.0$ kcal/mol) was obtained.

Run	Multicanonical	$1/k$	Simulated Tempering	Simulated Annealing
1	−11.6	−10.3	−11.8	−11.7
2	−12.0	−11.7	−11.7	−8.6
3	−10.2	−11.3	−11.5	−12.1
4	−10.1	−11.6	−11.4	−8.8
5	−11.9	−12.1	−11.1	−7.4
6	−12.0	−11.8	−11.1	−8.9
7	−11.9	−10.9	−11.6	−12.1
8	−11.7	−10.4	−11.5	−12.2
9	−11.8	−11.7	−11.4	−7.1
10	−11.9	−11.4	−11.5	−7.5
11	−12.0	−11.3	−11.3	−9.9
12	−12.1	−11.9	−10.5	−7.3
13	−12.0	−11.4	−12.0	−8.4
14	−11.8	−12.0	−9.0	−10.6
15	−11.3	−11.6	−11.8	−10.3
16	−11.9	−11.5	−10.9	−12.2
17	−12.0	−11.7	−11.4	−12.2
18	−11.9	−9.3	−10.7	−9.1
19	−11.9	−11.6	−11.1	−11.9
20	−11.6	−10.5	−11.7	−12.1
n_{GS}	18/20	15/20	16/20	8/20

distribution of energies is flat:

$$P(E) \propto n(E)w(E) = \text{const.} \tag{1}$$

Here, $n(E)$ is the spectral density. A free random walk in the energy space is realized which allows the simulation to escape from any local minimum, and even regions with small $n(E)$ can be explored in detail. Similarly, $1/k$-sampling yields a uniform distribution in (microcanonical) entropy and simulated tempering to an uniform distribution in temperature. From such a generalized ensemble simulation one can not only locate the global minimum in potential energy but also obtain the canonical distribution at any temperature by the reweighting techniques [7]. However, unlike in the canonical ensemble, the weights are not a priori known for simulations in these ensembles. For instance, in the multicanonical algorithm $w_{mu}(E) \propto n^{-1}(E)$, and the knowledge of the exact weights is equivalent to obtaining the density of states $n(E)$, i.e., solving the system. Hence, one needs their estimators for a numerical simulation.

Over the last three years we applied the generalized ensemble approach to the protein folding problem. Here, Met-enkephalin, which has the amino acid sequence Tyr-Gly-Gly-Phe-Met, has become a often used benchmark system to examine new algorithms. We used this simple peptide to compare the numerical performance of the above mentioned three algorithms with each other and with standard techniques like simulated annealing [8]. Most recent results are published in Ref. [9]. Here we show in Tab. 1 the number

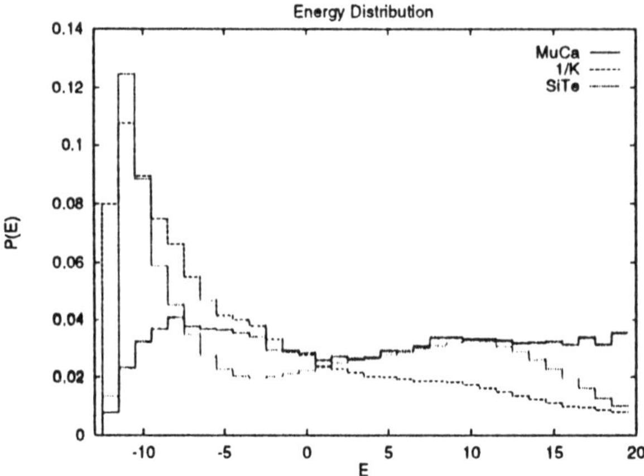

Figure 1: Probability distribution of energy E (kcal/mol) from simulations by multicanonical algorithm (MuCa), $1/k$-sampling (1/k) and simulated tempering (SiTe).

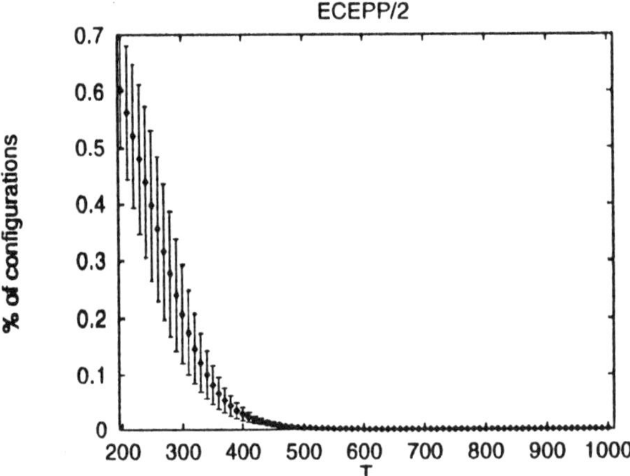

Figure 2: Relative weight of ground-state like configurations as a function of temperature.

of times the ground-state was found by the different methods. The generalized ensemble algorithms differ little in their ability to locate ground-state configurations, but their performance is better by at least a factor two than simulated annealing. However, the major advantage of the three algorithms over simulated annealing or other optimization methods is that they allow calculations of thermodynamic quantities at any temperature from just one simulation run, once the weights are determined. This is because all three algorithms sample over a large range of energies (temperatures). We show in Fig. 1 the histograms of energy distribution to underline this point. Our data were obtained from a simulation of 1,000,000 MC sweeps for each of the three methods, which followed 10,000 sweeps for thermalization. As an example for thermodynamic quantities which can be calculated from these runs the relative weight of ground-state configurations is displayed in Fig. 2. It is interesting to observe that ground-state configurations appear at room temperature only with about 20% probability.

3. A Generalized Ensemble with *A Priori* known weights

Despite many successful applications use of existing generalized ensemble algorithms is handicapped by the problem that one needs to find estimators for the weights. Here, we propose another ensemble where the problem of determining estimators for the weights is reduced to that of finding an estimator for a single quantity [10].

We are interested in an ensemble where not only the low-energy region can be sampled efficiently but also the high-energy states can be visited with finite probability. The latter feature ensures that energy barriers can be overcome and that the simulation can escape from local minima. The probability distribution of energy should resemble that of an ideal low-temperature Boltzmann distribution, but with a tail to higher energies. To obtain such an ensemble we propose the following weights:

$$w(E) = (1 + \beta \frac{E - E_0}{n_F})^{-n_F} , \qquad (2)$$

where E_0 is an estimator for the ground-state energy and n_F the number of degrees of freedom of the system. Obviously, the new weights reduce in the low energy (and hence low-temperature) region to canonical weights for $\frac{\beta(E-E_0)}{n_F} \ll 1$. On the other hand, high energies are no longer exponentially suppressed but only according to a power law. We remark that our ansatz is closely related to the Tsallis formulation of generalized mechanics [11].

In contrast to other generalized ensembles the weights of the new ensemble are explicitly given by Eq. (2). However, one needs to find an estimator for the ground-state energy which can be done by a procedure described in Ref. [10].

Our approach was again tested for Met-enkephalin. In Fig. 3 we display the "time series" of energy as a function of Monte Carlo sweep for both a regular canonical Monte Carlo simulation at temperature $T = 50$ K (dotted curve) and those from a generalized ensemble simulation of the new algorithm (solid curve). Here, the weight we used for the latter simulation is given by Eq. (2) with $n_F = 19$ and $E_0 = -12.2$ kcal/mol. For the

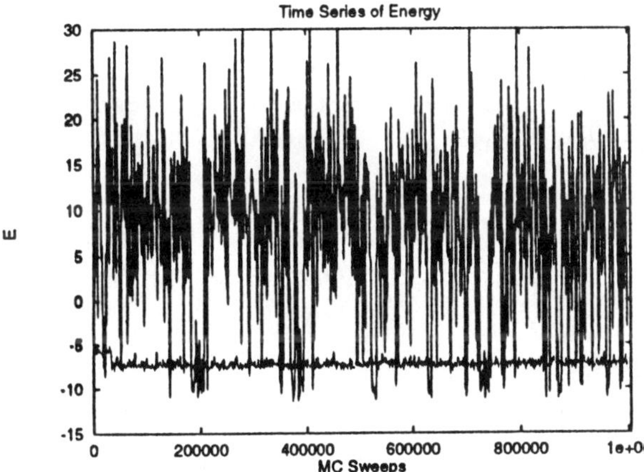

Figure 3: "Time series" of energy as a function of Monte Carlo sweeps for both a regular canonical Monte Carlo simulation at temperature $T = 50$ K (dotted curve) and that from a generalized ensemble simulation of the new algorithm (solid curve).

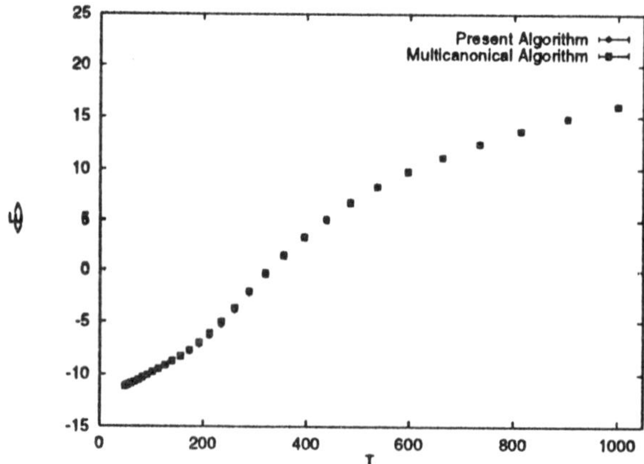

Figure 4: Average energy as a function of temperature. Shown are the results of a simulation in our new ensemble and of a multicanonical simulation.

canonical run the curve stays around the value $E = -6$ kcal/mol with small thermal fluctuations, reflecting the low-temperature nature. The run has apparently been trapped in a local minimum, since the mean energy at this temperature is $< E > = -11.1$ kcal/mol as found by a multicanonical simulation in Ref. [3]. On the other hand, the simulation based on the new weight covers a much wider energy range than the canonical run. It is a random walk in energy, which keeps the simulation from getting trapped in a local minimum. It indeed visits the ground-state region several times in 1,000,000 Monte Carlo sweeps. Since the simulation samples a large range of energies, we can use reweighting techniques [7] to calculate thermodynamic quantities over a wide temperature range. As an example we display in Fig. 4 the average energy as a function of temperature.

Conclusion

We presented results from generalized ensemble simulations of simple peptides. While these techniques allow for improvement, our results show that they are a promising approach for tackling the protein folding problem.

References

[1] M.J. Sipple, G. Némethy, and H.A. Scheraga, J. Phys. Chem. **88**, 6231 (1984), and references therein.

[2] R.H. Swendsen and J.-S. Wang, Phys. Rev. Lett. **58**, 86 (1987).

[3] U.H.E. Hansmann and Y. Okamoto, J. Comp. Chem. **14**, 1333 (1993).

[4] B.A. Berg and T. Neuhaus, Phys. Lett. B **267**, 249 (1991); Phys. Rev. Lett. **68**, 9 (1992).

[5] B. Hesselbo and R.B. Stinchcombe, Phys. Rev. Lett. **74**, 2151 (1995).

[6] A.P. Lyubartsev, A.A.Martinovski, S.V. Shevkunov, and P.N. Vorontsov-Velyaminov, J. Chem. Phys. **96**, 1776 (1992); E. Marinari and G. Parisi, Europhys. Lett. **19**, 451 (1992).

[7] A.M. Ferrenberg and R.H. Swendsen, Phys. Rev. Lett. **61**, 2635 (1988); Phys. Rev. Lett. **63** , 1658(E) (1989), and references given in the erratum.

[8] S. Kirkpatrick, C.D. Gelatt, Jr., and M.P. Vecchi, *Science*, **220**, 671 (1983).

[9] U.H.E. Hansmann and Y. Okamoto, *Numerical Comparisons of Three Recently Proposed Algorithms in the Protein Folding Problem*, to appear in *J. Comp. Chem.*

[10] U.H.E. Hansmann and Y. Okamoto, *New Generalized-Ensemble Monte Carlo Method for Systems with Rough Energy Landscape*, preprint IMS-TH-96/28, submitted for publication.

[11] E.M.F. Curado and C. Tsallis, *J. Phys. A: Math. Gen.* **27**, 3663 (1994).

Monte Carlo Studies of Surface-Induced Ordering in Cu₃Au-Type Alloy Models

W. Schweika and D.P. Landau*

Center for Simulational Physics, University of Georgia,
Department of Physics and Astronomy, Athens, GA 30602-2451, USA

* Permanent address: Institut für Festkörperforschung, Forschungszentrum Jülich GmbH, 52425 Jülich, Germany.

Using Monte Carlo simulations we study order-disorder phenomena near the surface to compare with experimental results for Cu_3Au alloys. Our simulations reveal possible new ordering cases near the surface which could lead to surface induced ordering.

I. INTRODUCTION

Surfaces may modify the order-disorder transitions with respect to their bulk counterparts, *e. g.* due to broken bonds at the surface, the long range order may be reduced near the surface. In particular, at a discontinuous bulk transition $T_{c,b}$ one would expect that the order parameter near the surface vanishes continuously as the interface between the less ordered "near surface" layer and the bulk becomes delocalized. A continuous order-disorder transition has been observed with LEED and SPLEED at the (100) surface of Cu_3Au [1] [2]. Surface induced disordering occurs when the interface delocalization occurs as the temperature is raised to $T_{c,b}$. This behavior, interpreted as a wetting phenomenon, has been theoretically studied by Lipowsky [3] and found in Monte Carlo studies [4] and in x-ray experiments [5] on Cu_3Au.

An observed pecularity is that the 50:50 composition of the uppermost layer in Cu_3Au remains fairly unaltered across the order-disorder transition [6]. More recently, x-ray experiments [7] revealed an oscillating concentration profile with a decay length that seems to diverge at a (spinodal) temperature below $T_{c,b}$. This observation is confirmed by a recent mean-field study [8]. The component of the Cu_3Au order parameter which is parallel to the surface causes the surface induced disordering, while the perpendicular component is stable above $T_{c,b}$.

The order in the surface layer can be enhanced because of (*i*) the possible removal of frustrations, as shown in a Monte Carlo study [9,10] of a (100) surface for a simple AB binary alloy model of $L1_0$ type of order (*e. g.* $CuAu$), (*ii*) effective surface fields which stem from any differences in the binding energies for the pure components for a given rigid lattice [11]. For $L1_0$ order (*i*) leads to a preference of AB-type surfaces with a simple 2×2 ordering parallel in the

Springer Proceedings in Physics, Volume 83
Computer Simulation Studies in Condensed-Matter Physics X
Eds.: D. P. Landau, K.K. Mon, H. -B. Schüttler
© Springer-Verlag Berlin Heidelberg 1998

surface plane, whereas (*ii*) induces an ABAB.. layering perpendicular to the surface. (Another example of ordering effects due to a free surface was found for FeAl alloys. [13]) Depending on the orientation of the ordered structure with respect to the surface, the frustration effects due to the missing bonds can vary, and in general different surface universality classes should exist [12].

II. MODEL AND SIMULATION METHOD

The A_3B alloy, *e. g. Cu_3Au*, ordering in the $L1_2$ structure, is depicted below. Ideally, one of the four simple cubic sublattices of the fcc lattice is occupied by a minority atom, B, each of which is thus solely surrounded by nearest neighbors of type A.

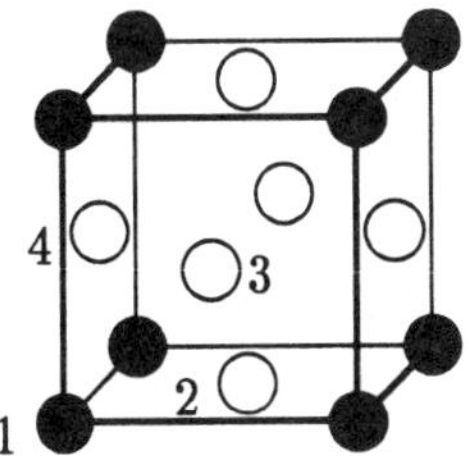

$L1_2$ structure

The complete order parameter is non-scalar and needs, in principle, four components. For simplicity, we emphasize two independent components, one for the 2×2 ordering parallel to the surface and a second, perpendicular component given by the concentration in the layers.

We consider the following configurational Hamiltonian of an alloy system

$$\mathcal{H} = \sum_{i>j} \sum_{\alpha,\beta} V_{\alpha,\beta}(\vec{R}_i - \vec{R}_j) c_\alpha(\vec{R}_i) c_\beta(\vec{R}_j) - \sum_\alpha \mu_\alpha \sum_i c_\alpha(\vec{R}_i), \qquad (1)$$

where α (β) labels the atomic species and i (j) the lattice sites, and μ and V are single - and two particle energies respectively. For a binary AB alloy system, $c_i \equiv c_A(\vec{r}_i) = 1 - c_B(\vec{r}_i)$, the Ising model is introduced *via* spin variables $s_i = 2c_i - 1$ and the interactions $J_{ij} = -\frac{1}{2}V_{ij} = -\frac{1}{4}(V_{AA} + V_{BB} - 2V_{AB})_{ij}$.

$$\mathcal{H} = -\sum_{i>j} J_{ij} s_i s_j - h \sum_i s_i - \sum_n h_n \sum_{i \epsilon n} s_{i,n}. \qquad (2)$$

For the fcc structure and only nearest neighbor interactions, the effective magnetic bulk field is $h = -3(V_{AA} - V_{BB}) + \frac{1}{2}(\mu_A - \mu_B)$ whereas the effective field acting only on the surface layer (n=1) is $h_1 = V_{AA} - V_{BB}$. Note that any difference in the binding energies of the pure components becomes apparent at free surfaces, while this is not the case for the bulk properties, see Eq. (1).

We used a fairly standard Metropolis, single spin-flip method where we included a preferential site selection which was determined by the nature of the order parameter profile. The program has been highly vectorized for a CRAY-YMP computer. We have chosen a slab geometry with two free (100) surfaces, which should behave independent of each other for a sufficient thickness of the slab. Screw type boundaries were used within (each of) the layers. For the case of surface induced ordering or disordering at the bulk transition the required slab thickness increases. Consideration of surface order-disorder transitions above $T_{c,b}$, for the $L1_2$ case was different than for the $L1_0$ structure studied earlier [10]: Here, the coupling of near surface effects to the perpendicular component of the $L1_2$ order parameter required the use of very large systems ($\approx$ 100 to 200 layers with cross-sections of typically 60×60) to investigate the ordering parallel to the surface in the first and second surface layers.

III. RESULTS AND CONCLUSION

For the nearest neighbor interaction model with $L1_2$ stoichiometry we found a strong preference for an A type surface, if we do not include explicit surface field h_1, see Fig. 1. This tendency is even sufficiently strong to stabilize order parallel to the surface in the second layer, and with exponentially decreasing amplitude also in the following even layers, well above $T_{c,b}$. For a sufficiently strong field h_1 $(-7 < h_1 < -3)$, the surface displays an AB composition and stable order at the surface above $T_{c,b}$.

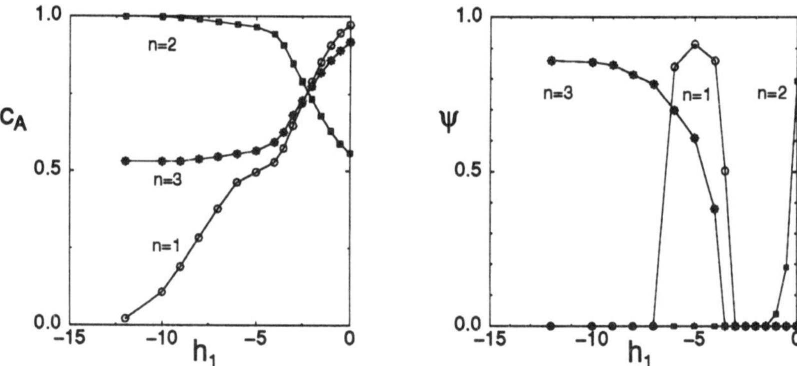

FIG. 1. Concentration of A-atoms (*left*) and order parameters (*right*) in the first three layers versus surface field h_1 for the nearest neighbor model $J_{nn} < 0$ at $h = 8.4|J_{nn}|$ and $k_B T = 1.875|J_{nn}|$ (which is slightly above $T_{c,b} \approx 1.85|J_{nn}|$).

There is another interesting situation for very strong (negative) surface field h_1 which leads to a surface covered mainly by minority atoms (B). Due to the interaction the second layer will be purely of A-type while the 3rd layer is then of AB-type; further layers alternate between A and AB type. Above $T_{c,b}$ in our nearest neighbor model there is a buried sheet of two-dimensional order parallel to the surface in the 3rd layer which can extend further into the bulk as T approaches $T_{c,b}$, while there will be no parallel component of order in the upper two layers.

According to the experimental observation for a free (100) surface, the uppermost surface layer is of AB type rather than of A-type. This is consistent with appropriately strong effective "near surface" fields. For our nearest neighbor model and for the effective surface field h_1 which is necessary to achieve the reported oscillating concentration profile, order will again be induced parallel to the surface, now in the uppermost layer. This is contrary to the confirmed experimental findings that surface induced *disordering* occurs in the parallel component of the $L1_2$ order parameter near and below $T_{c,b}$. This disagreement remains qualitatively the same, when one considers the *second nearest neighbor model*, which has been used previously for Cu_3Au [14,4] (providing a consistent description of this part of the phase diagram and of the temperature dependence of the first short range order parameter [14]). The perpendicular component of the order parameter, *i. e.* the measured so-called "near surface segregation profile", can be reproduced, as shown in Fig. 2.

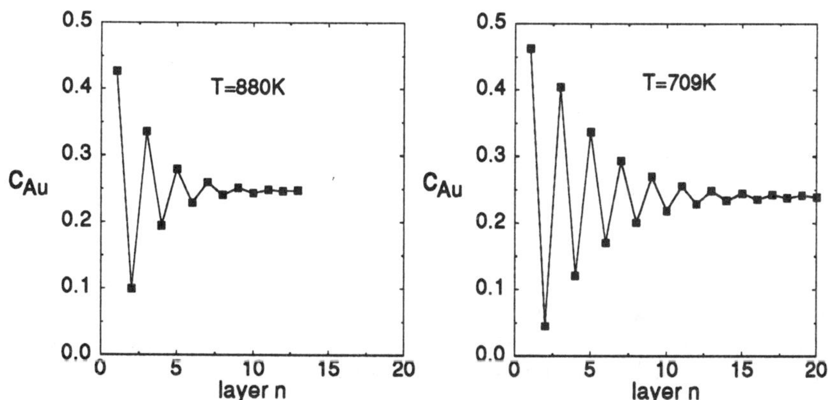

FIG. 2. Simulated Au concentration profile for $T_c + 217K = 880K$ and $T = 709K$, for $J_{2nn} = -0.2 J_{nn}$, $h_1 = -4|J_{nn}|$ and $h_2 = 1.25|J_{nn}|$.

However, as for the nearest neighbor model, near the surface the parallel order does not show surface induced disorder but is more stable than in the bulk. Our simulations provide an estimate for the strength of the effective surface fields which are supposed to originate from the asymmetry ($V_{AA} \neq V_{BB}$) in the pair interaction model.

A *third nearest neighbor interaction*, J_{3nn}, couples the four different sublattices but does not contribute to the ordering energy *in* the (100) surface layer. Hence, frustration due to missing bonds at the surface should reverse the trend

189

towards a less stable parallel component of the order parameter. It would be cumbersome (and is not the aim of the present paper) to achieve a realistic description by fitting a hamiltonian to the experimental results. Due to other possible effects, such as modified "near surface" interactions, relaxation- and many body terms, any result would not be unambiguous. However, if we wish to achieve a more realistic description of the Cu_3Au alloy we might use the interaction parameters [15] as determined from the measured short range order in diffuse scattering experiments [16]. Therefore, interactions to further distant neighbors are not negligible, and in some preliminary simulations the inclusion of J_{3nn} decreases the order near the surface as expected.

The present results can be of interest for other alloys which form the ordered $L1_2$ structure. We conclude three different kinds of possible surface order above the bulk transition, which could lead to possible surface induced ordering phenomena. They do not occur in Cu_3Au, and none of these have been observed yet. The reason probably is that these surface ordering phenomena are more likely to occur for alloys having very short range interactions and that they are rather demanding to verify experimentally.

This research was supported in part by NSF.

References

[1] V.S. Sundaram, B. Farrell, R.S. Alben, and W.D. Robertson, Phys. Rev. Lett. **31**, 1136 (1973); Surf. Sci. **46**, 653 (1974).

[2] E.G. McRae and R.A. Malic, Surf. Sci. **148**, 551 (1984); S.F. Alvarado, M. Campagna, A. Fattah and W. Uelhoff, Z. Phys. **B66**, 103 (1987).

[3] R. Lipowsky, Ferroelectrics **73**, 69 (1987).

[4] G. Gompper and D.M. Kroll, Phys. Rev. **B38**, 459 (1988).

[5] H. Dosch, L. Mailänder, A. Lied, J. Peisl, F. Grey, R.L. Johnson, and S. Krumnacher, Phys. Rev. Lett. **60**, 2382 (1988); H. Dosch, L. Mailänder, H. Reichert, J. Peisl, R.L. Johnson, Phys. Rev. **B43**, 13172 (1991)

[6] T.M. Buck, G. M. Wheatley, and L. Marchut, Phys. Rev. Lett. **51**, 43 (1983).

[7] H. Reichert, P.J. Eng, H. Dosch, and I.A. Robinson, Phys. Rev. Lett. **74**, 2006 (1995).

[8] K.R. Mecke and S. Dietrich, Phys. Rev. B **52**, 2107-16 (1996)

[9] W. Schweika, K. Binder, and D.P. Landau, Phys. Rev. Lett. **65**, 3321 (1990).

[10] W. Schweika, D.P. Landau, and K. Binder, Phys. Rev. B. **53**, 8937 (1996).

[11] W. Schweika, D.P. Landau, and K. Binder, in *Stability of Materials*, NATO-ASI Series B, eds. A. Gonis, P.E.A. Turchi, and J. Kudrnovsky, Plenum Press, New York 1996.

[12] A. Drewitz, R. Leidl, T. Burkhardt, and H.W. Diehl, Phys. Rev. Lett. **78**, 1090 (1997).

[13] F. Schmid, Z. Phys. B **91**, 77 (1993).

[14] K. Binder, W. Kinzel, and W. Selke, J. Magn. Mater. **31**, 1445 (1983).

[15] L. Reinhard and S.C. Moss, Ultramicroscopy **52**, 223 (1993).

[16] B. D. Butler and J. B. Cohen, J. Appl. Phys. **65**, 2214 (1989).

An Inherent-Structures Study of Two-Dimensional Melting

F.L. Sommer, Jr[1,2], *G.S. Canright*[1,2], *and T. Kaplan*[2]

[1]Department of Physics and Astronomy, The University of Tennessee,
 Knoxville, TN 37996-1200, USA
[2]Solid State Division, Oak Ridge National Laboratory, Oak Ridge, TN 37831, USA

Abstract. Preliminary results of a computational study of the "inherent structures" (IS) associated with equilibrium two-dimensional Lennard-Jones systems are presented. For the system sizes studied, the melting transition is first order, proceeding directly from the crystal to the isotropic liquid. The IS found clearly distinguish between these two equilibrium phases, thus providing strong confirmation of the inherent-structures concept of Stillinger and Weber, for these systems, and laying a foundation for similar studies of larger systems, where there is evidence of *three* condensed phases.

1. Introduction

Some years ago, Stillinger and Weber [1] developed a theory of condensed phases, based on the partitioning of the configuration space into potential-energy (PE) basins defined by steepest-descents paths to the nearest local PE minimum. These minima were coined "inherent structures" (IS) and all other configurations are considered to be vibrational excitations of them. Among the implications of this *inherent-structures theory* (IST) is that the IS associated with different equilibrium phases should differ in consistently reproducible ways. This has been confirmed for certain three-dimensional systems [2], but in 2D [3], there are only limited results. The purpose of the present study, then, is to test the applicability of IST to 2D atomic systems. In order to examine a reasonably large number of configurations, we have limited the present study to system sizes (4096 particles) for which the melting transition is first order, proceeding directly from the crystal to the isotropic liquid. A successful result at the present system size will provide a useful stepping stone for proceeding to much larger systems (an order of magnitude larger than the present systems), where simulations [4,5] have shown evidence of *three* condensed phases, including the novel *hexatic* phase [6].

2. Computational Procedure and Results

As mentioned above, finding the IS associated with any arbitrary configuration requires a steepest-descents minimization of the potential energy. For all but the smallest systems, however, this is not computationally feasible [2]. Instead, we employ a highly damped molecular-dynamics approach (herein, referred to as "quenching"), in which we insist that the potential energy strictly decrease,

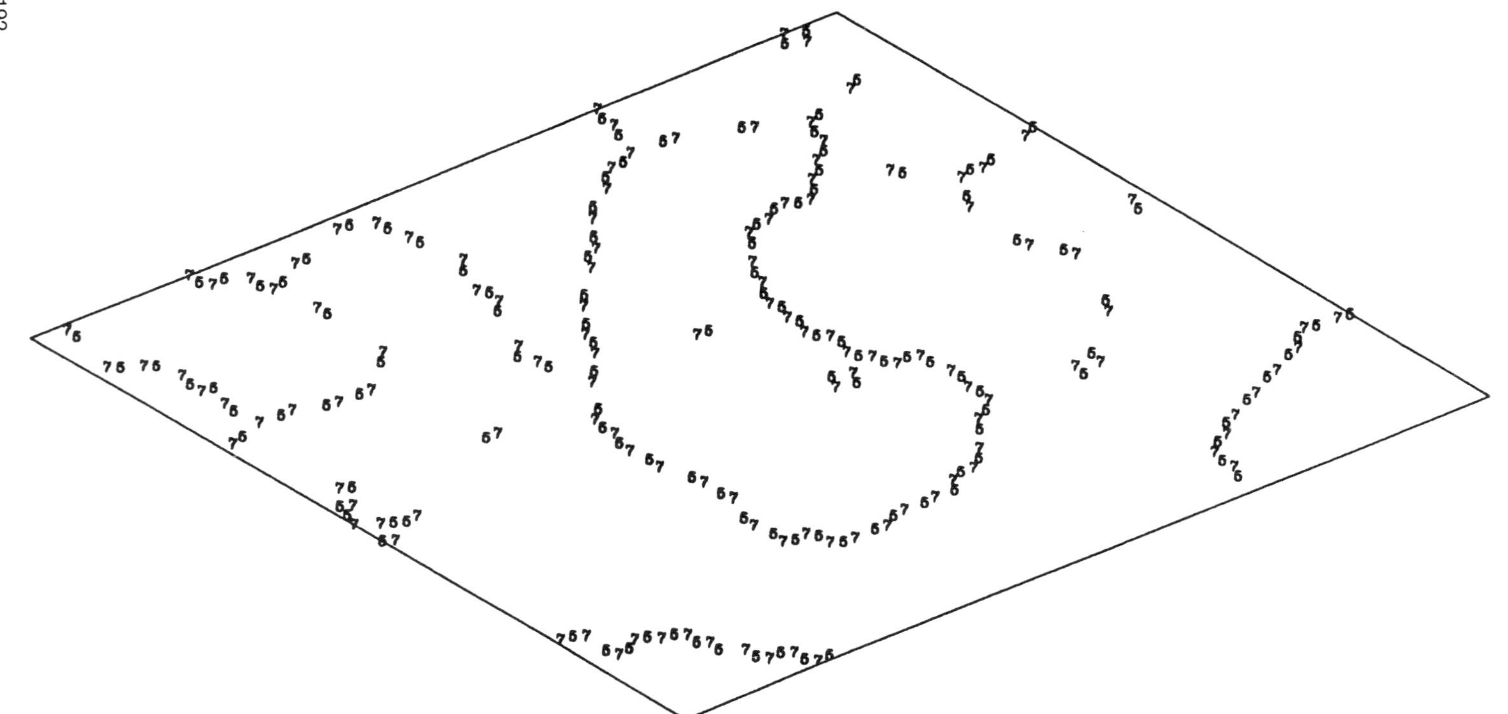

Figure 1. Inherent (mechanically stable, or 'quenched') structure for a system of 4096 particles with periodic boundary conditions, obtained by relaxing a configuration from an equilibrium liquid phase. The relaxation is done at constant pressure $p = 20$, from an equilibrium snapshot at $T_0 = 2.327$ (units, interaction potential, and constant-(p, T) MD algorithm, are as in Ref. [4]). Only those atoms which are not 6-fold coordinated are marked. Dislocations appear as 5-7 pairs, and grain boundaries as closely spaced chains of dislocations.

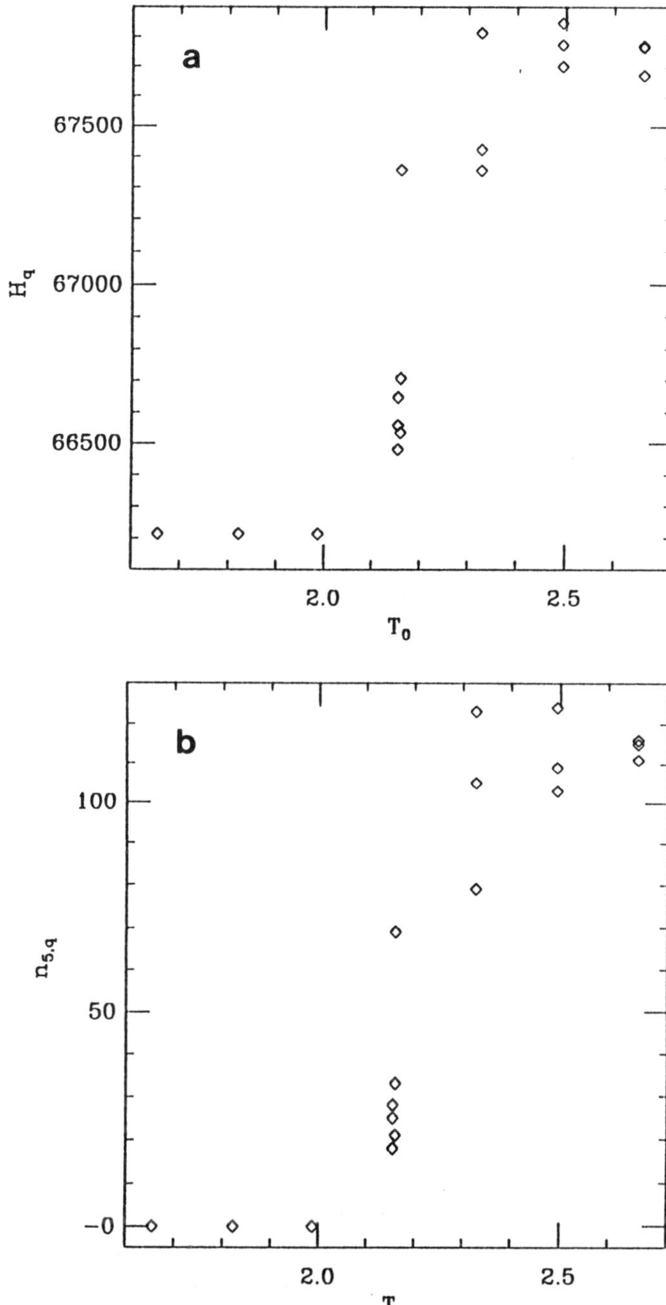

Figure 2. Plots of (a) the enthalpy and (b) the number of five-coordinated atoms for IS, as functions of the starting (equilibrium) temperature, T_0. A clear "phase transition" occurs in the IS, at the equilibrium melting point, $T_0 = 2.16$.

until a local minimum is reached. Of course, the structure so obtained will not, in general, be the true IS of the starting configuration, but since, according to IST, fluid systems at equilibrium transit frequently and exclusively between *thermodynamically equivalent* PE basins [7] (i.e. basins having the same values of structural parameters, such as defect densities, spatial correlations, etc.), we believe that our small deviation from steepest-descents minimization should have no significant effect on the results.

Starting with an equilibrium crystal at any temperature significantly below the melting temperature, our quenching procedure essentially always results in perfect-crystal IS. While this situation is not expected to persist at larger system sizes, it seems evident that we should not find any unbound dislocations or disclinations in the crystal IS, at any system size. With the isotropic liquid as the starting point, the quenching procedure results in such configurations as shown in Fig. 1. In this plot, only non-six-coordinated atoms are shown. Dislocations are identifiable as closely spaced 5-7 pairs, and free disclinations, if there were any, would show up as lone 5- or 7-coordinated atoms. The most obvious characteristic of these liquid IS is the presence of extensive large-angle grain boundaries, identifiable in the plot as chains of closely spaced dislocations. There also appear to be a few free dislocations, within the grains, but none of the unbound disclinations which might be expected on the basis of the theory of Kosterlitz, Thouless, Halperin, Nelson, and Young (KTHNY) [6,8]. Fig. 2 shows the enthalpy (both the equilibrations and quenches were done at constant pressure) and the number of 5-coordinated atoms (a useful structural parameter for 2D atomic systems) of quenched structures, as functions of the starting (equilibrium) temperature. The jump in each of these quantities, at the melting temperature, is clearly evident. Also, while the number of points is too small to make any strong statements concerning fluctuations, the fluctuations in the quenched properties do seem to increase as the transition is approached from above. (Since all configurations obtained significantly below the melting temperature have the same, perfect-crystal IS, there are no fluctuations in the quenched properties, in this regime.)

3. Discussion

The fact that each equilibrium phase of the present study has its own, qualitatively distinct type of IS provides strong confirmation of the applicability of IST to these systems. While the liquid IS lack the free disclinations which are characteristic of the KTHNY liquid, this situation might well be altered, on going to larger systems. For the present systems, the correlation length of the liquid is roughly equal to the system size, so that boundary effects might play a significant part in determining the liquid IS. The mechanical stability of free disclinations in these atomic systems is still in question, however, so that it might be impossible to "trap" them in IS, regardless of system size.

Another interesting difference, on going to larger systems, is the apparent existence of the hexatic phase, intermediate in energy to the crystal and liquid, and characterized, according to Halperin and Nelson [6], by the presence of free dislocations. Unlike free disclinations, free dislocations are well known to be mechanically stable—at least in some configurations. In fact, there are several, apparently free dislocations within the grains of the liquid IS of Fig. 1. An inherent-structures study of the hexatic phase, then, might provide direct, microscopic evidence for the Halperin-Nelson (HN) description, complementary to that obtained from previous studies [4,5]. In these studies, the correlation functions have been found to be consistent with the HN theory, but identification of the characteristic free dislocations has been difficult, due to the presence of many "virtual dislocations" [9] which are manifestations of vibrational distortions of the system. Of course, these would be removed on quenching, possibly facilitating the identification of the free dislocations of the HN picture.

Acknowledgments

We thank Kun Chen for the sharing of, and patient education regarding, his constant-(p, T) MD codes, and Mark Mostoller and Kun Chen for numerous helpful discussions. This work was supported in part by the U.S. Department of Energy through Contract No. DE-AC05-84OR21400 with Martin Marietta Energy Systems Inc. FLS and GSC were supported by the NSF under Grant # DMR-9413057.

References

1. F. H. Stillinger and T. A. Weber, Science **225**, 983 (1984).

2. F. H. Stillinger and T. A. Weber, Phys. Rev. A **28**, 2408 (1983); J. Phys. Chem. **87**, 2833 (1983); R.A. LaViolette and F.H. Stillinger, J. Chem. Phys. **83**, 4079 (1985).

3. F. H. Stillinger and T. A. Weber, Phys. Rev. A **25**, 978 (1982); T. A. Weber and F. H. Stillinger, Phys. Rev. E **48**, 4351 (1993).

4. K. Chen, T. Kaplan, and M. Mostoller, Phys. Rev. Lett. **74**, 4019 (1995).

5. K. Bagchi, H. C. Andersen, and W. Swope, Phys. Rev. Lett. **76**, 255 (1996); Phys. Rev. E **53**, 3794 (1996).

6. B. I. Halperin and D. R. Nelson, Phys. Rev. Lett. **41**, 121 (1978).

7. F. L. Somer and J. Kovac, J. Chem. Phys. **101**, 6216 (1994).; J. Chem. Phys. **102**, 8995 (1995).

8. J. M. Kosterlitz and D. J. Thouless, J. Phys. C **6**, 1181 (1973).; A. P. Young, Phys. Rev. B **19**, 1855 (1979).

9. D. S. Fisher, B. I. Halperin, and R. Morf, Phys. Rev. B **20**, 4692 (1979).

Monte Carlo Simulation on Aging Phenomena in the SK Spin-Glass Model: Temperature Dependence of the Time Evolution of Energy

H. Takayama, H. Yoshino, and K. Hukushima

Institute of Solid State Physics, University of Tokyo, 7-22-1 Roppongi, Minato-ku, Tokyo 106, Japan

Abstract. Aging processes after a rapid quench from $T = \infty$ to the spin-glass phase in the SK model have been studied by Monte Carlo simulation. Various quantities examined exhibit a rich variety of aging phenomena. Among them the temperature dependence of the energy evolution is further discussed in the present work.

1. Introduction

Since the first observation by Lundgren et al [1] aging phenomena in spin glasses have been extensively studied [2]. One of the key concepts in their recent studies is the *weak-ergodicity breaking* first introduced by Bouchaud [3]. It describes slow dynamics in such a system that has an unbounded, continuous distribution of relaxation times, and has been incorporated in analytic theories on aging phenomena in some mean-field spin-glass models and others [4-6].

Various simulations which reproduce some aspects of the experimentally observed aging phenomena have been already reported [7]. But they are yet far from satisfactory to be able to understand the phenomena in a unified way. Therefore we have started detailed numerical studies on aging phenomena in the SK model [8]. The results obtained [9,10] lead us to a scenario of the aging process in the model, which we call *growth of quasi-equilibrium domains*. One of the purposes of the present work is to examine further the energy evolution in the aging process.

2. Growth of Quasi-Equilibrium Domains (GQED) Scenario

It is well known that there are a huge number of solutions to the TAP equation of states in the spin-glass phase of the SK model [11]. Our GQED scenario concernes with how the system evolves stochastically in such a phase space after rapid quench. It has been originally derived from the analysis of relaxational modes, in which the Glauber dynamics of the model system of small sizes ($N \leq 12$, N being the total number of spins) has been solved numerically exactly [9].

As for an individual aging process simulated by the MC method, the GQED scenario is described in terms of Fig. 1. At time $t = 0$ the system is instanta-

Springer Proceedings in Physics, Volume 83
Computer Simulation Studies in Condensed-Matter Physics X
Eds.: D. P. Landau, K.K. Mon, H. -B. Schüttler
© Springer-Verlag Berlin Heidelberg 1998

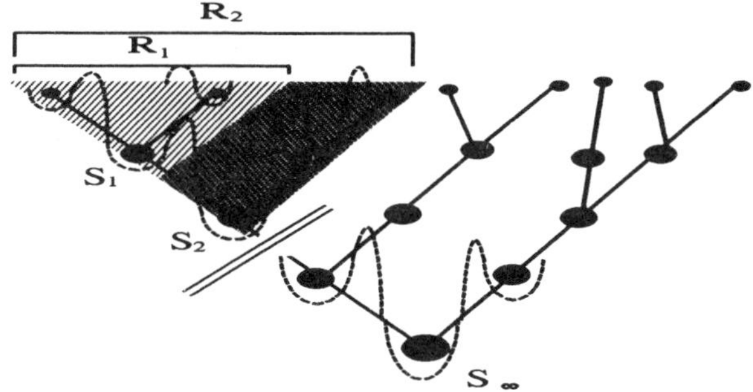

Figure 1: Schematic representation of the rugged energy structure in the phase space of the SK model. The dots represent the TAP solutions (or states $S_1, S_2, \cdots, S_\infty$) and the shaded areas quasi-equilibrium domains ($R_1, R_2, \cdots$).

neously quenched from $T = \infty$ to T below the spin-glass transition temperature (the latter is determined from the variance of interactions and is set unity in this work). After a certain initial transient time, the system is found in one of the huge number of states (which are solutions of the TAP equations of state) with a relatively higher energy, from which it starts to look for lower and lower energy states. At $t \simeq t_{w1}$ it reaches to a certain state, say S_1 in Fig. 1. In a time interval of $t = t_{w1} + \tau$ with $\tau \lesssim t_{w1}$ it is fluctuating in a quasi-equilibrium domain R_1 centered at S_1 (the shaded area in Fig. 1), thereby it is expected to visit states in the domain with frequencies proportionally to their relative Boltzmann weights. As time goes on further the system find a lower energy state with the bigger quasi-equilibrium domains, and finally at around $t = t_{erg}(N, T)$ it reaches to the lowest energy state S_∞, associated with which is a whole phase space with the same time reversal symmetry [10].

3. Time Evolution of Energy

The equilibration (or 'interruption' of aging processes [3]) is easily seen in the time evolution of energy. In Fig. 2a we show $\Delta E(t)$, the *extensive* energy of the system relative to its equilibrium value E_{eq} (which in turn is approximated by the value measured at the maximum observation time). Here the ergodic (equilibration) time $t_{erg}(N, T)$ is defined as the time, at which $\Delta E(t) \cong T$ holds. This $t_{erg}(N, T)$ coincides, within the logarithmic accuracy, with the time, above which the time translationally invariance is recovered, or the auto-correlation function defined by $C(t, t') \equiv N^{-1} \sum_i S_i(t) S_i(t')$ becomes a fnction only of difference $t - t'$ [10]. As shown in Fig. 2b we obtain $\ln t_{erg}(N, T) \propto N^{1/4}/T$. This N and T dependences agree with those of the relaxation time attributed to thermal activeted processes over free-energy barriers between the local minima with a common time-reversal symmetry [12].

The energy difference *per spin* exhibits a power-law decay, $\Delta E(t)/N \propto t^{-a}$, as demonstrated in Fig. 3. Its exponent a becomes independent of N for

197

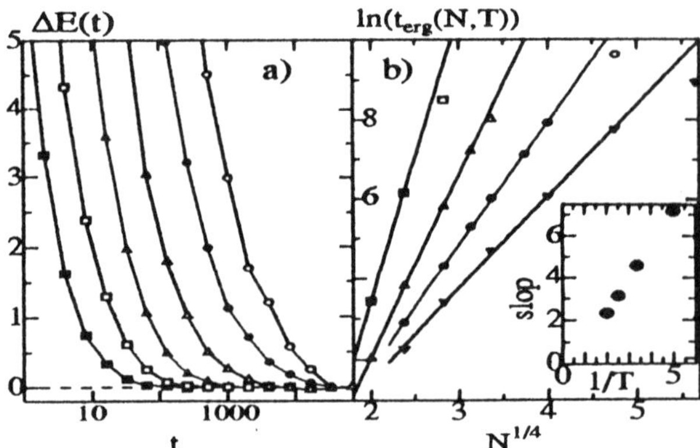

Figure 2: a) $\Delta E(t)$, the extensive energy relative to E_{eq}, at $T = 0.5$ in systems of various sizes; $N = 2^n$ with $n = 5 \sim 10$ (left to right). b) The plot of $\ln(t_{\mathrm{erg}}(N,T))$ versus $N^{1/4}$ at $T =$0.2, 0.3, 0.4, and 0.5 (left to right). The slopes of the linear fit (lines) are plotted aginst $1/T$ in the inset.

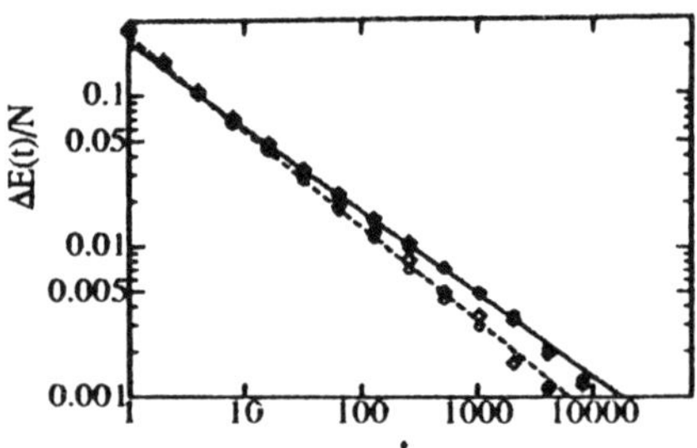

Figure 3: The double-logarithmic plot of $\Delta E(t)/N$ at $T =$0.4 (solid) and 0.5 (open) in systems with $N =$1024 (circles) and 2048 (diamonds). The lines represent the power-law fits described in the text.

large enough N ($\gtrsim$ 1024). The N-independent values of a is estimated as 0.55±0.005 and 0.63±0.01 at $T =$0.4 and 0.5, respectively. In contrast to the p-spin spherical spin glass model for which $a = 1$ independently of T [4], a of the present SK model depends on T similarly to the short-range spin-glass models [13].

3. Concluding Remarks

Beside the energy many aspects of our data obtained by the standard heat-bath MC simulation on systems of larger sizes ($N = 32 \sim 2048$) support the above GQED scenario [10]. A most typical result is the apparent fluctuation-dissipation theorem (FDT), $m(\tau;t_\mathrm{w})= h\{1-C(\tau + t_\mathrm{w}, t_\mathrm{w})\}/T$, where $m(\tau;t_\mathrm{w})$ is the magnetization induced by the field h applied after a waiting time t_w and $C(\tau+t_\mathrm{w}, t_\mathrm{w})$ is the auto-correlation function defined by $C(t,t') \equiv N^{-1} \sum_i S_i(t) S_i(t')$. It holds in the quasi-equilibrium time range, i.e., $\tau \lesssim t_\mathrm{w}$ ($\lesssim t_\mathrm{erg}(N,T)$). We expect that detailed analyses on aging phenomena will reveal nature of the low-temperature spin-glass phase which has not yet been settled in spite of nearly two decades of dispute.

Acknowledgments: Numerical calculations were mainly performed on Fujitsu VPP500 at the Institute for Solid State Physics, University of Tokyo.

References
1. L. Lundgren, P. Svedlindh, P. Nordblad and O. Beckman: Phys. Rev. Lett. **51**, 911 (1983).
2. E. Vincent, J. Hammann, M. Ocio, J.-P. Bouchaud and L.F. Cugliandolo: in *"Slow dynamics and aging"*, ed Rubi Siteges, Conference on Glassy Systems, 1996 (springer-Verlag, inpress) (cond-mat/9607224), and references therein.
3. J.-P. Bouchaud: J. Phys. I France **2**, 1705 (1992).
4. L.F. Cugliandolo and J. Kurchan: Phys. Rev. Lett. **71**, 173 (1993).
5. L.F. Cugliandolo and J. Kurchan: J. Phys. A: Math. Gen. **27**, 5749 (1994).
6. J.-P. Bouchaud, L.F. Cugliandolo, J. Kurchan and M. Mézard: preprint, cond-mat/9702070.
7. H. Rieger: in *"Annual Review of Computational Physics"*, vol.II, ed. D. Stauffer, (World Scientific, Singapore, 1995), and references therein.
8. A. Baldassarri: preprint, cond-mat/9607162.
9. H. Yoshino, K. Hukushima and H. Takayama, to appear in Prog. Theor. Phys. Supple No. 126 (1997) (cond-mat/9611230).
10. H. Takayama, H. Yoshino and K. Hukushima, submitted to J. Phys. A (cond-mat/9612071).
11. M. Mézard, G. Parisi and M.A. Virasoro: *"Spin Glass Theory and Beyond "* (World Scientific, Singapore, 1986).
12. N.D. Mackenzie and A.P. Young: Phys. Rev. Lett. **49**, 301 (1982).
13. E. Marinari, G. Parisi, J. Ruiz-Lorenzo and F. Ritort: Phys. Rev. Lett. **76**, 843 (1996).

Spontaneous Chiral Symmetry Breaking in 2D Aggregation

J. Sandler[1,2]*, G. Canright*[1,2]*, and Z. Zhang*[2]

[1]Department of Physics and Astronomy, The University of Tennessee,
 Knoxville, TN 37996, USA
[2]Solid State Division, Oak Ridge National Laboratory, TN 37831, USA

Abstract Recent experiments [1] using ionized-cluster-beam (ICB) deposition have revealed unusual "seahorse"-like growth patterns. These patterns possess a spontaneously broken chiral (left/right) symmetry, despite the fact that the particles constituting the aggregate are symmetric. We develop a continuum quasi-equilibrium growth model, assuming that the growing aggregate is charged, and that the incoming particles are polarizable, and hence attracted to regions of strong electric field. This model is used both for theoretical analysis and numerical simulation of the growth process. We find that our model possesses a *chiral instability*. That is, during the growth, the system amplifies enormously even tiny left-right asymmetry (which may arise due to noise). This instability leads to the formation of S-like patterns like those seen experimentally. The origin of this instability is the long-range interaction (competition and repulsion) among growing branches of the aggregate, such that a right or left side consistently dominates the growth process. We also show that the electrostatic interaction can account for the principal geometrical properties of the aggregates, such as the existence of only 2 main arms, and the "finned" external edge of the main arms.

Acknowledgements

This work was supported by the Oak Ridge National Laboratory, managed by Lockheed Martin Energy Research Corp. for the U.S. Department of Energy. IS and GC were supported in part by the NSF under grant # DMR-9413057,

References

1. Gao, H.J., Xue, Z.Q., Wu, Q.D. & Pang, S. *J. Mater. Res.* **9**, 2216 (1994).

Springer Proceedings in Physics, Volume 83
Computer Simulation Studies in Condensed-Matter Physics X
Eds.: D. P. Landau, K.K. Mon, H. -B. Schüttler
© Springer-Verlag Berlin Heidelberg 1998

Interactive Modeling of Granular Flow

D.C. Rapaport

Physics Department, Bar-Ilan University, Ramat-Gan, Israel

Abstract: The simulation of the flow of granular materials often leads to behavior that is difficult to quantify and that is therefore best described visually. In this paper we show several examples in which adding visualization to the calculations makes it possible to examine features which are not readily characterized by other means. The problems discussed are granular flow from a silo, the size-grading that occurs in inclined-chute flow, and the surface waves that appear in a thin, vertically vibrated layer.

1. Introduction

The ability to model the dynamical properties of dry granular materials [1] has important industrial consequences. The continuum theories and empirical approximations [2] that are widely used ignore the complex nature of the underlying dynamical processes, so that there is a need to develop reliable, quantitative techniques for predicting behavior in a variety of different kinds of granular flow problems. In recent years, techniques based on molecular dynamics (MD) simulation have achieved popularity [3,4], and considerable effort is being invested using this highly detailed modeling approach.

MD was originally developed for studying atomic and molecular systems, where much of the behavior can be expressed in terms of thermodynamic variables such as the pressure, structural features such as the radial distribution function, and transport coefficients such as the viscosity [5]. Because of the inherent self-averaging and the underlying statistical mechanics, many of the properties of such systems can be understood purely from such numerical "summaries". The situation with granular materials is very different since there is no underlying theory. Much of what happens in granular flow is far more complex, so that it is particularly important to be able to observe the details of what occurs during a simulation. Such observations could help explain laboratory measurements; moreover, they could also suggest other kinds of measurements that have no experimental equivalent which could lead to enhanced understanding of the behavior.

2. Silo flow

The first example considers the gravity-driven flow of granular material through a horizontal aperture. This is a simplified model for the outflow from a silo,

---

Springer Proceedings in Physics, Volume 83
Computer Simulation Studies in Condensed-Matter Physics X
Eds.: D. P. Landau, K.K. Mon, H. -B. Schüttler

or storage tank, with vertical walls; it is clearly a problem of considerable industrial importance, but engineering design is based on a continuous-medium treatment [2] that completely ignores the discrete nature of the grains. Here (and in the subsequent examples) the problem is treated in two dimensions for computational convenience; this follows the approach used in previous related work [6-8].

The grains themselves can be represented in a variety of ways, for example as disks or as rigid bodies of a more complicated shape. The interactions between grains can be expressed in terms of strongly repulsive short-range forces, as well as normal and (optional) tangential damping forces that provide the inelasticity and resistance to shear motion. Grains that exit through the hole at the bottom are reinjected into the system at the top, resulting in a system that eventually reaches a "steady" flow state. These and other details of the computational approach can be found in [9-12]. The simulation described here is controlled through a graphical user interface [13] that facilitates control

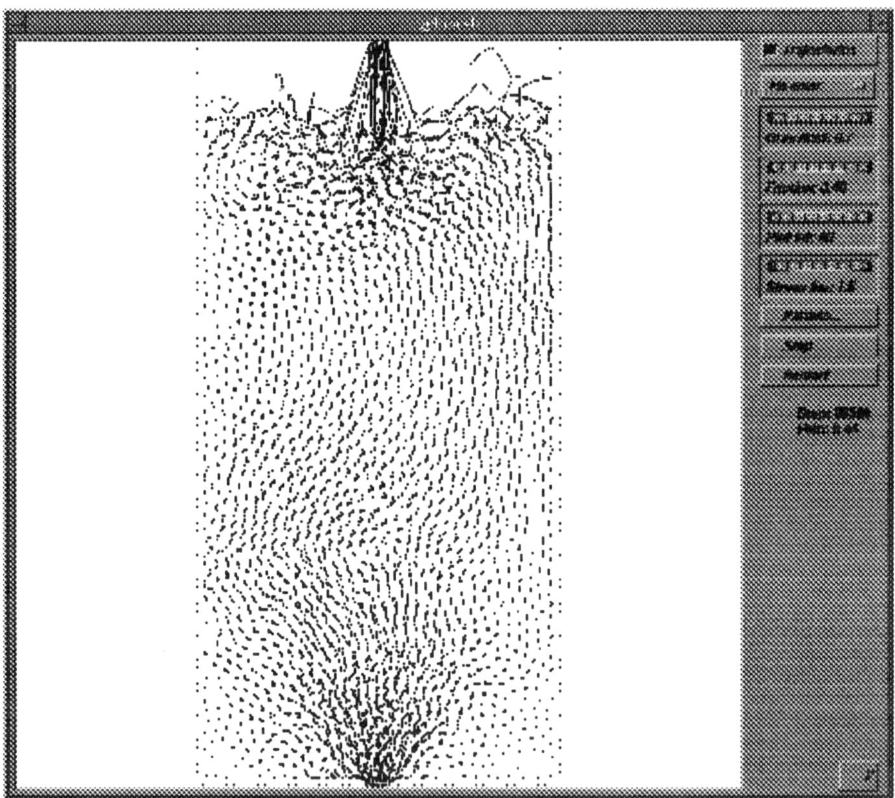

Fig.1 Silo flow: grain trajectories followed over a short time interval.

of the kind of problem being solved (for example, disks or triangular grains, vertical or slanting walls), the parameters of the problem (damping constants, hole width), the type of representation (the grains or their trajectories), and the display of other useful information (colored bands to allow following collective motion).

There are many aspects of the behavior that do not lend themselves to concise numerical summary. The actual motion of the grains turns out to be very complicated, with local regions undergoing shearing and rotation – very different from the uniformly sheared flow encountered in liquids. A sample plot of the trajectories is shown in Figure 1; the meandering patterns shown in this "time exposure" gradually change their form as the local behavior in different parts of the system changes with time. This variation becomes even more apparent after watching a real-time demonstration of this system; it is clearly difficult to convey such information by other than visual means. (Some of the controls available for the user to interact with the simulation are also shown.)

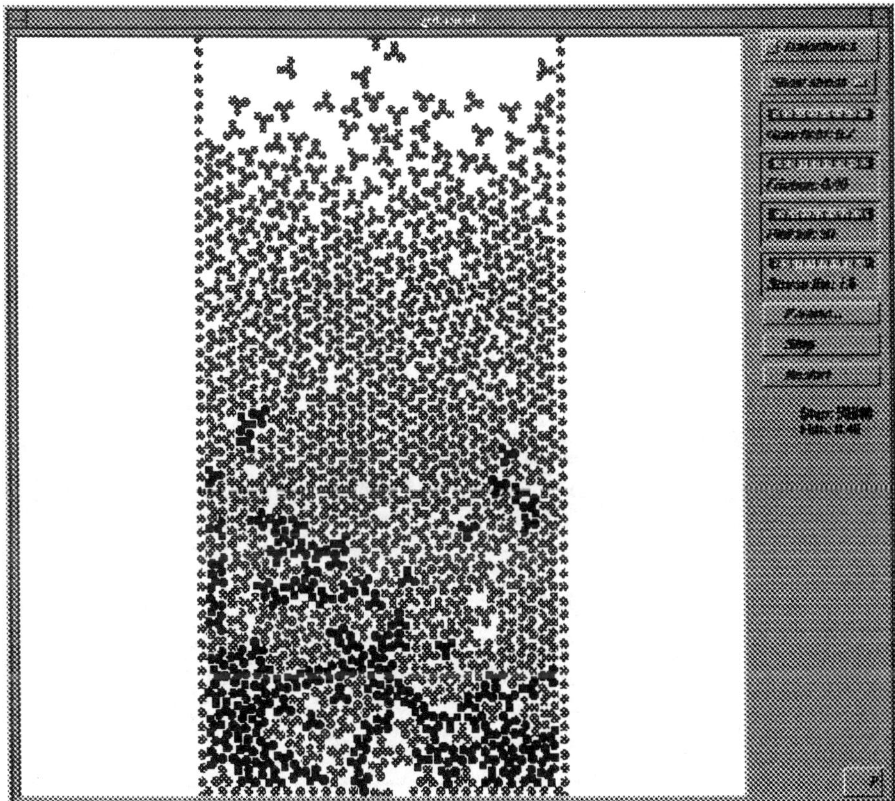

Fig.2 Snapshot showing regions of instantaneous high stress; grains subject to stress more than 60% above average are colored black.

The reason that the flow of granular materials differs from normal fluids is thought to be due to the way stress is transmitted throughout the system. Much of the vertical load in a silo is carried by the side walls due to the formation of "arches" that span the system; static friction is an important component of this effect. Despite the omission of static friction from the simulation, it is still possible to observe uneven stress distributions that vary rapidly with time. Figure 2 shows a typical case, with shading used to label grains that experience stress (computed from the virial) significantly above average.

3. Flow down an inclined chute

Size segregation is a widely observed consequence of sheared flow in granular media. Larger grains are found to rise to the upper surface of the bulk, despite the fact that all grains have the same material density [14]. In a simple MD simulation of this phenomenon [15] the grains are again represented by inelastic disks and the flow occurs down an inclined slope with a rough base. A simulation of this type readily reproduces the segregation effect.

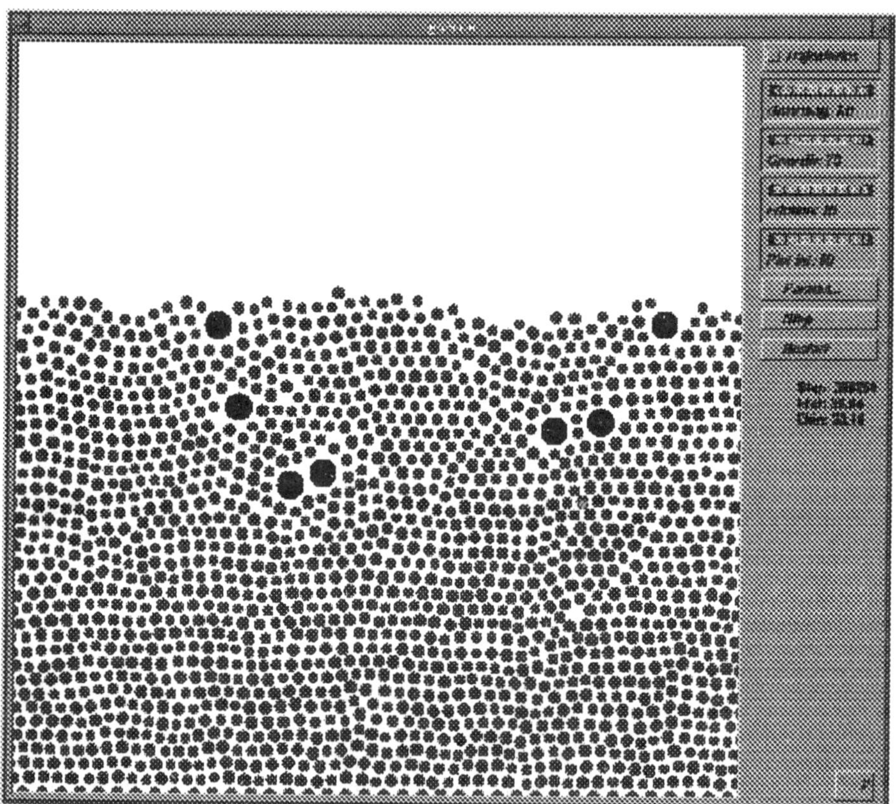

Fig.3 Chute flow: image of the system late in the run showing how the large grains have either reached or are close to the upper surface; gravitational acceleration is directed towards the lower-right corner.

While it is possible to analyze the heights of the larger grains as functions of time, the slope of the chute and the relative grain sizes, the ability to visually follow the segregation process has a far greater impact. At the beginning of the run all the large grains are positioned close to the base of the chute. Figure 3 shows the situation after the sheared flow has proceeded for a certain period of time; all the large grains have either reached the upper surface or are approaching it. (Note that the chute is drawn horizontally and the gravitational acceleration is directed towards the lower-right corner.) As the system evolves, the gradual rise of the large grains can be monitored and the organization of their local environments examined.

4. Vibrated layer

Another of the intriguing aspects of granular dynamics is the appearance of surface waves when a thin layer is vertically vibrated. Among the more novel features of this phenomenon reported recently are wave patterns having the

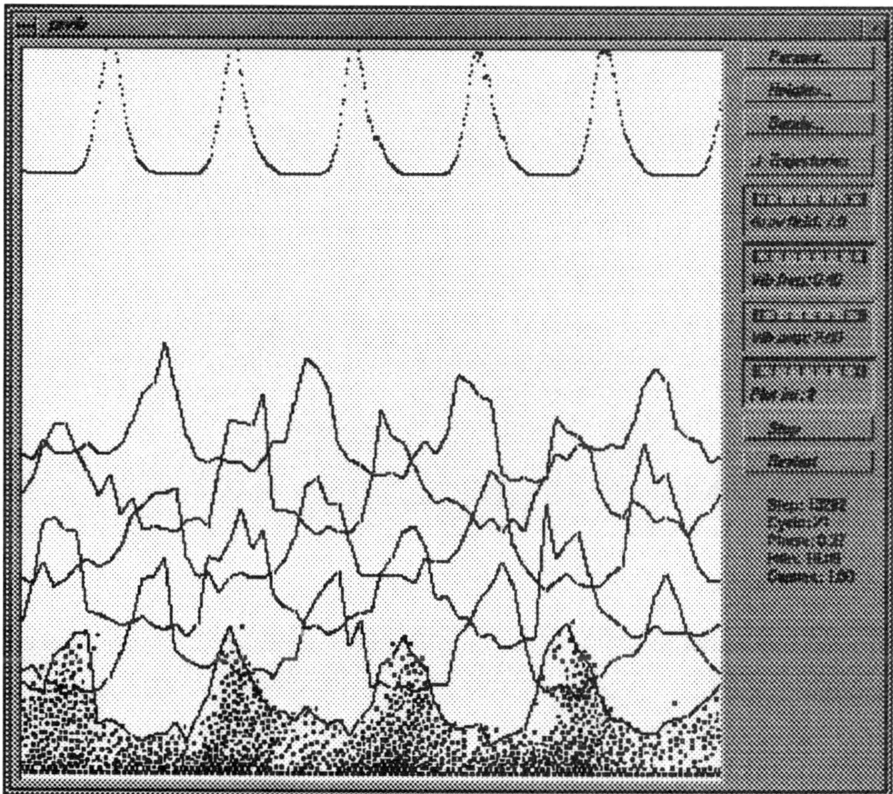

Fig.4 Vibrated layer: a typical surface wave pattern (other features are described in the text).

form of stripes and hexagons, period doubling of the waveform and localized excitations [16,17]. Certain aspects of this behavior have been modeled using two-dimensional MD [18,19].

The simulation shown here [20] allows detailed exploration of this phenomenon, and in fact reveals more than has been observed in [18,19]. The horizontal base undergoes vertical sinusoidal oscillations, and the disk-like particles are an even simpler version of the granular model than before in that they incorporate just the normal damping forces (but no transverse forces). By manipulating the parameters two period doublings are obtained, as observed experimentally. It is also possible to examine the mechanism responsible for the surface waveform; this turns out to be a stationary horizontal density wave, in which the highest points on the surface appear above the density maxima. How the vertical base oscillation is converted into a horizontal density wave still requires explanation.

A typical image from the simulation is shown in Figure 4. Several results are included in the figure. The grains themselves are shown of course. The vertically shifted series of waveforms show the recent shapes of the granular surface as if illuminated by a stroboscope on successive cycles when the base is at its lowest position. The periodic pattern at the top is a scrolling display of the time-dependent pressure.

5. Summary

The examples shown here are just a few of the fascinating granular systems whose modeling can be enriched by the addition of visualization to the simulation program. The ability to actually see the detailed motions of the grains themselves has the potential for significantly enhancing the understanding of the underlying phenomena. Full animation, rather than just static images, is even better [21], but this is not suitable for the printed medium. One challenge is to extend this approach to other granular systems where the behavior demands a full three-dimensional simulation; the problem then is not merely the additional computation required, but how to represent the large amount of information produced in a comprehensible manner.

Acknowledgment

This research was supported in part by a grant from the Israel Science Foundation.

References

[1] H. M. Jaeger, S. R. Nagel, and R. P. Behringer, Rev. Mod. Phys. **68**, 1259 (1996).

[2] R. M. Nedderman, *Statics and Kinematics of Granular Materials* (Cambridge University Press, Cambridge, 1992).

[3] G. C. Barker, in *Granular Matter: An Interdisciplinary Approach*, A. Mehta editor (Springer, Heidelberg, 1994), p.35.

[4] H. J. Herrmann, in *3rd Granada Lectures in Computational Physics*, P. L. Garrido and J. Marro editors (Springer, Heidelberg, 1995), p.67.

[5] D. C. Rapaport, *The Art of Molecular Dynamics Simulation* (Cambridge University Press, Cambridge, 1995).

[6] G. H. Ristow, J. Phys. I (France) **2**, 649 (1992).

[7] P. A. Langston, U. Tüzün, and D. L. Heyes, Chem. Eng. Sci. **49**, 1259 (1994).

[8] G. H. Ristow and H. J. Herrmann, Physica A **213**, 474 (1995).

[9] P. A. Cundall and O. D. L. Strack, Géotechnique **29**, 47 (1979).

[10] O. R. Walton, in *Mechanics of Granular Materials*, J. T. Jenkins and M. Satake editors (Elsevier, 1983) 327.

[11] D. Hirshfeld and D. C. Rapaport, submitted for publication (1996).

[12] Y. Radzyner and D. C. Rapaport, submitted for publication (1996).

[13] D. C. Rapaport, Computers in Physics (in press, 1997).

[14] S. B. Savage and C. K. K. Lun, J. Fluid Mech. **189**, 311 (1988).

[15] D. Hirshfeld and D. C. Rapaport, submitted for publication (1996).

[16] F. Melo, P. B. Umbanhowar and H. L. Swinney, Phys. Rev. Lett. **75**, 3838 (1995).

[17] P. B. Umbanhowar, F. Melo and H. L. Swinney, Nature **382**, 793 (1996).

[18] K. M. Aoki and T. Akiyama, Phys. Rev. Lett. **77**, 4166 (1996).

[19] S. Luding *et al*, Europhys. Lett. **36**, 247 (1996).

[20] D. C. Rapaport, to be published (1997).

[21] Animations are to be found on the author's web site: http://www.biu.ac.il/~rapaport.

Numerical Study of a Random Gauge XY Model

J.M. Kosterlitz and M.V. Simkin

Department of Physics, Brown University, Providence, RI 02912-1843, USA

Abstract. An XY model with random gauge as a model for a superconducting glass is studied in two and three dimensions by a zero temperature domain wall renormalization group which allows one to follow the flows of both the coupling constant and the disorder strength with increasing length scale. Weak disorder is found to be marginal in two and probably irrelevant in three dimensions. For strong disorder the flow is towards a non-superconducting gauge glass fixed point in $2d$ and a superconducting glass in $3d$. Our results are in agreement with recent analytic theory and are inconsistent with earlier predictions of a re-entrant transition to a disordered phase at very low temperature and with the loss of superconductivity for any finite amount of disorder.

The classical random gauge XY model described by the Hamiltonian [1]

$$H = - \sum_{<ij>} J_{ij} cos(\theta_i - \theta_j - A_{ij}) \tag{1}$$

has been the subject of much interest over the past decade. Here, θ_i is the phase of the order parameter at the ith site of a square lattice in $2d$ or a simple cubic lattice in $3d$ and the sum is over all nearest neighbor bonds. The coupling constants J_{ij} are uniform $J_{ij} = J > 0$ and the quenched random phase shifts A_{ij} are uncorrelated from bond to bond and uniformly distributed over the range $-\alpha\pi \leq A_{ij} \leq \alpha\pi$ with $0 \leq \alpha \leq 1$ so that $< A_{ij} >= 0$ and $< |A_{ij}| >= \alpha\pi/2$ where $<>$ means an average over disorder. In $d = 2$, the model of eq.(1) is a description of an array of Josephson junctions in a magnetic field perpendicular to the plane of the array when θ_i represents the phase of the superconducting order parameter of the ith grain and $A_{ij} = (2\pi/\Phi_0) \int_i^j \vec{A} \cdot d\vec{l}$ with $\vec{A}$ the vector potential of the external magnetic field and $\Phi_0 = hc/2e$ is the quantum of flux. The A_{ij} become independent quenched random variables when the average flux through an elementary plaquette is an integer multiple of Φ_0 but the superconducting grains are randomly displaced from their ideal lattice positions [2]. It is also a model for an XY magnet with random Dzyaloshinskii-Moriya interactions [3].

Whatever the physical origin of the quenched random phase shifts, the system described by eq.(1) has been a theoretical challenge for a decade. Early work concluded that weak disorder ($\alpha \ll 1$) does not destroy the superconducting phase at intermediate temperature T but at low T there is a re-entrant transition to a normal phase [3,2]. However, this was not confirmed either numerically nor experimentally [4]. Some later theoretical work [5] suggested that the bound vortex (KT) phase [6] is destroyed at any finite T by arbitrarily

Springer Proceedings in Physics, Volume 83
Computer Simulation Studies in Condensed-Matter Physics X
Eds.: D. P. Landau, K.K. Mon, H. -B. Schüttler
© Springer-Verlag Berlin Heidelberg 1998

small disorder and that the experimentally observed KT phase is a finite size effect. The more recent theoretical work [7–11], on the other hand, argue for a more conventional phase diagram in which there is a superconducting phase for $T < T_c(\alpha)$ where $T_c(\alpha) \geq 0$ for $\alpha \leq \alpha_c$. The case of maximum disorder ($\alpha = 1$) when the A_{ij} are uniformly distributed between $-\pi$ and $+\pi$ is called the gauge glass [12,13] which has been studied numerically at $T = 0$ by a domain wall renormalization group (DWRG) [14–16] in both $d = 2$ and $d = 3$ and the conclusions from these studies are that the stiffness to distortions of the phase vanishes in $d = 2$ so that the glass is not superconducting and that in $d = 3$ the gauge glass is probably superconducting, but the evidence is not conclusive. In view of the three conflicting scenarios in two dimensions, (i) re-entrant transition [3,2], (ii) destruction of superconductivity for any finite disorder [5] and (iii) superconductivity for $T < T_c(\alpha)$ with $T_c(\alpha) > 0$ for $0 \leq \alpha < \alpha_c$ [7,9,10], this system is an ideal candidate for study by a numerical DWRG at $T = 0$ as this will distinguish scenario (iii) from the others as only scenario (iii) predicts a finite stiffness at $T = 0$ for $0 \leq \alpha < \alpha_c$.

The standard DWRG [17] at $T = 0$ consists of computing the lowest energies of a set of systems of several linear sizes L with periodic and antiperiodic boundary conditions (BC) in one direction with some fixed BC in the other $d-1$ directions. The difference $\Delta E(L) =< |E_{ap}(L) - E_p(L)| >$ is the domain wall energy and $\Delta E(L)/2$ is interpreted as an effective coupling constant $J(L)$ at length scale L which one expects to scale as $J(L) \sim L^\theta$ at large L. The stiffness exponent θ is a crucial quantity as its value will distinguish between an ordered superconducting phase at small but finite T ($\theta \geq 0$) and a disordered phase ($\theta < 0$). If $\theta < 0$, then the energy of an excitation of size L is $\Delta E(L)$ which vanishes as $L \to \infty$ and the probability of such a phase unwinding excitation $P(\Delta E(L)) \sim exp(-\Delta E(L)/kT)$ is large at any $T > 0$ so the stiffness to twists in the phase will vanish. If $\theta > 0$, the converse is true and the stiffness will be finite for $T < T_c$ and the glass will be superconducting.

For the problem of interest with a variable disorder strength, this version of the DWRG, which considers only the scaling of the effective coupling $J(L)$ needs considerable modification as we also want to know how the disorder strength scales with L. For a single junction, the Hamiltonian is $H = -Jcos(\phi - A)$ where ϕ is the phase difference across the junction and the usual comparison of the energies with periodic and antiperiodic BC gives $\Delta E(1) = 2JcosA$ which does not separate the disorder strength from the coupling constant. At length scale L, the interaction is $V_L(\phi - A(L))$ where $A(L)$ is the phase shift at scale L and $V_L(\phi)$ is a 2π-periodic function with a minimum when its argument is zero. Thus, if we impose BC with phase shifts Δ_μ across the boundaries in the d directions $\mu = 1, 2, ...d$, minimizing the energy with respect to the phases θ_i will give the ground state energy of a system of linear size L as a function of Δ_μ which is 2π periodic in each of the d directions

$$E_L(\Delta_\mu) = E_L(\Delta_\mu + 2\pi) \tag{2}$$

with a minimum at some Δ_μ^0 which depends on the precise realization of disorder. The key observation is that Δ_μ^0 is exactly the phase shift $A_\mu(L)$ which minimizes the energy at scale L. A measure of the strength of disorder at this scale is

$$|A(L)| \equiv < |\Delta^0| > \tag{3}$$

with $|A(1)| = \alpha\pi/2$. The coupling constant $J(L)$ at scale L is found by first finding $E_L(\Delta^0_\mu)$, changing Δ^0 by π in one of the d directions and then finding the energy minimum $E_L(\Delta^0 + \pi)$ with these BC. As discussed above, the coupling constant $J(L)$ at scale L is

$$J(L) \equiv < (E_L(\Delta^0 + \pi) - E_L(\Delta^0)) > \tag{4}$$

and measuring $J(L)$ and $|A(L)|$ for several sizes L gives renormalization group flows for both the coupling constant and disorder strength. Of interest are the stable fixed point values $J^* \equiv J(L = \infty)$ and $A^* \equiv |A(L = \infty)|$ as these determine the nature of the phases. There are several possibilities of which the simplest are $[J^* = \infty,\ A^* = 0]$, $[J^* = \infty,\ A^* = \pi/2]$, $[J^* = 0,\ A^* = \pi/2]$ corresponding respectively to a superconducting state with long range order, a superconducting glass and a non-superconducting glass. There are other possibilities such as a state with quasi long range order corresponding to a flow to a fixed line with finite J^* and A^* whose values depend on the initial values of coupling and disorder. This is the scenario in $d = 2$ predicted by recent analytic work [9,10].

We used simulated annealing [18] to estimate the ground state energies which is considerably more efficient than simple repeated quenches to $T = 0$ [19]. Also, we imposed periodic $\Delta_\mu = 0$ BC in $d - 1$ directions and twisted $\Delta \neq 0$ BC in the remaining direction and minimized the energy with respect to the the phases θ_i and to the twist Δ to find Δ^0. To obtain the domain wall energy ΔE_L the twist is changed to $\Delta^0 + \pi$ and kept fixed while the energy is minimized with respect to the θ_i only. According to our earlier discussion, the energy should be minimized with respect to global phase shifts in all d directions and the domain wall energy ΔE_L obtained by increasing the phase shift by π in one direction. To within the errors of our simulations, ΔE_L is independent of the choice of BC in the $d - 1$ transverse directions so, for simplicity, we imposed periodic or $\Delta_\mu = 0$ BC in these directions. As a consistency check [19], we simulated two identical copies of each system with different random number sequences to obtain two estimates E_1, E_2 of the ground state energy. In the event that the simulation finds the exact ground state, then $\delta E = E_1 - E_2 = 0$, which often occurs for our small L values. If the simulation does not reach the exact minima, $< (\delta E)^2 >$ is a measure of the error. To minimize the errors caused by failure to reach the true energy minimum, we adjust the annealing schedule and the number of annealing attempts until $\delta E/E < N^{-1/2}$ where $N = 10^3$ in $2d$ and 10^4 in $3d$ is the number of realizations of disorder. This consistency check makes the error due to not reaching the true ground state no worse than the statistical error in the averaging over disorder. For repeated simulated annealings of N different samples the CPU time becomes prohibitive for $L > 8$ in $2d$ and $L > 4$ in $3d$. We therefore chose sizes $L = 2, 4, 8$ in $2d$

and $L = 2, 3, 4$ in $3d$ and the results are summarized in Fig.(1) for $2d$ and in Fig.(2) for $3d$.

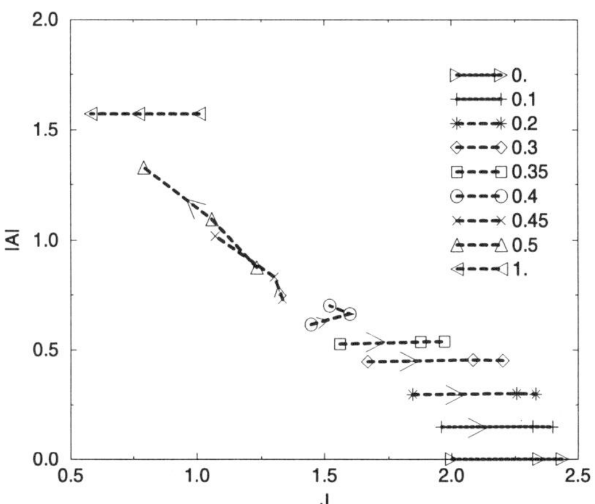

FIG. 1. DWRG flows for the $2d$ superconducting glass model in the coupling constant-disorder strength $(J, |A|)$ plane. The initial value of J is $J(1) = 1$ and the initial disorder strength α is indicated near corresponding symbols.

In $2d$, for small disorder $\alpha < \alpha_c \approx 0.37$, $J(L)$ increases more slowly than a power of L and seems to flow to a finite disorder dependent value $J^*(\alpha)$ and the disorder strength $|A(L)|$ does not change with L, at least for our small sizes, both of which are completely consistent with analytic RG calculations [9,10]. At larger disorder strength $\alpha > \alpha_c$, $|A(L)|$ increases and $J(L)$ decreases as L increases, in agreement with the analytic theory [9,10]. In this range of disorder strengths, the system is probably flowing to the non-superconducting glass fixed point at $J^* = 0, A^* = \pi/2$. When the disorder is maximal ($\alpha = 1$), $|A(L)|$ remains fixed at $\pi/2$ and $J(L) \sim L^\theta$ with the stiffness exponent $\theta \approx -1/2$, in agreement with other simulations of the gauge glass in $2d$ [15,14]. The flows shown in Fig.(1) may be regarded as RG flows in a higher dimensional parameter space projected on to the $(J(L), |A(L)|)$ plane, so the crossing of the two trajectories for $\alpha = 0.45$ and $\alpha = 0.5$ does not violate the non-crossing rule. For the $2d$ system, the RG flows are in the three parameter space of $J, |A|$ and the vortex fugacity y [7,9,10]. From the data of Fig.(1) and assuming that the trends for small L continue when L is large, one would conclude that the most likely scenario for the disordered system in $2d$ is, for weak disorder $\alpha < \alpha_c$, $J(L) \rightarrow J^*(\alpha)$ and $|A(L)| = |A(1)|$, so that the coupling constant scales to a finite but disorder dependent value while the disorder is marginal. For larger disorder $\alpha > \alpha_c$, $J(L) \rightarrow J^* = 0$ and $|A(L)| \rightarrow \pi/2$ which is a non-superconducting disordered state. Our results are consistent with recent analytical theory [9,10] and inconsistent with the re-entrant [3,2] and complete destruction of superconductivity [5] scenarios. However, they are consistent with a re-entrant scenario in which the ordered phase extends to $T = 0$ for a range of α [20]. The flows for $2d$ of Fig.(1) are consistent with a discontinuous jump in J^* at α_c as predicted analytically [9].

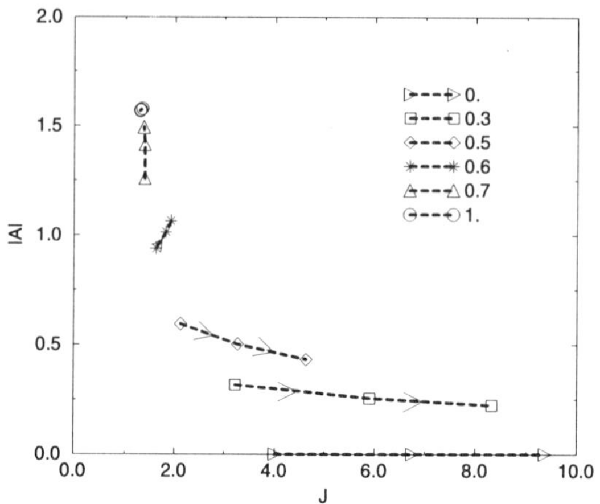

FIG. 2. DWRG flows for the model in $3d$.

The results of the simulations of the random gauge model in $3d$ are shown in Fig.(2). The system sizes are very small ($L = 2, 3, 4$) because of the CPU time needed to get close to the minimum energies so that irrelevant variables are giving large corrections to scaling. Nevertheless, some qualitative features are apparent, assuming that the small L trends continue. For small disorder $\alpha < \alpha_c \approx 0.55$, $J(L) \sim L^{d-2}$ as expected and the disorder strength seems to decrease. It is impossible to say if $|A(L)| \to 0$ as one expects but the data is consistent with this. One is tempted to conclude that, in this regime of weak disorder, the DWRG flows are to a stable fixed point at $J^* = \infty$, $A^* = 0$ corresponding to a true superconducting phase. For larger disorder $\alpha_c < \alpha \leq \pi/2$, the disorder increases with L and seems to flow to its maximum value of $\pi/2$. The coupling $J(L)$ seems to flow to a finite value which corresponds to a stiffness exponent $\theta = 0$. Although this is consistent with other simulations on the $3d$ gauge glass [14,16], our use of the phase representation of eq.(1) in the simulations together with the very small sizes may introduce large corrections to scaling. Nevertheless, the data is consistent with a superconducting glass phase which survives at finite T. These considerations suggest that the phase diagram for the model in $3d$ is similar to that of the corresponding infinite range model [21].

Our conclusions from our new $T = 0$ DWRG which follows the flows in two parameter space are in $2d$, the recent analytic theory which predicts a quasi long range ordered state for $T < T_c(\alpha)$ is the correct scenario and earlier suggestions of a re-entrant transition to a disordered phase at low T or no superconductivity at any finite disorder are ruled out. In $3d$, weak disorder has little or no effect on the superconducting phase and there is a critical disorder strength parameterized by $\alpha = \alpha_c$, above which the system is a superconducting glass at low T. After this work was finished we learned that similar conclusions in $2d$ have been reached by Maucort and Grempel [22] in a finite temperature Monte Carlo study of the model.

This work was supported by the NSF under Grant No. DMR-9222812. Computations were performed on the Cray EL98 at the Theoretical Physics Computing Facility at Brown University. JMK thanks B. Grossman and A. Vallat for invaluable discussions about gauge glasses. MVS is grateful to M. Cieplak, M.J.P. Gingras and A.V. Vagov for correspondence.

References

[1] C. Ebner and D. Stroud, Phys. Rev. B **31**, 165 (1985).

[2] E. Granato and J.M. Kosterlitz, Phys. Rev. B **33**, 6533 (1986); Phys. Rev. Lett. **62**, 823 (1989).

[3] M. Rubinstein, B. Shraiman, and D.R. Nelson, Phys. Rev. B **27**, 1800 (1983).

[4] M.G. Forrester, H.J. Lee, M. Tinkham, and C.J. Lobb, Phys. Rev. B **37**, 5966 (1988); M.G. Forrester, S.P. Benz, and C.J. Lobb, Phys. Rev. B **41**, 8749 (1990); A. Chakrabarty and C. Dasgupta, Phys. Rev. B **37**, 7557 (1988).

[5] S.E. Korshunov, Phys. Rev. B **48**, 1124 (1993).

[6] J.M. Kosterlitz and D.J. Thouless, J. Phys. C**6**, 1181 (1973).

[7] T. Natterman, S. Scheidl, S.E. Korshunov, and M.S. Li, J. Phys. it I France **5**, 555 (1995).

[8] S.E. Korshunov, and T. Natterman, Phys. Rev. B **53**, 2746 (1996).

[9] S. Scheidl. Phys. Rev. B**55**, 457 (1997)

[10] L.-H. Tang. Phys. Rev. B**54**. 3350 (1996)

[11] M.-C. Cha and H.A. Fertig, Phys. Rev. Lett. **74**, 4867 (1995)

[12] M.P.A. Fisher, Phys. Rev. Lett. **62**, 1415 (1989)

[13] D.A. Huse and H.S. Seung, Phys. Rev. B**42**, 1459 (1990)

[14] M.J.P. Gingras, Phys. Rev. B **45**, 7547 (1992)

[15] M.P.A. Fisher, T.A. Tokuyasu, and A.P. Young, Phys. Rev. Lett. **66**, 2931 (1991);

[16] J.D. Reger, T.A. Tokuyasu, A.P. Young, and M.P.A. Fisher, Phys. Rev. B **44**, 7147 (1991)

[17] J.R. Banavar and M. Cieplak, Phys. Rev. Lett. **48**, 832 (1982); W.L. McMillan, Phys. Rev. B **29**, 4026 (1983).

[18] S. Kirkpatrick, C. D. Gellat Jr, and M.P. Vecchi, Science, **220**, 671 (1983).

[19] M.V. Simkin, cond-mat/9609213, to appear in Phys. Rev. B.

[20] A.V. Vagov, unpublished.

[21] D. Sherrington and M. Simkin, J. Phys. A **26**(1993), L1201; Corr. *ibid* **27** (1994) 2237.

[22] J. Maucort and D.R. Grempel, cond-mat/9703109.

Perpendicular Order in Frustrated Magnetic Layers

M. Enjalran[1], *S.M. Kauzlarich*[2], *and R.T. Scalettar*[1]

[1]Physics Department, University of California, Davis, CA 95616, USA
[2]Chemistry Department, University of California, Davis, CA 95616, USA

Abstract. We report preliminary results of Monte Carlo simulations of coupled two dimensional, square lattice, classical, antiferromagnetic Heisenberg systems. We find that when the two layers are offset so that frustrating interactions are present, the magnetic order in the planes can be made orthogonal, even in the absence of any explicit symmetry breaking terms on the Hamiltonian. Connections are made to the magnetic behavior of recently sythesized pnictide–oxide materials.

1. Introduction

Studies of magnetic behavior at surfaces and in layered materials have a long history.[1] On the theoretical side, such systems offer the opportunity to explore the effect of anisotropy and reduced dimensionality on phase transitions. On the experimental side, advances in materials preparation techniques have enormously increased the ability to synthesize layered systems. Magnetic multilayers and high temperature superconducting oxides are two examples of classes of compounds with large anisotropy and remarkable physical and chemical properties.

The experimental motivation of the simulations we will describe in this paper is the recent improvement in synthesis techniques and characterization of the magnetic properties of the "pnictide–oxide" compounds, $A_2Mn_3Pn_2O_2$. (A=Sr,Ba;Pn=P,As,Sb,Bi).[2,3] The structure of these materials is such that the Mn is found in two different, alternating, types of layers. (See Fig. 1.) In both the MnO_2^{2-} and $Mn_2Sb_2^{2-}$ planes, the Mn atoms form a square array with relatively strong, antiferromagnetic, *intralayer* coupling. (Ordering transitions occur at room temperature in the $Mn_2Sb_2^{2-}$ planes, and at 50-100 K for the MnO_2^{2-} planes.) The square arrays which characterize these two sublattices are offset from each other as one moves perpendicular to the sheets, so that a Mn atom of a $Mn_2Sb_2^{2-}$ plane lies at the midpoint of a bond of the MnO_2^{2-} planes above and below it. Hence the *interlayer* coupling, which is also antiferromagnetic, is frustrated. The separation of the layers, and their magnetic coupling, can be controlled by changing the alkaline earth cation and the pnictogen, allowing for the possibility of studying systematic relations between structure and magnetic behavior.

At high temperatures the layers can be treated as independent, and the susceptibility is well described by that of the isotropic 2D Heisenberg model.[2,3] As T is lowered, the layers begin to interact. The two different types of layers then order such that the spins of each sublattice are perpendicular to each

Springer Proceedings in Physics, Volume 83
Computer Simulation Studies in Condensed-Matter Physics X
Eds.: D. P. Landau, K.K. Mon, H. -B. Schüttler
© Springer-Verlag Berlin Heidelberg 1998

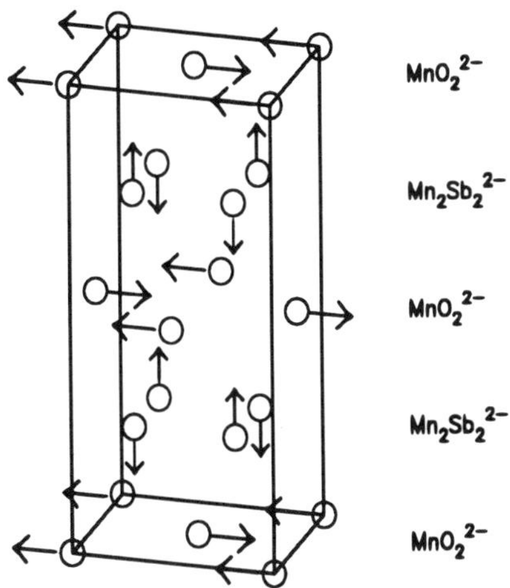

FIG. 1. Magnetic structure of $Sr_2Mn_3As_2O_2$. (From Ref. 3.)

other. (See Fig. 1). In this paper we would like to explore simple models which incorporate some of the basic features of these materials. The specific question we will address is whether the perpendicular order between the layers must originate in some explicit terms in the underlying Hamiltonian which force orthogonal arrangements of spins in different sheets, or whether such patterns can arise in isotropic models as a generic consequence of frustration effects.

2. Model and Simulation Method

We will study a Hamiltonian which consists of two, coupled, square arrays of classical Heisenberg spins,

$$H = J \sum_{\langle i,j \rangle \lambda} \vec{S}_{i\lambda} \cdot \vec{S}_{j\lambda} + J_\perp H_{\text{inter}}. \tag{1}$$

Here $\vec{S}_{i\lambda}$ is the value of the spin at spatial site $i = (i_x, i_y)$ of layer λ. For simplicity, we choose the intralayer exchange constant J to be the same in the two layers, though, as mentioned above, experimentally the ordering temperatures of the layers are different, suggesting that further work with $J \to J_\lambda$ should be done. We will consider three possible forms for the interlayer term:

- No offset. In this case the two square arrays will be arranged so that spins of the second layer lie immediately above spins in the first layer. The interlayer coupling is then between S_{i1} and S_{i2}. There is no frustration.

- The second layer is offset by one half a lattice constant in the $\hat{x}$ direction relative to the first layer. That is, spins of the second layer lie above the midpoints of bonds of the first layer. The interlayer coupling is between S_{j1} and S_{i2} where j takes on the two values given by the sites at the ends of the bond above which i sits. There is frustration.

- The second layer is offset by one half a lattice constant in the $\hat{x}$ *and* $\hat{y}$ directions relative to the first layer. That is, spins of the second layer lie above the midpoints of square plaquettes of the first layer. The interlayer coupling is between S_{j1} and S_{i2} where **j** takes on the four values given by the sites at the corners of the square above which **i** sits. There is frustration.

As we shall see, these represent three generic types of interlayer interaction.

Notice that our Hamiltonian does not include any terms which explicitly drive the perpendicular ordering of spins in the two layers seen experimentally. We could force such ordering by replacing the intralayer Heisenberg couplings with an xy planar anisotropy in one layer, and a uniaxial z axis anisotropy in the other. We could also put a tendency towards orthogonal order in "by hand" by including a biquadratic exchange interaction.

We can perform an analytic calculation of the ground state configuration of the classical models given by Eq. 1. In case (i), the ground state energy per spin is $E_0/N = -2J - J_\perp$, and, since there is no frustration, there is perfect Neel order, both in and between planes. Thus spins in the two different planes tend to point parallel or antiparallel, not, as observed experimentally, at right angles to each other. Case (ii) is more interesting. Here the frustration makes the ground state nontrivial. It turns out that $E_0/N = -2J - J_\perp^2/8J$. Interestingly, the spin configuarations are now such that spins in the two layers do arrange themselves perpendicular to each other. This calculation is similar to that considered by White *et. al.* in their studies of coupled spin–1/2 Heisenberg chains,[4] and also has analogies to calculations of the energy of isotropic Heisenberg models in external magnetic fields.[5] Finally, case (iii) has $E_0/N = -2J$, independent of $J_\perp$. In this case, the low temperature spin correlations are essentially those of uncoupled sheets, with no evidence for orthogonal, or any other type, of interlayer ordering.

3. Results

We simulate the Hamiltonian Eq. 1 using a standard single spin change Metropolis algorithm.[6] At high temperatures, the spins are disordered. The spin–spin correlation function, $c(l) = \langle S_i S_{i+1} \rangle$ decays rapidly to zero from its value $c(l = 0) = 1/3$ as l increases. As the temperature is lowered, and for $J > J_\perp$ correlations begin first to develop between spins of the same layer, and $c(l)$ begins to approach the value $(-1)^l$ characteristic of intralayer Neel order. As T is lowered further, interlayer spin correlations have the potential to develop.

In Fig. 2 we show the interlayer spin–spin correlations

$$c_\perp(\mathbf{i} - \mathbf{j}) = \langle |\vec{S}_{i1} x \vec{S}_{j2}|^2 \rangle$$
$$c_\parallel(\mathbf{i} - \mathbf{j}) = \langle |\vec{S}_{i1} \cdot \vec{S}_{j2}|^2 \rangle \tag{2}$$

as a function of temperature for case (ii) of partially offset layers. Here **i** and **j** are two sites connected by an interlayer bond $J_\perp$. At high temperatures the

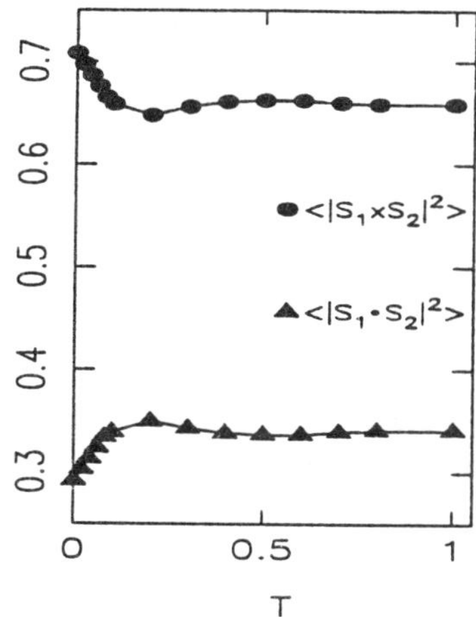

FIG. 2. Nearest neighbor spin-spin correlations as functions of temperature for partially offset layers. The interlayer coupling is $J_\perp = 0.8J$, and the lattice size is two 4x4 layers.

expectation values in Eq. 2 factorize, and reduce to a product of individual expectation values for the individual layers, and $c_\perp \to (2/3)$, and $c_\parallel \to (1/3)$. As the temperature is lowered below $T = 0.5$, the ground state energy per spin crosses the value $E/N = -2$ which represents the result for uncoupled layers. Roughly at this point, the interlayer correlations differ from their noninteracting values. At first, the parallel correlations are enhanced relative to the perpendicular ones, but at temperatures below $T = 0.1$, the perpendicular correlations become dominant, showing evidence that the spins in the two layers tend to orient orthogonal to each other.

Meanwhile, consistent with the analytic analysis, we have determined that perpendicular ordering does not take place in the fully offset case.

In this contribution, we have considered whether the interlayer perpendicular magnetic order observed in the pnictide oxides can originate generically in isotropic models in which interlayer couplings are frustrated. Our conclusion is that for some forms of frustration geometries, orthogonal order can naturally arise without being explicitly introduced into the couplings. We still need to extend the analysis to the experimental geometry of the pnictide oxides, where the spins of one layer lie on the midpoint of the bonds of the next.

There are a number of other ways we plan to extend the present work. Studies of a quantum mechanical spin–1/2 Heisenberg layers with a long range interplanar interaction have shown that perpendicular order can exist at low temperatures.[7] It would be interesting to extend this work to more reasonable, finite range, models. Perhaps the fully offset model can recover orthogonal order when the spins are quantum mechanical.[8]

Acknowledgements

We gratefully acknowledge the support of the Campus–Laboratory Collaboration Program of the University of California.

REFERENCES

1. D.L. Mills, J. Mag. and Mag. Phen. **100**, 515 (1991) and references cited therein; and A.J. Freeman and R.-q. Wu, J. Mag. and Mag. Phen. **100**, 497 (1991) and references cited therein.

2. S.L. Brock and S.M. Kauzlarich, J. Alloys and Compounds **241**, 82 (1996); and S.L. Brock and S.M. Kauzlarich, Comments Inorg. Chem. **17**, 213 (1995). 82 (1996); and

3. S.L. Brock and S.M. Kauzlarich, Chemtech **25**, 18 (1995).

4. S.R. White and I. Affleck, Phys. Rev. **B54**, 9862 (1996).

5. R.T. Scalettar, E.Y. Loh, Jr., J.E. Gubernatis, A. Moreo, S.R. White, D.J. Scalapino, R.L. Sugar, and E. Dagotto, Phys. Rev. Lett. **62**, 1407 (1989).

6. More complicated algorithms which introduce collective moves which beat critical slowing down will be used in future work.

7. J. Richter, C. Gros, S.E. Kruger, W. Wenzel, and J. Schulenberg, J. of Magnetism and Magnetic Materials, **156**, 433 (1996).

8. Note, however, that the Mn ions in the pnicide oxides are spin–5/2, so that a classical model might not be so inapropriate.

Simulation of Polymers Using the Ellipsoidal Model

G. Schöppe[1] and D.W. Heermann[2]

[1]Institut für Theoretische Physik*, Universität Heidelberg,
 Philosophenweg 19, D-69120 Heidelberg, Germany
[2]Interdisziplinäres Zentrum für wissenschaftliches Rechnen
 der Universität Heidelberg

Abstract. We present a recently developed model for polymer simulation, the ellipsoidal model. The geometric shape of the chemically realistic monomers is conserved by using ellipsoids as building units for the coarse grained model of the monomers. Static and dynamical properties are investigated using a parameterization for the Bisphenol-A-Polycarbonate. Our advantage employing this model is, that smaller chain lenths can be used to investigate physical properties than with other models. One also can reinsert the chemically realistic chain after a simulation to investigate properties of the atomistic scale.

1   Introduction

The interest in the physics of polymers was pushed by the need of new materials in all fields of everyday life. Prominent examples for this are plastic bags made out of polyethylene or compact discs which mainly consist out of Bisphenol-A-Polycarbonate (BPA-PC). The properties of these materials were investigated by experiments and for some a theory has been established. Nevertheless computer simulations can give a more detailed picture in this context. Polymers can be modeled for computational purposes in a variety of ways. Depending on the kind of question and the degree of abstraction, one has the basic choice between a lattice model and a model in continuum. If one wants to stay as close as possible to the chemically realistic chain it may be of advantage to remain in real space. The approach to the computational modeling we have undertaken is to retain as much as possible of the chemically realistic chain and still stay computationally efficient. The goal is to examinate the correlation between the chemical structure and the physical properties which is very time consuming at atomistic scale.

2   Model and Methods

We consider the interaction volume of chemical sequences to be of rotational symmetric ellipsoidal form. These ellipsoids are then connected at their focal

---

* http://wwwcp.tphys.uni-heidelberg.de/

Springer Proceedings in Physics, Volume 83
Computer Simulation Studies in Condensed-Matter Physics X
Eds.: D. P. Landau, K.K. Mon, H. -B. Schüttler
© Springer-Verlag Berlin Heidelberg 1998

points to form a chain. The interaction volume is taken as a con-focal force field and its hard core region, the connection between the focal points, is chosen to represent the mass of the building unit, which is assumed to be homogeneous. Thus in the ellipsoidal model linear polymers are simulated by rod chains which on the one hand give a realistic excluded volume region along the backbone of the chain and on the other hand are closer to the geometric form of the chemical sequence in contrast to a bead in e.g. united-atom model see M. Bishop et. al. (1979). The bonded interactions between neighboring units are harmonic angle and torsion potentials. For details see K.M. Zimmer and D.W. Heermann (1995). In this paper we present data obtained by molecular dynamics simulations. To estimate the force on the rods, we use a 10 point Gauss integration. At each time step we solve the equations of motion for a rod chain, which leeds to a tridiagonal matrix.

3 Parameterization of BPA-PC

In this investigation we focus on the simulation of BPA-PC. We use the parameterization of K.M. Zimmer et. al. (1996) in which each repeating unit is mapped to one ellipsoid. We identify the focal points of the modeling ellipsoids with the center of mass of the backbone atoms of the carbonate-groups. Thus the lengths of the ellipsoids are about $l \approx 11$ Å. This potential of the chemically detailed monomer shows us that in the case of polycarbonate the strength of the attractive part of the potential is very small. Thus we set the absolute value of the intermolecular potential to be a r^{-6} potential. The intra-chain parameters are optimized for each temperature, using lengths and angles distributions received form quantum chemical calculations as input.

4 Static Properties

First we want to present some data and considerations concerning the statics of a BPA-PC melt within the ellipsoidal model. We did compute the single chain structure factor, as described in P. G. de Gennes (1979). This structure function

$$S_s(q) = \left\langle \frac{1}{N} \left| \sum_{j=1}^{N} e^{i\mathbf{q}\cdot\mathbf{r}_j} \right|^2 \right\rangle_{|\mathbf{q}|} \tag{1}$$

of the individual chains provides a test on which length scale the self-avoiding interaction is screened. The index $|\mathbf{q}|$ indicates the spherical average over $\mathbf{q}$ vectors of the same absolute value. One expects good solvent properties $S_s(q) \sim q^{-2}$ on length scales $\langle l \rangle < 2\pi/q < \xi(\rho)$ and random walk behavior $S_s(q) \sim q^{-1/\nu}$ for $\xi(\rho) < 2\pi/q < \langle R^2 \rangle^{1/2}$, whereby ρ is the density of the polymer melt. Fitting straight lines with the slope of $1/\nu = 1.695$ and 2 to the data in fig. 1 we find a crossing of the two regimes at $q = 0.225\,\text{Å}^{-1}$.

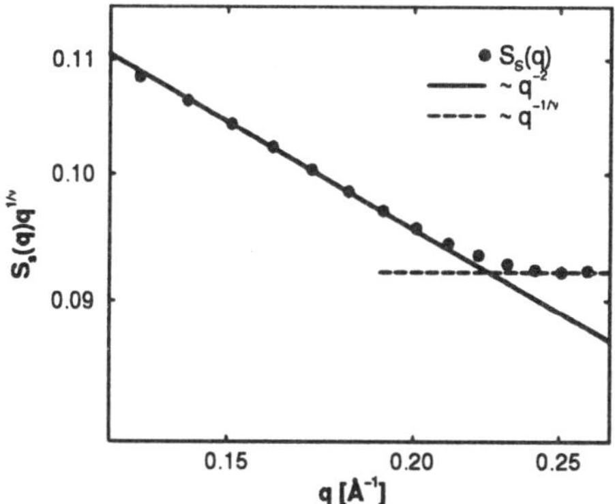

Fig. 1. Single chain structure factor for chains with length $N = 30$ at $\rho = 1.374\mathrm{g/cm}^3$ and $T = 800$ K

This gives a screening length $\xi(\rho = 1.374\mathrm{g/cm}^3) \approx 27.9\,\text{Å}$. Since our average monomer length is about 11 Å, the inter-molecular interaction is screened out at a distance of about 2.5 monomer lengths.

5 Dynamical Properties

The examination of various mean-square displacements leads to an understanding of the dynamical behavior. For this we define two different displacements: the mean-square displacement of the monomers in the center of the chains

$$g_1(t) \equiv \left\langle \left[\mathbf{r}_{N/2}(t) - \mathbf{r}_{N/2}(0) \right]^2 \right\rangle, \qquad (2)$$

and the mean-square displacement of the center of mass of the chains,

$$g_3(t) \equiv \left\langle \left[\mathbf{r}_{\mathrm{CM}}(t) - \mathbf{r}_{\mathrm{CM}}(0) \right]^2 \right\rangle. \qquad (3)$$

The crossover from Rouse behavior to reptation dynamics can be easily shown by the investigation of the diffusion constant D

$$D = \lim_{t \to \infty} \frac{g_3}{6t}. \qquad (4)$$

According to the reptation theory one would expect a crossover from $\sim N^{-1}$ to a $\sim N^{-2}$ regime (M. Doi and S.F. Edwards (1987)). Our data in fig. 2 show a crossover from a Rouse to a reptation regime which is consistent with the reptation theory. During the last years evidence for the reptation model was given by simulations, e.g. K. Kremer and G.S. Grest (1990) and W. Paul (1991). The study of the late time behavior of the mean-square

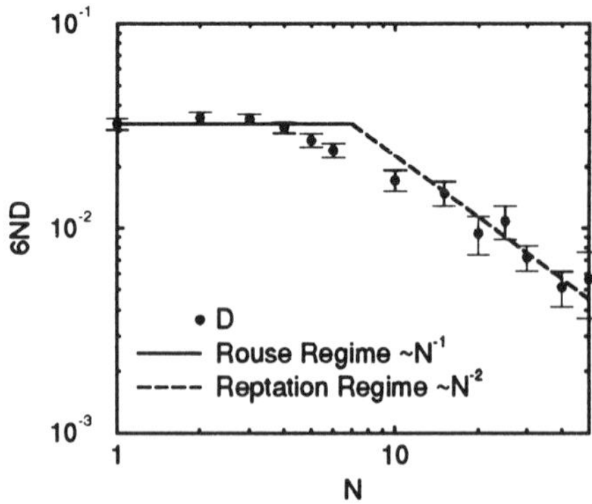

Fig. 2. Diffusion constants at $T = 800$ K and $\rho = 1.374\text{g/cm}^3$

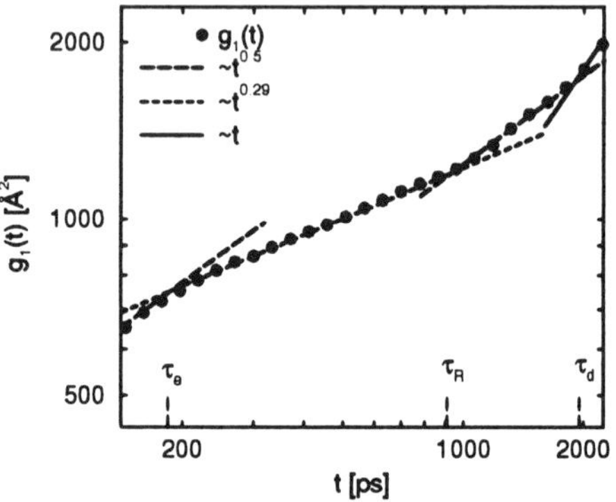

Fig. 3. $g_1(t)$ at $T = 1000$ K, $\rho = 0.687\text{g/cm}^3$, chain length $N = 40$

displacement $g_1(t)$ leads to the several crossovers predicted by the reptation theory, M. Doi and S.F. Edwards (1987). $g_1(t)$ shows Rouse behavior $\sim t^{1/2}$ at times $t < \tau_e = N_e^2/W$. At larger times $t > \tau_e$ the motion is restricted to a movement along a tube surrounding the chain. The exponents predicted for this region are only half of the Rouse values $g_1(t) \sim t^{1/4}$. This prediction is valid only for times $t < \tau_R = N^2/W$ (Rouse time). For later times we expect $g_1(t) \sim t^{1/2}$, while starting at $t = \tau_N = N^3/N_e W$ we have fully relaxed tube constraints and $g_1(t) \sim t$. Fig. 3 shows this predicted behavior of $g_1(t)$ very clearly in all of these four regimes. Only the theoretical $t^{1/4}$ law is not fully reached but displays an exponent of approximately 0.29. It coincides with the result presented in W. Paul (1991) (≈ 0.3). The reason for this deviation from the theoretical value is the influence by crossover effects due to the fact that our chain lengths are only six times the entanglement

222

length. Nevertheless it is interesting that we can see the predicted behavior in all of the four regimes with chains not longer than 40 monomers whereas simulations using the bond fluctuation model need to perform systems with 5 times larger chain lengths (up to $N = 200$). In spite of using long chain segments we still have a continuous chain, thus no bond crossing is possible. So we are able to take advantage of scaling laws and examinate rather long effective chain lengths. Looking at the mean-square displacements of the inner monomers of the chains one is able to determine the tube diameter d_T. Its square is of the order of the value of the mean-square displacement g_1 at the entanglement time τ_e. Our data for chains with chain length $N = 40$ at a mean temperature $T = 1000$ K and density of $\rho = 0.68\text{g/cm}^3$ yields a square tube diameter $d_T^2 \approx 745\text{Å}^2$. Thus we have $d_T \approx 27$ Å.

6 Conclusions

We have presented data obtained by molecular-dynamics simulations of the ellipsoidal model using a parameterization for Bisphenol-A-polycarbonate (BPA-PC). Some static and dynamical properties of BPA-PC were investigated. We estimate a screening length $\xi(\rho = 1.374\text{g/cm}^3)$ of ≈ 27.9 Å. For the entanglement length we get $N_e \approx 77$ Å. Looking at the mean-square displacements of inner monomers we could identify directly the four regimes of different dynamics according to the reptation model, even with rather short chain lengths. With our model it is possible to see effects at moderate chain lengths which can usually only be investigated at very large chain lengths.

7 Acknowledgments

Part of this work was supported by BMFT project 03 N 8008 D7 and the "Graduiertenkolleg Modellierung und Wissenschaftliches Rechnen in Mathematik und Naturwissenschaften" at the Interdisciplinary Center for Scientific Computing (IWR), as well as the EU Project No. CIPA CT 93-0105.

References

M. Bishop, M.H. Kalos, and H.L. Frisch (1979): *J. Chem. Phys.* **70**, 1299

M. Doi and S.F. Edwards (1987): *Theory of Polymer Dynamics* Wiley, New York

P. G. de Gennes (1979): *Scaling Concepts in Polymer Physics* Cornell University Press, London

K. Kremer and G.S. Grest (1990): *J. Chem. Phys.* **92**, 5057

W. Paul, K. Binder, D.W. Heermann, and K. Kremer (1991): *J. Chem. Phys.* **95** 10, 7726 *J. Chem. Phys.* **88**, 1407

K.M. Zimmer and D.W. Heermann (1995): *J. Computer-Aided Materials Design* **2**, 1

K.M. Zimmer, A. Linke, D.W. Heermann, J. Batoulis, and Th. Bürger (1996): *Makromol. Theory Simul.* **5**, 1065

Microcanonical Transfer Matrix and Yang-Lee Zeros of the Q-State Potts Model

R.J. Creswick and S.-Y. Kim

Department of Physics and Astronomy, University of South California, Columbia, SC 29208, USA

1 Introduction

The Q-state Potts model in two dimensions is very fertile ground for the investigation of phase transitions and critical phenomena. For $Q = 2$ and 3 there is a second order phase transition between Q ferromagnetic ordered states and a disordered state. For $Q = 4$ the transition is also second order, but the usual critical behavior is modified by strong logarithmic corrections. For $Q > 4$ the transition is first order, with $Q = 5$ exhibiting *weak* first order behavior and a very large correlation length at the critical point.

The Hamiltonian for the Potts model is

$$\mathcal{H} = J \sum_{<i,j>} (1 - \delta(q_i, q_j)), \tag{1}$$

where J is the coupling constant and $0 \leq q_i \leq Q - 1$ are the Potts spin variables. The energies (in units of J) are evenly spaced and take on interger values in the range $0 \leq E \leq N_b$ where N_b is the number of bonds on the lattice. Here we will consider simple square lattices with periodic, cylindrical, and self-dual boundary conditions.

In addition to the energy there are Q order parameters

$$M_q = \sum_k \delta(q_k, q), \tag{2}$$

which for the Ising model is simply related to the magnetization. The possible values of the order parameter are also intergers, $0 \leq M \leq N_s$, where N_s is the number of sites on the lattice.

If we denote the number of states with energy E by $\Omega(E)$, then the canonical partition function for the Q-state Potts model is

$$Z_Q(y) = \sum_E \Omega_Q(E) y^E, \tag{3}$$

where $y = e^{-\beta J}$. From (3) it is clear that Z is simply a polynomial in y, and the analytic structure of Z is completely determined by the zeros of this polynomial, as first discussed by Lee and Yang[2].

If we wish to study the partition function in an external field which couples to the order paramter, (2), then one needs to enumerate the number

Springer Proceedings in Physics, Volume 83
Computer Simulation Studies in Condensed-Matter Physics X
Eds.: D. P. Landau, K.K. Mon, H. -B. Schüttler
© Springer-Verlag Berlin Heidelberg 1998

of states with fixed energy E and fixed order parameter M, $\Omega(E, M)$. The partition function is again a polynomial given by

$$Z_Q(y, x) = \sum_E \sum_M \Omega_Q(E, M) x^M y^E, \qquad (4)$$

where $x = e^{-\beta h}$, and h is the external field. As discussed in the second[2] of Lee and Yang's two famous papers, the zeros of the partition function for the Ising model in the complex-x plane all lie on the unit circle.

For finite systems the analyticity of Z in both x and y ensures that no zeros lie on the real axis. However, according to Lee and Yang, in the thermodynamic limit the zeros of the partition function in either the complex-x or $-y$ planes approach arbitrarily close to the real axis at the critical point, leading to nonanalytic behavior in the partition function.

If the zeros lie on a one-dimensional locus in the thermodynamic limit one can define the density of zeros (per site) $g(\theta)$[3, 4] in terms of which the free energy per site is

$$f(y) = -\int g(\theta) log[y - y_0(\theta)]d\theta. \qquad (5)$$

In the critical region the singular part of the free energy is a homogeneous function of the reduced temperature, $y - y_c$, from which it follows[5] that $g(\theta)$ must also be a homogeneous function for small θ of the form

$$g(\theta) = b^{(-d + y_T)} g(\theta b^{y_T}). \qquad (6)$$

This in turn implies that $g(\theta)$ vanishes as θ^κ as θ goes to zero where $\kappa = (d - y_T)/y_T$. On the other hand, if the system has a first order transition, $\kappa = 0$ and the discontinuity in the first derivative of f, the latent heat, is given by $L = 2\pi g(0)$. Exactly parallel arguments hold in the complex-x plane if one replaces the temperature exponent y_T by the magnetic exponent y_h.

In a recent paper Creswick[6] has shown how the numerical transfer matrix of Binder[7] can be generalized to allow the evaluation of the density of states $\Omega(E)$ and the restricted density of states, $\Omega(E, M)$ for the Q-state Potts model. Similar calculations of $\Omega(E)$ have been carried out by Bhanot[8] in both two and three dimensions for the $Q = 2$ and $Q = 3$ Potts models, and Pearson[9] for the 4^3 Ising model. Beale[10] has used the exact solution for the partition function of the Ising model on finite square lattices to calculate $\Omega(E)$. Bhanot's method is far more complex than the μTM and requires essentially the same computer resources. Pearson's method is only applicable to lattices with very few spins (e.g. 64) and Beale's approach makes essential use of the exact solution for the Ising model and so can not be used for other values of Q or in three dimensions. The μTM is quite general and the algorithm itself requires less than 100 lines of code. In addition, it is straightforward to generalize the μTM to count states with fixed energy *and magnetization*, or any other function of the Potts variables.

2 Results

The partition function for the Potts model maps into itself under the dual transformation

$$u \to \frac{1}{u}, \tag{7}$$

where

$$u = \frac{y^{-1} - 1}{\sqrt{Q}}. \tag{8}$$

In the complex u-plane a subset of the zeros of the partition function tend to lie on a unit circle (which maps into itself under (7)); however, cylindrical and periodic boundary conditions are not self-dual, and this causes the zeros to move slightly off the unit circle. For this reason we have modified the μTM for the self-dual lattice introduced by Wu et $al.$[11], so that the zeros do indeed lie on the unit circle and therefore are simply parameterized by the phase θ.

In the complex x-plane the zeros of the partition function for the Ising model are guaranteed to lie on the unit circle by the circle theorem of Lee and Yang[2], irrespective of the boundary conditions.

Given that the zeros are well parameterized by a single variable, we can define the density of zeros for finite lattices as

$$g(\frac{1}{2}(\theta_{k+1} - \theta_k)) = \frac{1}{N} \frac{1}{\theta_{k+1} - \theta_k}. \tag{9}$$

In Fig.1 we show the density of zeros in the complex-x plane for the Ising model at $y = y_c$ and $y = 0.5y_c$. Note that at the critical temperature g tends to zero as one approaches the real axis, but below the critical temperature it approaches a constant

$$2\pi g(0, y) = m_0(y), \tag{10}$$

where $m_0(y)$ is the spontaneous magnetization. We have applied finite-size scaling to the density of zeros calculated in this way and find excellent agreement with the exact solution for the magnetization except close to the critical point where crossover complicates the FSS analysis. There is reason to hope that a more sophisticated FSS analysis will improve these results substantially.

Finally, in Fig.2 we show the density of zeros in the complex y-plane for the 3-state Potts model, which is known to have a second-order transition at the critical point.

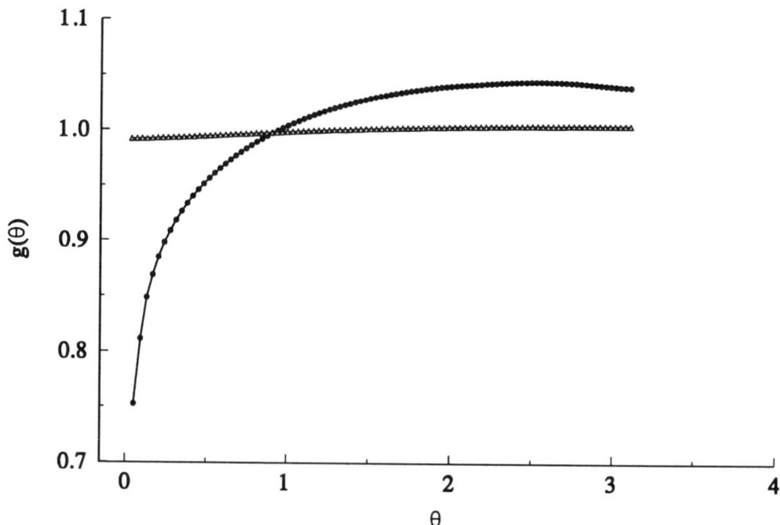

FIGURE 1. Density of zeros in the complex x-plane for the Ising model for $L = 14$, cylindrical boundary conditions, for $y = y_c$ (filled circles) and $y = 0.5y_c$ (open triangles).

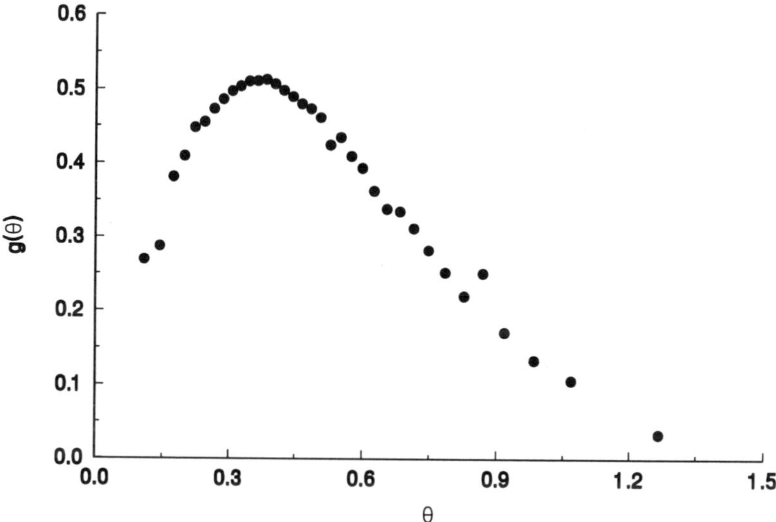

FIGURE 2. Density of zeros in the complex y-plane for the 3-state Potts model for $L = 10$, self-dual boundary conditions.

3 Conclusions

The μTM and its extensions offer a new way of obtaining exact information about finite two dimensional lattices. While the method is easily extended to three dimensions, memory requirements limit its use to the 2-state model on $4^2 \times L$ lattices. However, Monte Carlo techniques have been developed[12] which have no such limitations and show great promise in extending many of these results to larger lattices. Preliminary studies indicate that while most of the Yang-Lee zeros are very sensitive to the exact value of $\Omega(E)$, the edge singularity and the next two nearest the critical point are not.

In addition, it is possible to generalize the μTM to calculate the microcanonical distribution of any function of the Potts variables, and in particular the order parameter and correlation function.

4 References

[1] C. N. Yang and T. D. Lee, Phys. Rev. **87**, 404 (1952).

[2] T. D. Lee and C. N. Yang, Phys. Rev. **87**, 410 (1952).

[3] R. Abe, Prog. Theor. Phys. **38**, 72 (1967).

[4] M. Suzuki, Prog. Theor. Phys. **38**, 1225 (1967).

[5] R. J. Creswick and S.-Y. Kim, to be published.

[6] R. J. Creswick, Phys. Rev. E **52**, R5735 (1995).

[7] K. Binder, Physica **62**, 508 (1972).

[8] G. Bhanot, J. Stat. Phys. **60**, 55 (1990).

[9] R. B. Pearson, Phys. Rev. B **26**, 6285 (1982).

[10] P. D. Beale, Phys. Rev. Lett. **76**, 78 (1996).

[11] C.-N. Chen, C.-K. Hu, and F. Y. Wu, Phys. Rev. Lett. **76**, 169 (1996).

[12] G. Bhanot, R. Salvador, S. Black, P. Carter, and R. Toral, Phys. Rev. Lett. **59**, 803 (1987); K.-C. Lee, J. Phys. A **28** 4835 (1995); C. M. Care, *ibid.* **29** L505 (1996); R. J. Creswick and A. Pavel'yev, to be published.

Roughening Transitions in HCP Lattices

J. Adler, A. Hashibon, and S.G. Lipson

Department of Physics, Technion-IIT, 32000, Haifa, Israel

Abstract. Direct observation of three equilibrium roughening transitions (at about one degree Kelvin) for HCP He^4 was possible because equilibrium between the solid and superfluid phases of helium can be reached on the scale of seconds. While this provides an exciting opportunity for comparison between experimental data and models for the roughening transition, the non-Bravais nature of the HCP lattice severely complicates direct simulations. Using computer visualization to simplify the HCP sample preparation, and parallel algorithms to speed up the calculations, we have been able to evaluate roughening temperatures for three facets and directly estimate the surface tension for many temperatures. Substantial agreement with the experimental data was found.

1. Introduction

The Roughening Transition(RT) is a phase transition which corresponds to a morphological change in the interface between a crystal and its fluid or vapor in thermal equilibrium. There is a singularity in the interfacial tension at the RT, and an intrinsic width d_w can be associated with the interface as the distance over which the local physical properties change from those of phase A to phase B. Thermal excitations cause local interface fluctuations, so the interface may wander over a characteristic length w. If $w \sim d_w$ the interface is smooth but if $w \to \infty$ in the strict thermodynamic limit, the interface is rough.

The concept of the RT arose in the context of an abstract crystal growth model put forward [1] by Burton, Cabrera and Frank (BCF) in 1951. They made an analogy between a crystal/fluid interface and up and down spins. The interface in a $d = 3$ Ising model between two regions of spins mostly up and mostly down is $\sim d = 2$. In the BCF model the growing layer is described by a $d = 2$ Ising model with $T_R \sim T_c(\mathrm{d}=2) \sim \frac{1}{2}T_c(\mathrm{d}=3)$. The basic model of BCF, though naive, gives surprisingly good numerical estimates of T_R, and in fact can be shown to provide a rigorous lower bound to T_R. But the RT and the phase transition of the $d = 2$ Ising model are, of course, of a completely different nature and therefore a more appropriate model for the study of a crystal in equilibrium with its fluid or vapor is the $d = 3$ Ising model. Also appropriate are solid-on-solid (SOS) models which are [2] within the universality class of the Kosterlitz-Thouless (KT) transition, resemble Ising interfaces except that no

Springer Proceedings in Physics, Volume 83
Computer Simulation Studies in Condensed-Matter Physics X
Eds.: D. P. Landau, K.K. Mon, H. -B. Schüttler
© Springer-Verlag Berlin Heidelberg 1998

holes or overhangs are present and give similar critical results for cubic systems [3].

About 17 years ago the first equilibrium roughening transition on ^{4}He crystals was discovered at the Technion [4]. Helium is especially suited to experimental work on RTs because equilibrium between the solid and superfluid at the interface is achieved on the timescale of seconds for crystals that are large enough to be observed with suitable optical techniques. To date three roughening transitions in HCP He^4 have been seen: The first two facets appear [4,5] in the (0001) or $\mathbf{c}$ direction at $T_R(\mathbf{c}) = 1.28K$. The six facets in the directions equivalent to $(1\bar{1}00)$ or $\mathbf{a}$, start emerging [6] at $T_R(\mathbf{a}) \sim 1.0K$. Finally, twelve $(1\bar{1}01)$ or $\mathbf{s}$-facets begin to appear [7] at about $T_R(\mathbf{s}) \sim 0.35K$.

Despite the fact that the best RT data was measured on an HCP crystal most analytic and numerical work continued to be carried out on cubic crystals without higher neighbours. An exception was the study of Touzani and Wortis (TW), who developed [8] exact and mean-field results for HCP models of the BCF type. When comparing ratios of their T_R values to the experimental ratios, problems are observed with the T_R of the $\mathbf{s}$ facet. While the ratio of T_R for the $\mathbf{c}$ and $\mathbf{a}$ facets is within less than 20% of the measured ratio, ratios including the $\mathbf{s}$ facet are more than 50% off. Given the errors in the $\mathbf{a}$ and $\mathbf{s}$ measurements 20% is not too bad but 50% is obviously excessive. Several reasons for this discrepancy were proposed, including (i) lack of higher neighbour interactions and (ii) lack of quantum effects by TW, (iii) problems with the BCF type approximation, (iv) problems with the basic idea of lattice models for RTs, or (v) experimental problems associated with equilibration in [7].

We began a comprehensive program to develop models that would describe the experimental data, and report here on simulations of roughening transitions in an HCP crystal with both nearest(NN) and next-nearest neighbour(NNN) interactions, of strengths J_{NN} and J_{NNN} respectively. A discussion of earlier calculations for the case of positive NNN interactions, as well as quantum mechanical estimates showing that zero point motion may lead to negative NNN interactions was given in [9]. While we were carrying out the negative NNN HCP study, new (body-centred cubic) BCC He^3 experimental results [11] were published. Simulations [10] for BCC He^3 with NNN interactions in a BCF type of approximation were made and the results were in good agreement with the new measurements, greatly strengthening the idea that lattice models and even BCF-type approximations are reasonable and that (i), (ii) or (v) must be the cause of the discrepancy.

2. Calculations

We began by developing a SOS model that maps to a 12-vertex model for the c facet. Unfortunately this model does not appear to have an exact solution. In general, however a full three-dimensional Ising spin system was found to be easier to simulate than SOS models for the HCP system, since to change facet we merely needed to change the boundary conditions rather than the model details. To force an interface in a given direction in the three-dimensional Ising model, anti periodic boundary conditions (APBC), or fixed boundary conditions (FBC) in the orientation specified may be used. We sectioned the HCP lattice in the c, a, and s directions. In order to ascertain that the crystals had been correctly cut visualizations of the samples were made. The type of boundary condition errors that occasionally plague devlopment of simulations for cubic systems with higher-neighbour interactions are far more likely to occur in a non-Bravais lattice and direct visualliztion proved to be helpful in eliminating these. To facilitate this, and other projects of the Computational Physics group at the Technion, a system of visualization for crystal structures in OpenGL/mesa was developed by A. Hashibon and D. Saada and is described in [12]. A picture of the He^4 crystal drawn with this system was given in [12].

In our earliest studies [9] we had naively assumed that higher-neighbour interactions would be positive. The results of this assumption (which led to the prediction of additional and fascinating new facets which have never been observed) did not improve the agreement with experiment. Consideration of zero-point energy [9] led to the idea that the NNN interactions may be negative. For negative interactions no new facets appear that are not present in the NN case, and agreement with experimental ratio for s/c improves substantially. Sample sizes of $L \times L \times 10$ and $L \times L \times 20$ for $L = 5, 10, 20$, and 40 were simulated with FBC. The T_R estimates were made by assuming the KT theory and fitting w^2 near the critical temperature. Some interesting results are given in Table 1; we have results for many more J_{NNN} values and find a linear realtionship between J_{NNN} and T_R. A detailed account of these simulations will be given in [13]. Since systems with FBC usually suffer from boundary effects and systems with APBC give the same physics with less boundary effects, we also studied larger samples ($L \leq 160$) of the c facet for longer times on the SP2 power parallel machine and achieved comparable results.

Table 1 Experimental and calculated ratios of values of T_R, (the last column is an extrapolation).

Ratio	Expt	TW	Our Results $J_{NNN} =$			
			0.0	0.234	−0.4	−0.8
a/c	0.78	0.93	0.96	0.97	0.96	0.95
s/c	0.29	0.77	0.80	0.90	0.53	0.29

We have also calculated the interface tension for many temperatures. At zero temperature, the interface tension is the excess energy per unit area; and is easily calculated for crystal lattices for any $\hat{n}$ by enumeration of broken bonds across an interface normal to the surface $\hat{n}$; a visualization of this is given in [9]. At finite T the free energy (F_s) of the interface must be calculated. The excess interfacial energy is the energy difference between two systems with and without an interface. Two systems (one with PBC and one with APBC) were simulated for several sizes between $L = 4, 6, 8, 10, 16, 20, 30$ and 60: in the **c**, **a**, and **s** facet directions. The surface excess energy was calculated in steps of $0.05\frac{J_{NN}}{K_B}$. The finite size scaling is consistent with $F_s \simeq \text{Const} + \sigma L^2$. The results of the Monte-Carlo calculation of the interface tension may be compared directly to the experimental measurements if the units of the temperature are transformed between the model and the experiments. As a by-product we measured the magnetization for both systems and estimated the critical temperature of the $d = 3$ HCP Ising model to be $J_{NN}/K_B T_c = 0.10$.

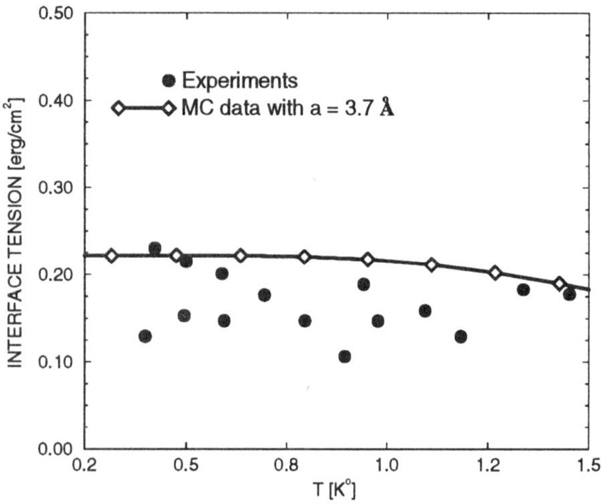

Figure 1. Comparison between experiments[4] and simulations of the interface tension for the c-facet with NN interaction only. The size of the system is 10x10x10, and the integration starts at $T = 0K$.

3. Conclusions

We have applied techniques well established on cubic lattice models for calculating the equilibrium roughening temperature, and the temperature dependence of the interface tension on a real non-Bravais lattice, with both NN and NNN interactions. From Table 1. we see that for $J_{NNN}/J_{NN} < 0.4$ reasonable agreement for the c/s ratio is obtained. Reviewing the list of reasons for the problems with the TW study we see that we have eliminated (i) and (ii); referring to Table 1 we can also see that the BCF approximation was not of itself the problem. We are not certain that J_{NNN}/J_{NN} as high as 0.8 is entirely reasonable, thus we cannot say that there are no problems of equilibration in the experiment, however large effects of this type are no longer required. Given the overall success for both the surface tension and T_R measurements we suggest that any corrections obtained by removing the lattice constraint (iv) will be a higher order effect. In summary, we have shown that careful work on a conceptually simple Ising system with complicated but realistic geomtry gives results in good agreement with experiment.

Acknowledgements: Support of the Israel Science Foundation was essential for both the experimental and simulational aspects of this project. We acknowledge the Technion Visualization Centre (and advice from B. Peri), Minerva Non-Linear Centre and the IUCC HPCC for support for the visualization and simulation. We thank D. Stauffer for advice on aspects of the simulations, and G. Baum and A. Gemintern for their earlier collaboration on this project.

References

1. W. K. Burton, N. Cabrera and F. C. Frank, Phil. Trans. R. Soc. **243**, 299 (1951).
2. S. T. Chui and J. D. Weeks, Phys. Rev. Lett. **40**, 733 (1978).
3. M. Hasenbuch, M. Marcu, K. Pinn, Physica A **208**, 208 (1994).
4. J. Landau, S. G. Lipson, L. M. Maatanen, L. S. balfour and D. O. Edwards, Phy. Rev. Lett. **45**, 31 (1980).
5. K. O. Keshishev, A. Ya. Parshin and A. V. Babkin, Sov. Phys. JETP **53**, 362 (1981).
6. J. A. Avron, L. S. Balfour, C. G. Kuper, J. Landau, S. G. Lipson and L. S. Schulman, Phys. Rev. Lett **45**, 814 (1980).
7. P. E. Wolf, S. Balibar and F. Gallet, Phy. Rev. Lett **51**, 1366 (1983).
8. M. Touzani and M. Wortis, Phys. Rev. B **36**, 3598, (1987).
9. S. G. Lipson, J. Adler, G. Baum, A. Gemintern and A. Hashibon, JLTP **101** 683 (1995).
10. A. Gemintern, S. G. Lipson and J. Adler, Phys. Rev. B, to appear.
11. R. Wagner, S. C. Steel, O. A. Andreeva, R. Jochemsen and G. Frosatti, Phys. Rev. Lett. **76**, 263 (1996).
12. J. Adler, Proceedings of the Scientific Visualization Conference, Jerusalem, 59 (1995).
13. A. Hashibon, J. Adler and S. G. Lipson, in preparation.

Spin Patterns in the Three-Dimensional Chiral Clock Model

P.D. Scholten[1] *and D.R. King*[2]

[1]Department of Physics, Miami University, Oxford, OH 45056, USA
[2]Abbott Laboratories, Abbott Park, IL 60064, USA

Abstract. A Monte Carlo study of the three-dimensional 4-state chiral clock model was performed. For values of the chiral parameter Δ close to 0.5 the Monte Carlo results were consistent with those obtained by low temperature series expansion. However, for Δ near 0.2 the computer simulations revealed the existence of a new type of chiral phase.

1. Introduction

Phase transitions in systems with spatially modulated phases have been studied theoretically for more than a decade. Although much of the theoretical analysis has been done with the axial next-nearest-neighbor Ising (ANNNI) model [1], the q-state chiral clock model provides another approach [2-10]. Here the standard clock model Hamiltonian is modified by introducing an anisotropic coupling, Δ, between spins of neighboring planes, i.e.

$$ H = -J \sum^{\perp} \cos\left[\frac{2\pi}{q}(m-n)\right] - J \sum^{z} \cos\left[\frac{2\pi}{q}(m-n+\Delta)\right] $$

where the first sum is over nearest neighbors in the x-y plane and the second sum is over nearest neighbors in the z direction. The integers m and n refer to the states (from 1 through q) of the neighboring spins. The chiral nature of this model comes from the fact that for certain choices of Δ the predominant direction of the spin in successive layers goes through cyclic patterns.

Yeomans studied the 3-dimensional chiral clock model using a low temperature series expansion technique, and produced, among other cases, the phase diagram for q=4.[11,12] For low temperatures and $\Delta<0.50$ the system will be in a ferromagnetic state. As the temperature increases, Yeomans predicted that the system will undergo a first-order phase transition and pass into the <4> state which is characterized by the following progression of spin states of the planes: 11112222333344441111122222... . At higher temperatures other, more complicated phases appear, e.g. the <3²2> phase which has the spin state pattern 11122233444111122333... . For Δ near 0.50 the <2> phase is stable over a relatively large portion of the phase diagram.

It was the purpose of our work to use Monte Carlo simulation to study both the three-dimensional 4-state chiral clock model over wider ranges of Δ and temperature than had previously been considered. Considering the complexity of the phase

Springer Proceedings in Physics, Volume 83
Computer Simulation Studies in Condensed-Matter Physics X
Eds.: D. P. Landau, K.K. Mon, H. -B. Schüttler

diagram published by Yeomans, the initial goal was to map the outer phase boundaries, i.e. those between the FM and <4> phases, between the PM phase and any ordered phase, and perhaps between the <2> phase and the more complicated phases. Given the relatively small sizes of the systems that could be simulated, there was no real possibility of resolving all the phase boundaries.

2. Monte Carlo Simulations

Simulations were done for simple cubic lattices consisting of D LxL planes stacked along the z-axis. Periodic boundary conditions were used in the x- and y-directions, whereas free boundaries were used in the z-direction to reduce frustration. The Metropolis algorithm was used to update the spin states. To obtain a better picture of the system's behavior during a simulation run instantaneous values ("snapshots") of system energy, order parameter and planar spin pattern along the z-axis were recorded every several hundred Monte Carlo steps per spin (MCSS). The spin patterns were determined by finding the dominant spin state of each plane and recording the resulting sequence of numbers. In order that several periods of any pattern could be observed D was set equal to either 2L or 4L. The initial search for the FM phase boundary in the q=4 model began with $\Delta=0.2$ and a system size of L=20, D=40. Each simulation consisted of between 3×10^5 and 9×10^5 MCSS. An example of the plane spin patterns at kT/J=1.95 is shown in Fig. 1. Each single number is the dominant spin state of a plane, and each row of 40 numbers represents an instantaneous state of the system. It is evident that one or more interfaces may form. Even as the locations of the interfaces fluctuate during the simulation, the proper chiral progression is maintained. The corresponding distribution function of the system order parameter (magnetization) M is shown in Fig. 2; this distribution includes data for every MCSS. Using the instantaneous order parameter information we identify the peak at M=0.72 with the ferromagnetic state (no interfaces) and the peak at M=0.50 with a state of one interface. The broader peak near M=0.27 corresponds to the system having more than one interface.

To study the transition from the FM to this chiral or interface phase in more detail, runs of 10^7 MCSS were done on systems of L=10, D=20. These smaller systems were used to ensure better statistics for the different system states. The adequacy of these run lengths was confirmed by examining the "snapshots" of the system order parameter to see that the system made numerous switches between the FM and chiral states during a simulation run.

For each of several different values of Δ complete two-dimensional histograms of energy and order parameter, P(E,M), were recorded at temperatures within a few hundredths of a kT/J of the FM transition point. These data were then used to compute the order parameter distributions P(M), the heat capacities, and the susceptibilities at nearby temperatures using the histogram methods described by Ferrenberg and Swendsen.[13] Each order parameter distribution was, in turn, used to create a free energy distribution using $F-F_0 = -kT \ln(P(M)\, dM)$, where P(M) dM is the probability that the order parameter will lie in the range M to M+dM.[14] The

MCSS #	Planar Spin Pattern	M
500	444444444444444444444444444411111111111	0.5585
1000	444444444444444444444444444441111111111111	0.5243
1500	444444444444444444444444444441111111111111	0.5236
2000	444444444444444444444444444441111111111111	0.5261
2500	444444444444444444444444444444111111111122	0.5069
3000	444444444444444444444444444441111111111112	0.5277
3500	444444444444444444444444441111111111111111	0.4907
48000	4444444111111111111111111111111111111111	0.5999
48500	4444441111111111111111111111111111111111	0.6373
49000	4444441111111111111111111111111111111111	0.6375
49500	1111111111111111111111111111111111111111	0.7212
50000	1111111111111111111111111111111111111111	0.7198
106500	1111111111111111111111111111111111111112	0.6976
107000	1111111111111111111111111111111111111111	0.7291
107500	1111111111111111111111111111111111111111	0.7131
108000	4444411111111111111111111111111111111111	0.6452
108500	4444441111111111111111111111111111111111	0.6222
109000	4444444111111111111111111111111111111111	0.6039
127500	444444444444444444444411111111111111111111	0.4918
128000	444444444444444444444441111111111111111111	0.5164
128500	444444444444444444444444441111111111111111	0.5244
129000	444444444444444444444444441111111111111111	0.5322
129500	3333444444444444444444441111111111111111112	0.4117
130000	333344444444444444444444411111111111111111	0.4233
130500	333333344444444444444444411111111111111111	0.3627
131000	333333333444444444444444411111111111111111	0.2955
131500	333333333334444444444444444111111111111111	0.2739
132000	233333333333334444444444444444411111111111	0.2661

1. Planar spin patterns for a system with q=4 and Δ=0.20 at kT/J=1.95. Each number in a pattern corresponds to the dominant state of a plane, and each row shows the instantaneous state of the 40-plane system. Also shown for each pattern are the Monte Carlo step number and system order parameter (M).

temperature at which the free energy minima associated with the ferromagnetic and single interface states were of equal depth was taken to be one measure of the transition temperature.

The temperatures of the maxima in the susceptibility and heat capacity were also used to estimate the transition temperature. Although differences in the estimates

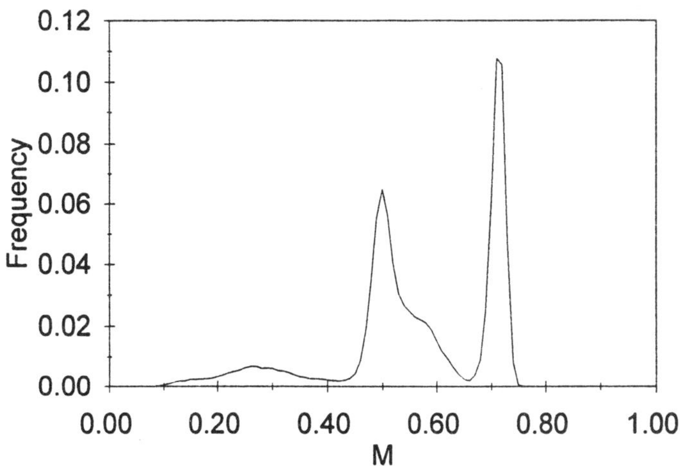

2. System order parameter (magnetization) distribution for q=4 and Δ=0.20 at kT/J=1.95. The system size was L=20, D=40.

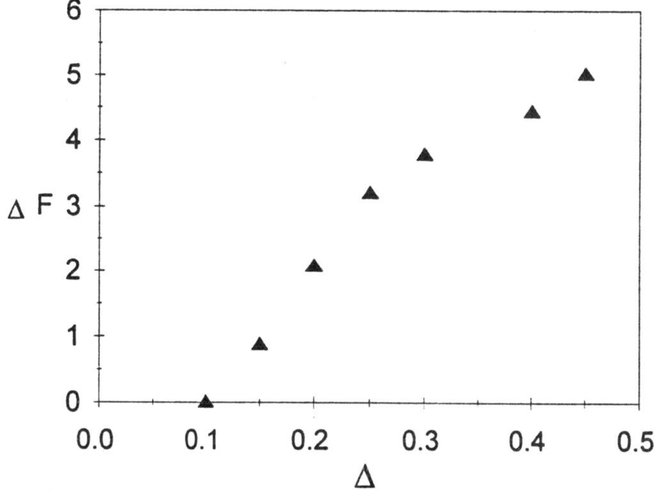

3. Depth of the free energy minima as a function of Δ.

were to be expected due to finite size effects, the transitions temperatures obtained by the three methods agreed to within 1% for given values of q and Δ.

The multi-peaked structure of the system order parameter distribution functions suggests that the transition from the ferromagnetic state to the chiral state is of first-order. This was examined in detail using the free energy analysis of Lee and Kosterlitz.[15] They showed that if the height of the free energy barrier ΔF between the two phases at kT_c/J increases with the system size, the transition is first-order. Although the system sizes were small, the data clearly showed the transition to be first-order. If ΔF is plotted as a function of Δ (Fig. 3), we see that the free energy barrier goes to zero near $\Delta = 0.10$. Lower values of Δ are too small to produce well-defined chirality, and the system as a whole is believed to undergo a second-order transition from the ferromagnetic state to the paramagnetic state. This would be consistent with the behavior of the simple three-dimensional clock model.[16]

Uncertainties in the values of the transition temperatures and free energy barrier heights were estimated by performing additional MC simulations for L=10 with Δ=0.30, 0.40, and Δ=0.45. The transition temperatures for duplicate runs agreed to within ±0.01 kT/J. The depths of the free energy minima, ΔF, were also in good agreement for the first two cases (ΔF=3.80 and 3.72 for Δ=0.30 and ΔF=4.46 and 4.44 for Δ=0.40). For Δ=0.45 the ΔF values were 5.04 and 4.54.

The boundary between the chiral and paramagnetic regions was explored for Δ=0.20 and 0.45. These critical temperatures were largely independent of Δ, and in fact were all found to be within the range 2.12 to 2.20 for system sizes ranging from L=10, D=20 to L=20, D=40.

3. Comparison with Earlier Work

It is interesting to compare the picture of the behavior of the 4-state chiral clock model provided by the Monte Carlo data to that of Yeomans' theory. Consider the case of Δ=0.2. Fig. 4 shows how the average the number of planes between interfaces, changes with temperature. We will refer to this number as the interface spacing. These spacings were computed from the snapshot data, such as those shown

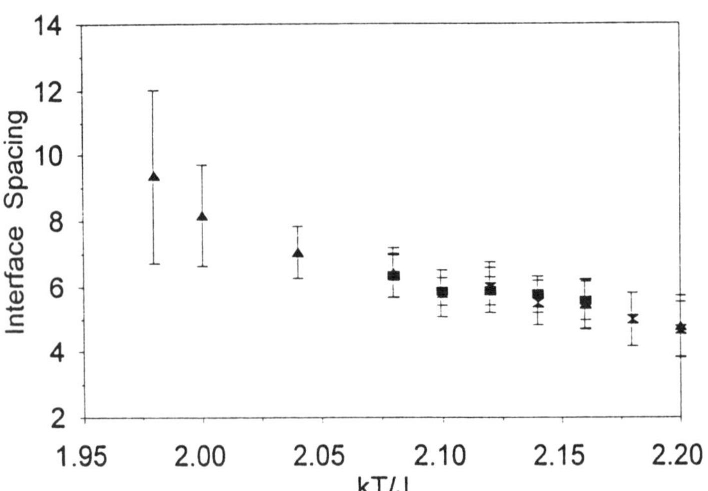

4. Average interface spacing as a function of temperature for q=4 and Δ=0.20. Data are shown for a system with D=80 initialized in the FM state (■), a system with D=40 initialized in the FM state (▲), and a system with D=40 initialized in the <4> state (✖). In each case L=20.

in Fig. 1. The patterns on both ends of the data string were not included in the analysis as their lengths were affected by the free boundaries. The results shown in Figs. 4 and 5 are from simulations with approximately 5×10^4 MCSS and 200 MCSS between recorded snapshots. Using this smaller number of MCSS was justified because at temperatures above the FM transition the fluctuations among states were much more frequent.

In Fig. 4 we see that, as expected, the interface spacing decreases with increasing temperature. But most significantly, for both systems with D=40 and D=80 the average interface spacing for the q=4 model does not reach 4 until the system is very close to the paramagnetic transition. Furthermore, the results were independent of whether the system was initialized in either the FM state or the <4> state. This shows that for $\Delta = 0.2$ the system does not exhibit a stable <4> phase, or if the phase does exist, it is confined to a very narrow temperature range.

As Δ increases, the average interface spacing decreases for all temperatures in the chiral phase. Tests showed that for $\Delta=0.45$ the spacings are all less than 2.7 for kT/J > 1.40, and in fact, the average interface spacings for all the previously predicted phases would fit within the bounds of the Monte Carlo results. Therefore, although these simulations cannot resolve the predicted phases, our results are not inconsistent with those of Yeomans.

Assuming that all of the predicted phases do exist, there is still the matter of the long spin patterns (what we have called the chiral phase) observed in the Monte Carlo results as the system leaves the FM phase. We do not believe these patterns to be just a finite size effect, for they were strongly evident in all q=4 systems studied including those with D=80. It should be noted, however, that the existence of this phase does not contradict the series expansion results as those were obtained for low temperatures.

4. Conclusion

Monte Carlo investigation of the 3-dimensional 4-state chiral clock model reveals a transition from the FM phase to a chiral phase not previously predicted. This phase is observed for Δ well below 0.50 and is characterized by large numbers of planes between interfaces. For Δ near 0.50 the Monte Carlo results are consistent with those obtained by low temperature series expansion.

Acknowledgements

The authors wish to thank Prof. J. Yeomans for helpful comments during the course of this study. The project was supported in part by Ohio Supercomputer Center Grant No. POS030-3.

References

1. For a review see Walter Selke, Phys. Rept. $\underline{170}$, 213 (1988).
2. S. Ostlund, Phys. Rev. B $\underline{24}$, 398 (1981).
3. Walter Selke and Julia M. Yeomans, Z. Phys. B $\underline{46}$, 311 (1982).
4. G. von Gehlen and V. Rittenberg, Nucl. Phys. B $\underline{230}$, 455 (1984).
5. David A. Huse, Anthony M. Szpilka, and Michael E. Fisher, Physica $\underline{121A}$, 363 (1983).
6. Phillip M. Duxbury, Julia Yeomans, and Paul D. Beale, J. Phys. A $\underline{17}$, L179 (1984).
7. David A. Huse and Michael E. Fisher, Phys. Rev. Lett. $\underline{49}$, 793 (1982).
8. F. D. M. Haldane and P. Bak and T. Bohr, Phys. Rev. B $\underline{28}$, 2743 (1983).
9. H. J. Schultz, Phys. Rev. B $\underline{28}$, 2746 (1983).
10. W. Scott McCullough, Phys. Rev. B $\underline{46}$, 5084 (1992).
11. Julia M. Yeomans, J. Phys. C $\underline{15}$, 7305 (1982).
12. Julia Yeomans, Physica $\underline{127B}$, 187 (1984).
13. Alan M. Ferrenberg and Robert H. Swendsen, Phys. Rev. Lett. $\underline{61}$, 2635 (1988).
14. James Glosli and Michael Plischke, Can. J. Phys. $\underline{61}$, 1515 (1983).
15. Jooyoung Lee and J. M. Kosterlitz, Phys. Rev. Lett. $\underline{65}$, 137 (1990).
16. P. D. Scholten and L. J. Irakliotis, Phys. Rev. B $\underline{48}$, 1291 (1993).

The Critical Region in Finite-Sized Systems

G.A. Baker, Jr.

Theoretical Division, Los Alamos National Laboratory,
University of California, Los Alamos, NM 87545, USA

Abstract The critical point in an infinite system is smeared out or "rounded" in a finite system into a "critical region." This region has some structure associated with it, which is difficult to study except by means of simulations, *i.e.*, Monte Carlo. I have begun a study of this structure in small two-dimensional Ising model systems, using the Markov property method.

There is still an area of critical phenomena that has not been nearly as thoroughly explored as it might be. The regions where the temperature is greater than the critical temperature and where the temperature is less than critical temperature in infinite size systems have by now been quite thoroughly studied. However, when the system is finite in size, we know that the thermodynamic quantities which would have been divergent in an infinite system are rounded over a critical region instead. There is some interesting structure in this region described by finite-size scaling functions. Other problems which are of the same character have recently come under investigation by Borgs *et al.*, Caracciolo *et al.*, and Kim *et al.* [1]. Some methods of scaling are much less sensitive to the system size than others. Put otherwise, we see better data collapse (independence of the system size) using some sets of variables rather than other sets which are asymptotically the same. The elucidation of which versions to use is of considerable practical importance, as one can get by with much smaller system sizes than would otherwise be required.

In this paper I will begin such a study, primarily for the two-dimensional Ising model on small system sizes. The method of computation of these results is the Markov property method [2]. I have results at various temperatures for the 2, 4, 6, 8 and 10 edged squares on a plane square lattice with periodic boundary conditions. In their study of the question of hyperscaling in the three dimensional Ising model, Baker and Kawashima [3], and Baker and Erpenbeck, and Kim [4] before them. noticed that the variable ξ_L/L where ξ_L is the correlation length computed from the system of size L is very useful in this regard. That quantity is zero in the limit $L \to \infty$ for temperatures above the critical point and infinite for temperatures below the critical point. Thus the infinite system limit where this quantity has any finite value is just the critical point, and so looking at the behavior of the system for fixed ξ_L/L in effect spreads out the critical point. We define the correlation length by,

Springer Proceedings in Physics, Volume 83
Computer Simulation Studies in Condensed-Matter Physics X
Eds.: D. P. Landau, K.K. Mon, H. -B. Schüttler
© Springer-Verlag Berlin Heidelberg 1998

$$\xi_L^{-2} = [2f(\Delta k, 0, 0) + 2f(0, \Delta k, 0) + 2f(0, 0, \Delta k)$$
$$- f(0, \Delta k, \Delta k) - f(\Delta k, 0, \Delta k) - f(\Delta k, \Delta k, 0)]/3, \quad (1)$$

where $\Delta k \equiv 2\pi/L$, $\chi = \chi(\vec{0})$ is the magnetic susceptibility, $\langle\ \rangle$ is the mean value over all the states of the system, weighted by the Gibbs factor, and

$$f(\vec{k}) \equiv 4\left(\sin^2\frac{k_x}{2} + \sin^2\frac{k_y}{2} + \sin^2\frac{k_z}{2}\right)\left(1 - \frac{\chi(\vec{k})}{\chi}\right)^{-1},$$

$$\chi(\vec{k}) \equiv \left\langle\left|\sum_{\vec{r}} \exp(-i\vec{k}\cdot\vec{r})S_{\vec{r}}\right|^2\right\rangle, \quad (2)$$

where $S_{\vec{r}}$ are the spin variables. In the region $T > T_c$, this definition of the correlation length is correct up to the second order in $1/L$. For $T < T_c$ we continue to use this definition. It is the analytic continuation (for finite L), and is *not to be confused with the proper low temperature quantities*. The same remark is true of the other thermodynamic quantities we use. If M is the total magnetization, then the Binder cumulant ratio U, and the renormalized coupling constant g are given by,

$$U = \frac{\langle M^4\rangle}{(\langle M^2\rangle)^2} - 3, \quad g = -\left(\frac{L}{\xi_L}\right)^d U. \quad (3)$$

where d is the spatial dimension.

Next I remark that the argument usually used in finite scaling discussions instead of ξ_L/L is $L^{1/\nu}(K_c - K)$, where $K = J/kT$ with J the exchange energy and k Boltzmann's constant. (For the 2-D Ising model, $\nu = 1$.) There is a considerable difference in these two variables for small values of L as is shown in Fig. 1.

The point of the following graphs is that they are appropriate for quantities which differ on the high and low sides of the the critical point. They will show how the transition occurs. The critical point for an infinite system is a point of non-uniform approach in the system-size, temperature plane. The functions described in these graphs are basically finite-size scaling functions. It has been argued by Privman and Fisher [5] that they should be universal, at least in the high-temperature region. The most successful data collapse that I have found so far is for the renormalized coupling constant. These results are shown in Fig. 2. The limiting value for $\xi_L/L \to 0$ is marked by a large tic mark on the left hand margin. Note is taken that this limit is not the same as the value at $K = 0$, but rather is the limit obtained by first holding $\xi_L/L > 0$ fixed and taking $L \to \infty$ followed by the limit $\xi_L/L \to 0$. Note is taken that as the Binder cumulant ratio differs from g by a factor of $(\xi_L/L)^d$, a similar data collapse is also found for it as well.

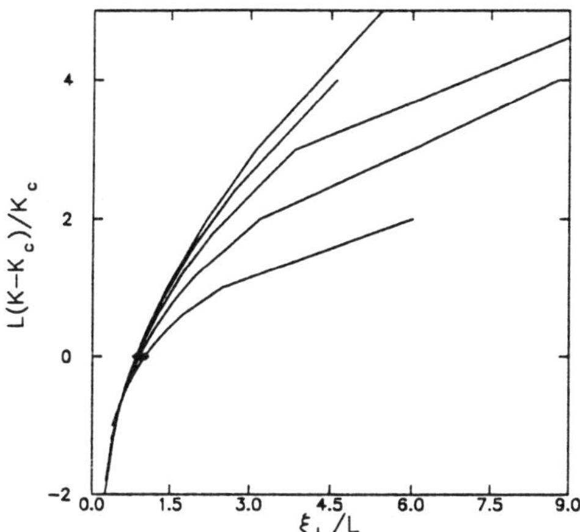

Fig. 1. $L(K_c - K)/K_c$ versus ξ_L/L for the 2D Ising model. The curves from left to right are for $L = 10, \ldots 2$, and the diamonds mark the value at the critical temperature.

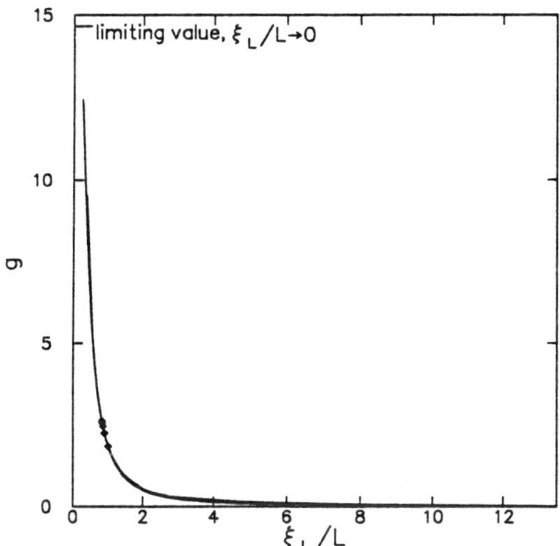

Fig. 2. g versus ξ_L/L for the 2-D Ising model. The curves are for $L = 10, \ldots 2$, and the diamonds mark the value at the critical temperature, with $L = 10$ the highest diamond.

Similar results were obtained by Baker and Kawashima [3] for the three dimensional Ising model. We show them in Fig. 3.

A further example is shown in Fig. 4. Here we plot $(K/K_c)^{7/8}\chi/\xi_L^{7/4}$. The factor of $\xi^{-7/4}$ reflects the standard scaling behavior in the two dimensional Ising model, and is chosen so the plotted quantity should be finite at the critical

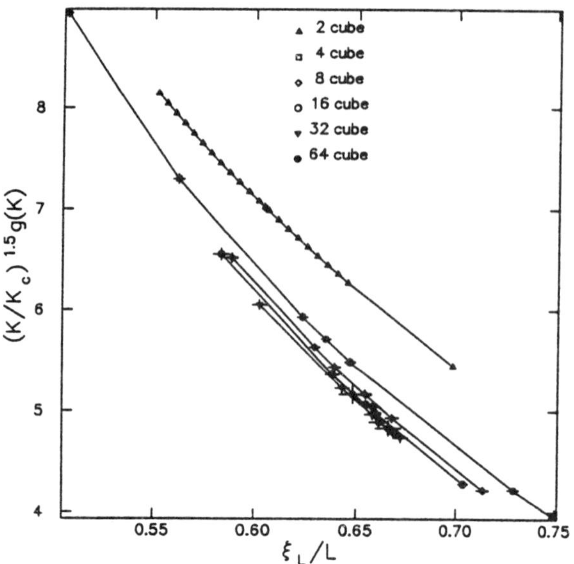

Fig. 3. $(K/K_c)^{3/2}g(K)$ versus ξ_L/L for the 3-D Ising model. The curves are for cubes with edge length $L = 2$, 4, 8. 16. 32. and 64, with periodic boundary conditions.

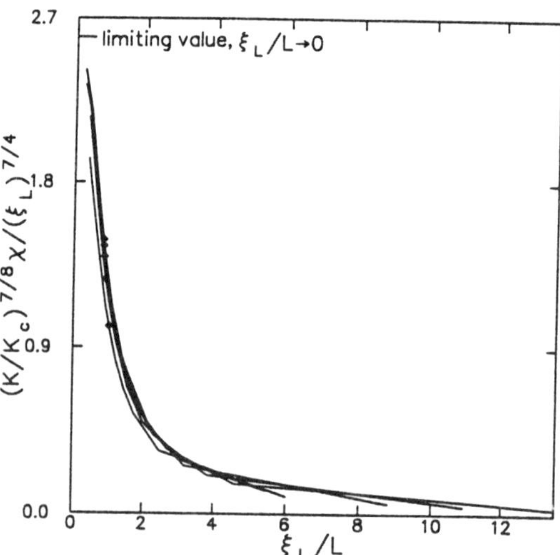

Fig. 4. $(K/K_c)^{7/8}g(K)/\xi_L^{7/4}$ versus ξ_L/L for the 2-D Ising model. The curves from left to right are for squares with edge lengths $L = 2, \dots, 10$. The diamonds mark the values for the critical temperature, with $L = 10$ having the largest value.

point. The factor of $(K/K_c)^{7/8}$ is included to compensate for the fact that $\xi \propto K^{1/2}$ as $K \to 0$. Thus the plotted quantity will also be finite at $K = 0$. Again the limit $\xi_L/L \to 0$ is marked with a large tic mark on the left margin. The same remarks about this limit apply as those made with respect to Fig.3.

References

1. C. Borgs, J. T. Chayes, H. Kesten and J. Spencer, oral communication, expected preprints, "Uniform boundedness of crossing probabilities implies hyperscaling." —, "Birth of the infinite cluster: finite-size scaling in percolation." S. Caracciolo, R. G. Edwards, S. J. Ferreira, A. Pelissetto, and A. D. Sokal, *Phys. Rev. Letts.* **74**, 2969 (1995). J.-K. Kim, A. J. F. de Souza, and D. P Landau. *Phys. Rev. E* **54**, 2291 (1996).
2. G. A. Baker, Jr. *Computer Simulation Studies in Condensed Matter Physics* **6**, eds. D. P. Landau, K. K. Mon and H.-B. Schüttler, (Springer, New York, 1993) pg. 178. —, *J. Stat. Phys.* **77**, 955 (1994).
3. G. A. Baker, Jr. and N. Kawashima, *Computer Simulation Studies in Condensed Matter Physics* **8**, eds. D. P. Landau, K.K. Mon and H.-B. Schüttler, ((Springer, New York, 1995) pg. 112. —, *J. Phys. A* **29**, 7183 (1996).
4. G. A. Baker, Jr. and J. J. Erpenbeck, *Computer Simulation Studies in Condensed Matter Physics* **7**, eds. D. P. Landau, K.K. Mon and H.-B. Schüttler, ((Springer, New York, 1994) pg. 213. J.-K. Kim, *Phys. Rev. D* **50**, 4663 (1994).
5. V. Privman and M. E. Fisher. *Phys. Rev. B* **30**, 322 (1984).

Projected Dynamics for Metastable Decay in Ising Models

M. Kolesik[1,2], M.A. Novotny[1,3], P.A. Rikvold[1,4], and D.M. Townsley[1]

[1]Supercomputer Computations Research Institute, Florida State University,
  Tallahassee, FL 32306-4052, USA
[2]Institute of Physics, Slovak Academy of Sciences, Dúbravská cesta 9,
  84228 Bratislava, Slovak Republic
[3]Department of Electrical Engineering, 2525 Pottsdamer Street, Florida A&M
  University-Florida State University, Tallahassee, FL 32310-6046, USA
[4]Center for Materials Research and Technology and Department of Physics,
  Florida State University, Tallahassee, FL 32306-3016, USA

Abstract. The magnetization switching dynamics in the kinetic Ising model is projected onto a one-dimensional absorbing Markov chain. The resulting projected dynamics reproduces the direct simulation results with great accuracy. A scheme is proposed to utilize simulation data for small systems to obtain the metastable lifetime for large systems and/or for very weak magnetic fields, for which direct simulation is not feasible.

In simulations of metastable decay one faces the problem of measuring the lifetime of the metastable phase, which is by definition very long. For example, in Monte Carlo modeling of magnetization switching in ferromagnets the physically relevant simulation time scales are on the order of $10^{12} - 10^{15}$ Monte Carlo Steps per Spin (MCSS). Even with sophisticated algorithms [1,2] such simulations have not yet been feasible, and one has to resort to extrapolation of the results into the physical time-scale regime.

In this article, a scheme is presented to map the system under study onto a simpler one, which is faster to simulate but still gives accurate results for most important physical quantities. This idea is not new. For example, recently Lee et al. introduced a macroscopic mean-field dynamics [3] which semiquantitatively describes magnetization switching in Ising systems. The aim of the present work is to construct a scheme which is very similar in spirit, but is intended to create a practical computational tool applicable to the simulation of ferromagnets in the physical time regime.

To simplify our notation and reasoning, we explain the proposed Projected Dynamics (PD) as applied to the isotropic kinetic Ising model on a square or cubic lattice. The Hamiltonian has the standard form,

$$\mathcal{H} = -J \sum_{\langle ij \rangle} s_i s_j - H \sum_i s_i , \tag{1}$$

with the ferromagnetic nearest-neighbor spin-spin interaction $J > 0$ and an external magnetic field H. In what follows, we denote by V the total number of spins in the system. To study the magnetization reversal, we initialize all spins in the state $+1$, fix the temperature T well below its critical value T_c, and apply

Springer Proceedings in Physics, Volume 83
Computer Simulation Studies in Condensed-Matter Physics X
Eds.: D. P. Landau, K.K. Mon, H. -B. Schüttler
© Springer-Verlag Berlin Heidelberg 1998

a negative magnetic field. Then we apply Metropolis or Glauber dynamics with updates at randomly chosen sites to measure the time the system needs to reach a configuration with a given stopping magnetization. Repeating this procedure many times, we obtain the mean lifetime, τ, of the metastable state and its standard deviation, $\Delta\tau$.

To speed up the simulations as much as possible, we use a rejection-free algorithm [1,2] which uses the notion of spin classes. By a spin class we mean the state of the spin itself and its neighbors. There are ten classes for the given model on a square lattice. Classes $i = 1, \ldots, 5$ correspond to spins in the state $+1$ which have exactly $i - 1$ neighbors in the state $+1$. Similarly classes $i = 6, \ldots, 10$ are assigned to those -1 spins which have $i - 6$ neighbors in the state $+1$. All spins in class i have the same flipping probability, p_i.

Rejection-free algorithms keep track of the number c_i of spins in each class, and it is not computationally expensive to measure the growth and shrinkage rates of the stable phase, which are defined as

$$g(n) = \sum_{i=1}^{5} \langle c_i \rangle_n p_i \quad , \quad s(n) = \sum_{i=6}^{10} \langle c_i \rangle_n p_i \; , \tag{2}$$

respectively. The angular brackets mean the average taken over the configurations generated during the Monte Carlo lifetime experiment, conditionally on the number n of overturned spins. Thus, $g(n)/V$ is the probability that a $+1$ spin will be flipped in the next Monte Carlo step, and $s(n)/V$ is the probability to flip one of the -1 spins, both conditionally on n. Flipping a $+1$ spin increases the volume fraction of the stable phase, hence the names $g(n)$ and $s(n)$.

The main idea of the proposed method is to make use of the observed growth and shrinkage rates to map the switching dynamics onto a one-dimensional absorbing Markov chain. We assign to all configurations with n overturned spins a single state n in the chain. The one-dimensional dynamics is given by the above probabilities: From the state n we have the probability $g(n)/V$ of jumping to the state $n + 1$, the probability $s(n)/V$ of jumping to $n-1$, and the probability $1-g(n)/V-s(n)/V$ of remaining in the current state. This random walk starts at $n = 0$ and terminates when it reaches $n = N + 1$ where $M = V - 2N - 2$ corresponds to the stopping magnetization. Using standard methods from the theory of absorbing Markov chains [2,3], we obtain the mean lifetime τ and the total average time $h(n)$ (in MCSS) spent by the random walker in the state n in terms of the growth and shrinkage rates:

$$\tau = \sum_{n=0}^{N} h(n) \quad , \quad h(n) = \frac{1 + s(n+1)h(n+1)}{g(n)} \quad , \quad h(N) = \frac{1}{g(N)} \; . \tag{3}$$

Higher moments of the lifetime distribution can be obtained in a similar way.

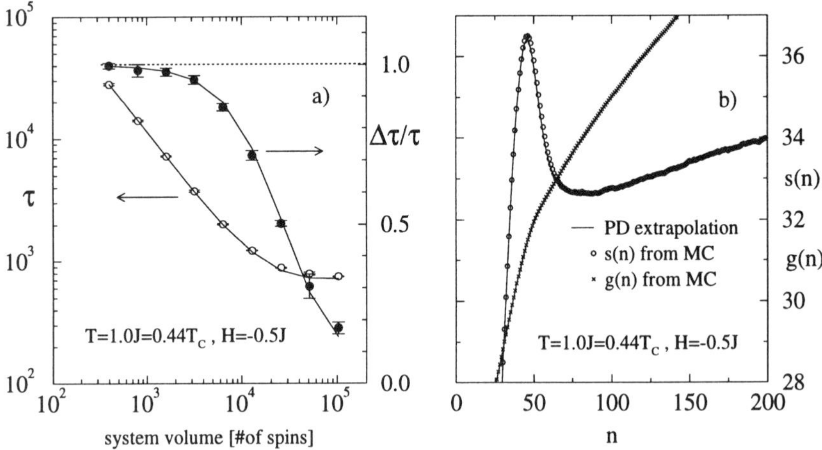

Fig. 1 Illustration of the Projected Dynamics [Eqs. (2,3)] and of the "size extrapolation" scheme [Eq. (4)]. The lines in both panels are Projected Dynamics extrapolations based on $\langle c_i \rangle_n$ sampled during lifetime measurements on a 20×20 lattice. Symbols are direct simulation results. a) The lifetime and its relative standard deviation vs. the system volume. The agreement is near-perfect in the single-droplet regime (where $\Delta\tau/\tau \approx 1$), and deviations are only observed in the crossover region to the multidroplet regime (where $\Delta\tau/\tau \to 0$). b) Extrapolation from the 20×20 lattice reproduces the measured growth and shrinkage rates, $g(n)$ and $s(n)$, of the 160×160 lattice very well. The crossings of the two curves correspond to the metastable magnetization (left) and the critical fluctuation (right).

As shown in Fig. 1, the proposed scheme reliably reproduces the direct Monte Carlo results. The lifetimes obtained from Eq.(3) are usually only slightly different from their counterparts from the direct measurements. The projected dynamics also gives the standard deviation of the lifetime distribution, which agrees with the measured values within the error bars.

Suppose we have measured $g(n)$ and $s(n)$ in a system of volume V. How are they related to their counterparts in a system of volume $2V$? Since the relevant configurations typically contain many small droplets of the stable phase, it is a reasonable approximation to view the larger system as consisting of two "independent" copies of the smaller system. Then, the growth rate of the large system can be approximated as a weighted sum of the sub-system contributions,

$$g(2V, n) \approx \frac{\sum_{i=0}^{n} h(V, n-i)h(V, i)[g(V, n-i) + g(V, i)]}{\sum_{i=0}^{n} h(V, n-i)h(V, i)} \, . \tag{4}$$

Here, we have explicitly shown the dependence on the volume V. An analogous formula is proposed for the shrinkage rate $s(n)$. Provided the smallest system is in the so-called single-droplet regime, in which the nucleation is triggered

248

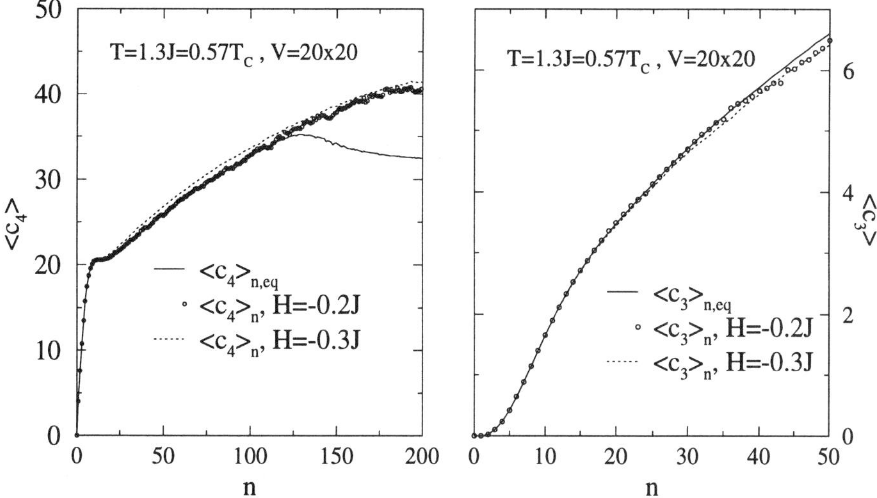

Fig. 2 Populations of two spin classes vs. the number of overturned spins n. Solid lines are $\langle c_i \rangle_{n,\mathrm{eq}}$ measured in the equilibrium fixed-n ensemble; circles and dotted lines show $\langle c_i \rangle_n$ measured during the lifetime measurement for two different fields. Note that as the field decreases the region in which $\langle c_i \rangle_n \rightarrow \langle c_i \rangle_{n,\mathrm{eq}}$ increases.

by a single critical fluctuation smaller than the system size (for the theoretical description of different regimes of the magnetization switching, see Refs. [4-8]), one can repeat the extrapolation $V \rightarrow 2V$ several times without finding a big discrepancy between growth and shrinkage rates calculated from the small-lattice data and those directly measured on large lattices (see Fig. 1b).

Now we return to our main goal, namely predicting lifetimes in very weak fields. What we need is to calculate $g(n)$ and $s(n)$ as functions of the magnetic field. The naive choice would be to replace the class spin-flip probabilities p_i by their values calculated for the desired field, and keep the values $\langle c_i \rangle_n$ entering Eq. (2) unchanged. Such a scheme works quite well when extrapolating to stronger fields, but the lifetimes are systematically overestimated for fields weaker than the one at which the $\langle c_i \rangle_n$ were sampled. The reason lies in the nonequilibrium character of the configurations with n larger than the critical droplet volume (see Fig. 2). As it tries to escape from the metastable free-energy minimum, when n is small the system passes through configurations which are very close to "equilibrium". Farther from the free-energy minimum, the configurations which appear in the system are increasingly different from "equilibrium" configurations. Although these configurations do not contribute significantly to the lifetime, if we use them to estimate the lifetime in a *weaker* field their contribution becomes more important because the system then spends more time in them. However, the actual configurations are closer to "equilibrium," because the free-energy minimum is deeper in a weaker field.

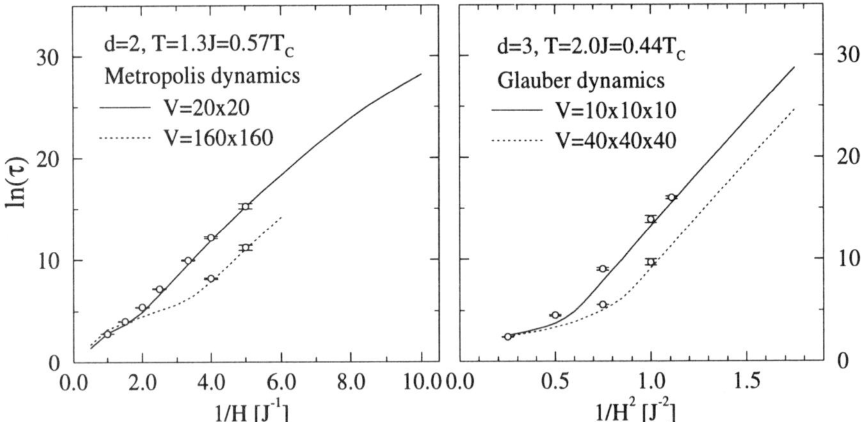

Fig. 3 The metastable lifetime of a kinetic Ising model as a function of the magnetic field. Points are conventional Monte Carlo measurements, and lines are Projected Dynamics calculations based on the "equilibrium" $\langle c_i \rangle_{n,\mathrm{eq}}$ data sampled on the smaller systems. Predictions of the Projected Dynamics improve with decreasing field.

Thus, we effectively replace the actual weak-field configurations by ones which appear slightly "overheated." This enhances the shrinkage probabilities more than the growth probabilities, thus leading to an overestimate of the metastable lifetime. Understanding the cause of this problem also offers a remedy: we can do better by using the equilibrium configurations. Then we expect to see underestimation of relatively short lifetimes (by reversing the above argument). On the other hand, such an approximation will improve with decreasing field, which is exactly what we need to extrapolate towards zero field. Thus, we define the equilibrium growth and shrinkage rates as follows

$$g_{\mathrm{eq}}(n) = \sum_{i=1}^{5} \langle c_i \rangle_{n,\mathrm{eq}} p_i \quad , \quad s_{\mathrm{eq}}(n) = \sum_{i=6}^{10} \langle c_i \rangle_{n,\mathrm{eq}} p_i \ . \tag{5}$$

The only difference from Eq. (2) is in the sampling. Here, $\langle c_i \rangle_{n,\mathrm{eq}}$ are sampled in an equilibrium ensemble with a fixed number of overturned spins n. Thus, instead of sampling the configurations during the lifetime measurement, we perform a static measurement for each value of n needed (up to the stopping magnetization). The $\langle c_i \rangle_{n,\mathrm{eq}}$ depend on temperature, but the only dependence of g_{eq} and s_{eq} on the magnetic field comes from the spin-flip probabilities p_i. In that way, a single measurement provides sufficient data to obtain a good approximation for the lifetime as a function of the external field.

Figure 3 shows the lifetime calculated from the projected dynamics for two pairs of lattice sizes in two and three dimensions. Comparison with the direct simulation results (points) corroborate our expectation that the lifetime is underestimated in the crossover region between the single-droplet and multidroplet regimes, and that the approximation provides progressively better results as the external field strength decreases. Measurements on small lattices, for which we can obtain reasonable statistics in zero field, show that the Projected Dynamics based on $\langle c_i \rangle_{n,\mathrm{eq}}$ reproduces the "lifetime" in zero field.

In conclusion, our Projected Dynamics maps the complex Monte Carlo dynamics onto a much simpler one-dimensional absorbing Markov chain. The Projected Dynamics is computationally much easier to study than the full underlying Monte Carlo dynamics, while it provides reliable results for metastable lifetimes and their standard deviations. Data from small systems can be utilized to predict lifetimes for large systems. When based on "equilibrium" class populations $\langle c_i \rangle_{n,\mathrm{eq}}$, Projected Dynamics yields results as functions of the field. Such estimates are most reliable in the experimentally relevant weak-field region, which it is currently not feasible to investigate with direct simulations. The method also offers a deeper insight into the mechanism of metastable decay. For example, the histogram $h(n)$ provides a simple method to measure the metastable magnetization and susceptibility, and the rates $g(n)$ and $s(n)$ contain information about the size of the critical fluctuations.

For other types of models, the applicability of the projected dynamics will depend on the existence of a "single channel" for the decay. The projected parameter (n in the present case) need not necessarily be related to the magnetization, but it should parameterize the optimal path for escape from metastability in such a way that the configurations at a fixed value of the parameter will exhibit similar growth and shrinkage rates.

We dedicate this paper to Professor Masuo Suzuki on his 60th birthday. This research was supported by FSU-MARTECH, FSU-SCRI (DOE Contract No. DE-FC05-85ER25000), NSF Grants No. DMR-9315969, DMR-9520325, and DMR-9634873.

References

[1] A. B. Bortz, M. H. Kalos and J. L. Lebowitz, J. Comput. Phys. **17**, 10 (1975).

[2] M. A. Novotny, Phys. Rev. Lett. **74**, 1 (1995); Erratum **75**, 1424 (1995).

[3] J. Lee, M. A. Novotny and P. A. Rikvold, Phys. Rev. E **52**, 356 (1995).

[4] P. A. Rikvold, H. Tomita, S. Miyashita and S. W. Sides, Phys. Rev. E **49**, 5080 (1994).

[5] H. L. Richards, S. W. Sides, M. A. Novotny, and P. A. Rikvold, J. Magn. Magn. Mater. **150**, 37 (1995).

[6] H. L. Richards, S. W. Sides, M. A. Novotny, and P. A. Rikvold, J. Appl. Phys. **79**, 5479 (1996).

[7] H. L. Richards, M. A. Novotny, and P. A. Rikvold, Phys. Rev. B **54**, 4113 (1996).

[8] H. L. Richards, M. Kolesik, P.-A. Lindgård, P. A. Rikvold, and M. A. Novotny, Phys. Rev. B, **55** in press.

Short-Time-Scaling Behavior of Growing Interfaces

M. Krech

Fachbereich Physik, BUGH Wuppertal, D-42097 Wuppertal, Germany

Abstract

The short-time evolution of a growing interface is studied analytically and numerically for the Kadar-Parisi-Zhang (KPZ) universality class. The scaling behavior of response and correlation functions is reminiscent of the "initial slip" behavior found in purely dissipative critical relaxation (model A). Unlike model A the initial slip exponent for the KPZ equation can be expressed by the dynamical exponent z. In 2+1 dimensions z is estimated from the short-time evolution of the correlation function for ballistic deposition and for the RSOS model.

1. Introduction

Interface formation and growth are typical processes in nonequilibrium systems. Two important examples are fluid flow in porous media [1] and deposition of atoms during molecular beam epitaxy (MBE) [1, 2]. It is expected that at times much later than typical aggregation times and on macroscopic length scales these interfaces develop a characteristic scaling behavior, where the scaling exponents fall into certain dynamic *universality classes* [1, 2, 3]. In certain cases, however, interfaces can also show turbulent, i.e., spatial multiscaling behavior [4]. Usually a d-dimensional interface is embedded in $d + 1$-dimensional space such that the interface position at time t can be described by a height function $h(\mathbf{x}, t)$, where $\mathbf{x}$ denotes the lateral position in a d-dimensional reference plane given by the surface of a substrate. Complete information about the scaling behavior is contained in the dynamic structure factor, which is related to the time displaced height-height correlation function $C(\mathbf{x} - \mathbf{x}', t, t') \equiv \langle h(\mathbf{x}, t)h(\mathbf{x}', t') \rangle - \langle h(\mathbf{x}, t) \rangle \langle h(\mathbf{x}', t') \rangle$, where a laterally translational invariant system is assumed. For $t, t' \to \infty$ and finite $|t - t'|$ the correlation function displays the asymptotic scaling behavior

$$C(\mathbf{x} - \mathbf{x}', t, t') = |\mathbf{x} - \mathbf{x}'|^{2\alpha} F_C(|t - t'|/|\mathbf{x} - \mathbf{x}'|^z), \qquad (1.1)$$

where α denotes the *roughness* exponent and z is the *dynamic* exponent [1, 2]. For a laterally translational invariant system the interfacial width $w^2(t) \equiv \langle h^2(\mathbf{x}, t) \rangle - \langle h(\mathbf{x}, t) \rangle^2$ is only a function of t and displays the scaling behavior $w(t) \sim t^\beta$ for late times, where $\beta = \alpha/z$ is the *growth* exponent. For MBE as an example the scaling behavior displayed in Eq.(1.1) gives access to the exponents α and z both experimentally by reflection high energy electron diffraction (RHEED) (see, e.g., chaper 16 of Ref.[1]) and by direct imaging using a surface tunneling microscope [6] and theoretically by continuum models [1, 2] and Monte-Carlo simulations [2, 5].

Continuum descriptions of interfacial growth processes can be obtained from general symmetry principles and conservation laws obeyed by the growth process [1]. For a wide class of growth processes the resulting continuum model is given by the well-known Kadar-Parisi-Zhang (KPZ) equation [7], which reads

$$\tfrac{\partial}{\partial t} h(\mathbf{x}, t) = \nu \nabla^2 h(\mathbf{x}. t) + \tfrac{\lambda}{2}(\nabla h(\mathbf{x}, t))^2 + \eta(\mathbf{x}, t). \qquad (1.2)$$

The noise $\eta(\mathbf{x}, t)$ has a Gaussian distribution with $\langle \eta(\mathbf{x}, t) \rangle = 0$ and

$$\langle \eta(\mathbf{x}, t)\eta(\mathbf{x}', t') \rangle = 2D\delta(\mathbf{x} - \mathbf{x}')\delta(t - t'). \qquad (1.3)$$

Springer Proceedings in Physics, Volume 83
Computer Simulation Studies in Condensed-Matter Physics X
Eds.: D. P. Landau, K.K. Mon, H. -B. Schüttler
© Springer-Verlag Berlin Heidelberg 1998

The parameters ν, D, and λ are assumed to be constants and averages $\langle\ldots\rangle$ are taken over the noise distribution. In the long time limit Eq.(1.2) has a global symmetry which is commonly denoted as Galileian invariance [1, 7]. An important consequence is that the exponents z and α of the KPZ equation fulfill the scaling relation $\alpha + z = 2$. The exponents of the KPZ equation are exactly known only in $d = 1$, where $z = 3/2$ and $\alpha = 1/2$ due to the existence of a dissipation fluctuation theorem [8, 9]. In $d = 2$ numerical investigations indicate $z \simeq 1.6$ and $\alpha \simeq 0.4$ [1]. For $d > 2$ the asymptotic scaling behavior is either governed by linear theory ($\lambda = 0$, weak coupling regime) or by another set of exponents inaccessible by analytical methods (strong coupling regime) depending on the value of the effective coupling constant $g \equiv D\lambda^2/(4\nu^3)$ [1, 8]. Furthermore, it is interesting to note that the nonlinearity in Eq.(1.2) renders all other possible nonlinearities irrelevant in the renormalization group sense in the long-time limit. For intermediate times, however, the presence of other nonlinearities in the growth equation gives rise to various crossover phenomena [1, 10].

2. Analytic Theory

In Fourier space the KPZ equation Eqs.(1.2) and (1.3) is equivalent to the dynamic functional $\mathcal{J} = \mathcal{J}_0 + \mathcal{J}_1$ [8, 11, 14] which consists of the Gaussian part

$$\mathcal{J}_0[\tilde{h}, h] = \int \frac{d^d q}{(2\pi)^d} \int_0^\infty dt \left\{ D\tilde{h}(\mathbf{q}, t)\tilde{h}(-\mathbf{q}, t) - \tilde{h}(\mathbf{q}, t) \left(\frac{\partial}{\partial t} h(-\mathbf{q}, t) + \nu \mathbf{q}^2 h(-\mathbf{q}, t) \right) \right\} \qquad (2.1)$$

and the interaction part

$$\mathcal{J}_1[\tilde{h}, h] = -\frac{\lambda}{2} \int \frac{d^d q_1}{(2\pi)^d} \int \frac{d^d q_2}{(2\pi)^d} \int_0^\infty dt\, \mathbf{q}_1 \cdot \mathbf{q}_2\, \tilde{h}(-\mathbf{q}_1 - \mathbf{q}_2, t)h(\mathbf{q}_1, t)h(\mathbf{q}_2, t), \qquad (2.2)$$

where $\tilde{h}(\mathbf{q}, t)$ is the Fourier transform of the response field [12]. The initial condition $h(\mathbf{q}, 0) = 0$, which is implicitly assumed in Eqs.(2.1) and (2.2), breaks the temporal translational invariance of the KPZ dynamics. The analytic treatment of Eqs.(2.1) and (2.2) is based on the identities

$$\tfrac{\partial}{\partial t} h(\mathbf{q}, t = 0) = 2D\tilde{h}(\mathbf{q}, t = 0) \quad \text{and} \quad G(\mathbf{q} = 0, t, t' < t) = 1, \qquad (2.3)$$

where the response function G is given by the formal average $\langle h(-\mathbf{q}, t)\tilde{h}(\mathbf{q}, t')\rangle$ with respect to the dynamic functional $\mathcal{J}[\tilde{h}, h]$ [15]. For details of the field-theoretic treatment of Eqs.(2.1) and (2.2) we refer to Refs.[11, 15] and only quote the main results for later reference.

The correlation function C and the response function G are found to obey the scaling relations

$$C(\mathbf{q}, t, t' \ll t) = (t'/t)^\theta |\mathbf{q}|^{-d-2\alpha} f_C(|\mathbf{q}|^z t) \quad \text{and} \quad G(\mathbf{q}, t, t' \ll t) = (t'/t)^{\bar{\theta}} |\mathbf{q}|^{-d} f_G(|\mathbf{q}|^z t), \qquad (2.4)$$

respectively, where in contrast to model A [11] the short-time exponents θ and $\bar{\theta}$ are given by

$$\theta = (d + 4)/z - 2 = (d + 2\alpha)/z \quad \text{and} \quad \bar{\theta} = 0 \qquad (2.5)$$

at the nontrivial fixed point ($\lambda \neq 0$) of Eq.(1.2). In $d = 1$ the exact value $\theta = 4/3$ can be obtained from the exact value $z = 3/2$ and is confirmed by a Monte-Carlo simulation of ballistic deposition [15]. From numerical estimates for z in $d = 2$ one obtains $\theta \simeq 1.7$. The exponent relation given by Eq.(2.5) simply means that the short-time and the long-time scaling behavior of the correlation function are *identical*, i.e., the short-time scaling behavior can be obtained by extrapolating the t'-dependence of $C(\mathbf{q}, t, t')$ from $t' \sim t$ to $t' = 0$. In fact, the scaling relation given by Eq.(2.5) can be derived independently by analyzing the fluctuation spectrum of the interface displacement velocity averaged over a macroscopic portion of the interfacial area [16]. It should be noted, however, that the perturbative analysis of Ref.[15] only consitutes a rigorous proof of Eq.(2.5) for $d = 1$. For $d \geq 2$ one encounters the strong coupling regime of Eq.(1.2) which can no longer be treated analytically.

Finally, we remark that an alternative scaling form for C can be obtained from the definition of the growth exponent β which leads to $\theta = d/z + 2\beta$. The scaling behavior displayed in Eq.(2.4) can then be written in the simplified form $C(\mathbf{q}, t, t' \ll t) = t'^{\theta} g_C(|\mathbf{q}|^z t)$, where $g_C(y) = y^{-\theta} f_C(y)$.

3. Monte-Carlo Results

The scaling behavior of $C(\mathbf{q}, t, t' \ll t)$ according to Eq.(2.4) can be tested numerically by a Monte-Carlo simulation of simple deposition models on lattices [1]. The continuum description underlying Eqs.(1.2) and (1.3) is replaced by a discretized description according to

$$h(\mathbf{x}, t) = h(\mathbf{x} = a\mathbf{j}, t = n/(FL^d)) \equiv a\, h_{\mathbf{j}}(n), \tag{3.1}$$

where the lattice constant a is assumed to be the same both in the plane of the substrate and perpendicular to it and $\mathbf{j} = (j_1, \ldots, j_d)$. The lattice has L^d sites, F is the incoming particle flux, and n is the number of deposited particles. Furthermore, the incoming particle flux F has been normalized to unity, so that t in Eq.(3.1) is dimensionless and given by the number of deposited layers. Finally, $h_{\mathbf{j}}(n)$ defined by Eq.(3.1) is also dimensionless and denotes the number of particles deposited at lattice site $\mathbf{j}$ after n particles have been deposited on the lattice. Ballistic deposition on a two-dimensional substrate is defined by the *deterministic* growth rule

$$h_{j,k}(n+1) = \max(h_{j-1,k}(n), h_{j,k-1}(n), h_{j,k}(n)+1, h_{j+1,k}(n), h_{j,k+1}(n)), \tag{3.2}$$

(see, e.g., Ref.[1]), where the site (j, k) in Eq.(3.2) has been selected randomly from the $L \times L$ sites of the lattice and periodic boundary conditions are applied.

In order to measure the short time exponent given by Eq.(2.5) it is sufficient to probe the *integrated* time displaced correlation function, i.e., one probes $C(\mathbf{q} = 0, t, t')$ as described in Ref.[15]. Like a real deposition process the simulation is characterized by an a priori unknown microscopic aggregation time t_a. The scaling behavior of C according to Eq.(2.4) can only be observed for $t' \gg t_a$. On the other hand $t' \ll t$ is required for Eq.(2.4) to hold, so that short-time scaling is restricted to the time window $t_a \ll t' \ll t$. Furthermore, the lattice size L must be chosen sufficiently large in order to avoid the onset of finite-size crossover effects if $t'^{1/z} \sim L$ when t' is still much smaller than t. For the simulation described here $t = 1000$ and $L \geq 200$ fulfill the above requirements. In order to cope with the very small signal to noise ratio in each measurement of $C(0, t, t')$ for $t' \ll t$ averages are taken over 2×10^4 realizations. These are distributed over 20 individual runs at every point in time so that the jackknife method can be applied for the data analysis. The result for ballistic deposition according to Eq.(3.2) and for the RSOS model in $d = 2$ according to Ref.[17] is displayed in Fig.1, where $C(0, t, t')$ is shown as a function of t'/t for $t = 1000$ and $L = 300$. According to Fig.1 the interval $0.02 \leq t'/t \leq 0.2$ is available to determine the short-time exponent θ. From Eq.(2.5) we obtain the estimate $z = 1.608 \pm 0.013$ by averaging over the two values for θ given in Fig.1. Finally, we note that according to the scaling relation $\alpha + z = 2$ one has $\alpha = 0.392 \pm 0.013$ for the roughness exponent. These values are in agreement with other numerical data for z and α in $d = 2$ (see chapter 8 of Ref.[1] for a collection of recent estimates) and they therefore provide some support for the general validity of Eq.(2.5).

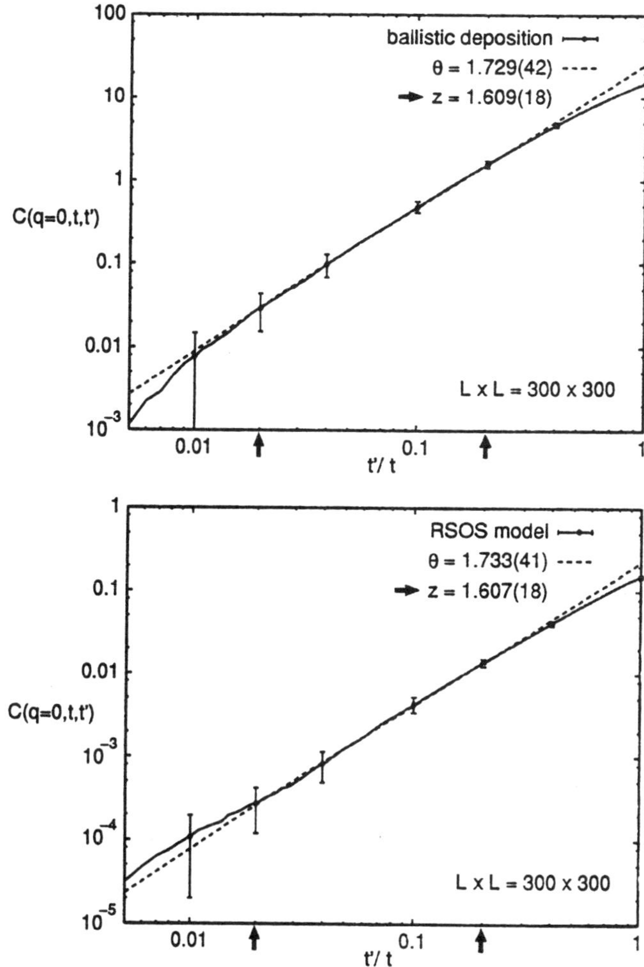

Fig.1: Correlation function $C(0, t, t')$ in $d = 2$ for ballistic deposition (top) and the RSOS model (bottom) as a function of t'/t for $0.005 \leq t'/t \leq 1$ and $L \times L = 300 \times 300$ (solid line). Error bars are shown only at a few selected points in time and represent one standard deviation. The dashed lines display power laws with the measured short-time exponents $\theta = 1.729 \pm 0.042$ and $\theta = 1.733 \pm 0.041$, respectively. The data follow this power law rather accurately in the interval $0.02 \leq t'/t \leq 0.2$ (bold arrows).

Acknowledgment

The author gratefully acknowledges partial financial support of this work through the Heisenberg program of the Deutsche Forschungsgemeinschaft.

References

[1] A.-L. Barabasi and H.E. Stanley, *Fractal Concepts in Surface Growth* (Cambridge University press, New York, 1995) and references therein.

[2] S. Das Sarma, C.J. Lanczycki, R. Kotlyar, and S.V. Ghaisas, Phys. Rev. E **53**, 359 (1996) and references therein.

[3] J. Krug and H. Spohn, Phys. Rev. A **38**, 4271 (1988); J. Krug, J. Phys A **22**, L769 (1989); J. Krug, M. Plischke, and M. Siegert, Phys. Rev. Lett. **70**, 3271 (1993).

[4] J. Krug, Phys. Rev. Lett. **72**, 2907 (1994).

[5] S. Pal and D.P. Landau, Phys. Rev. B **49**, 10597 (1994) and references therein.

[6] J. Krim, I. Heyvaert, C. Van Haesendock, and Y. Bruynseraede, Phys. Rev. Lett. **70**, 57 (1993).

[7] M. Kadar, G. Parisi, and Y.-C. Zhang, Phys. Rev. Lett **56**, 889 (1986).

[8] E. Frey and U.C. Täuber, Phys. Rev. E **50**, 1024 (1994).

[9] U. Deker and F. Haake, Phys. Rev A **11**, 2043 (1975).

[10] K. Sneppen, J. Krug, M.H. Jensen, C. Jayaprakash, and T. Bohr, Phys. Rev. A **46**, R7351 (1992).

[11] H.K. Janssen, B. Schaub, and B. Schmittmann, Z. Phys. B **73**, 539 (1989).

[12] P.C. Martin, E.D. Siggia, and H.H. Rose, Phys. Rev. A **8**, 423 (1973).

[13] P.C. Hohenberg and B.I. Halperin, Rev. Mod. Phys. **49**, 435 (1977).

[14] M. Lässig, Nucl. Phys. **B448**, 559, (1995).

[15] M. Krech, Phys. Rev. E **55**, 668 (1997).

[16] J. Krug, Phys. Rev. A **44**, R801 (1991).

[17] J. M. Kim and J. M. Kosterlitz, Phys. Rev. Lett. **62**, 2289 (1989).

Scaling Behavior of the 2D *XY* Model Revisited

W. Janke

Institut für Physik, Johannes Gutenberg-Universität, D-55099 Mainz, Germany

Abstract. Using two sets of high-precision Monte Carlo data for the two-dimensional XY model in the Villain formulation on square $L \times L$ lattices, the scaling behavior of the susceptibility χ and correlation length ξ in the vicinity of the Kosterlitz-Thouless phase transition is analyzed with emphasis on multiplicative logarithmic corrections $(\ln \xi)^{-2r}$ in the high-temperature phase and $(\ln L)^{-2r}$ in the finite-size scaling region, respectively.

1   Introduction

In the past two decades systems belonging to the two-dimensional (2D) XY model universality class have been the subject of extensive experimental, analytical and numerical investigations [1]. Examples are layers of superconducting materials, films of liquid helium, Josephson-junction arrays, and some magnetic systems [2]. Theoretically the peculiar behavior of the Kosterlitz and Thouless (KT) phase transition [3, 4], which is believed to be driven by the unbinding of defect pairs, still poses many interesting questions and the details of the transition are not yet fully understood.

In the past year in particular multiplicative logarithmic corrections [4, 5] to the leading scaling behavior have attracted much interest. In a Monte Carlo (MC) simulation study of Lee-Yang partition function zeros on square lattices of size $L \times L$, Kenna and Irving [6] found a multiplicative correction to the leading finite-size scaling (FSS) behavior of $(\ln L)^{-2r}$ with $r = -0.02(1)$, while the standard KT theory would predict quite a different exponent of $r = -1/16 = -0.0625$ [4, 5]. Moreover, by reanalyzing "thermodynamic" MC data [7] for lattices satisfying $L > 7\xi$, where ξ is the correlation length, Patrascioiu and Seiler [8] arrived at a completely different estimate of $r = 0.077(46)$, and by analyzing long high-temperature series expansions, Campostrini *et al.* [9] also obtained positive values in the range $r = 0.042(5) - 0.05(2)$, depending on the quantity considered.

All numerical estimates quoted above were obtained in the cosine formulation of the XY model. In view of the severe inconsistencies I found it therefore worthwhile to reanalyze the logarithmic corrections in the Villain formulation [10, 11] of the XY model as well [12], which is actually (sometimes implicitly) the starting point of most if not all theoretical investigations.

2   Scaling Predictions

In the XY Villain model [10] the Boltzmann factor of the cosine formulation, $B_{\cos} = \prod_{\boldsymbol{x},i} \exp\left[\beta_{\cos} \cos(\nabla_i \theta(\boldsymbol{x}))\right]$, is replaced by the periodic Gaussian

$$B_{\mathrm{vil}} = \prod_{\boldsymbol{x},i} \sum_{n=-\infty}^{\infty} \exp\left[-\frac{\beta}{2}(\nabla_i \theta - 2\pi n)^2\right], \tag{1}$$

Springer Proceedings in Physics, Volume 83
Computer Simulation Studies in Condensed-Matter Physics X
Eds.: D. P. Landau, K.K. Mon, H. -B. Schüttler
© Springer-Verlag Berlin Heidelberg 1998

where β is the inverse temperature in natural units, and $\nabla_i\theta = \theta(x+i) - \theta(x)$ are lattice gradients. A discussion of the relation between the two formulations as well as numerical comparisons can be found in Refs. [10, 11].

The two-point correlation function, $G(x) \equiv \langle \vec{s}(x) \cdot \vec{s}(0)\rangle = \langle \cos(\theta(x) - \theta(0))\rangle$ with $\vec{s} = (\cos(\theta), \sin(\theta))$, is predicted to behave at the critical temperature $T_c = 1/\beta_c$ as [5]

$$G(x) \propto \frac{(\ln|x|)^{-2r}}{|x|^{\eta}} \left[1 + \mathcal{O}\left(\frac{\ln(\ln|x|)}{\ln|x|}\right)\right], \tag{2}$$

with $r = -1/16$ and $\eta = 1/4$. In the high-temperature phase near criticality, i.e. $0 < t \equiv (T - T_c)/T_c \ll 1$, this implies for the magnetic susceptibility, $\chi = \sum_x G(x)$, a scaling behavior

$$\chi \propto \xi^{2-\eta}(\ln\xi)^{-2r}\left[1 + \mathcal{O}\left(\ln(\ln\xi)/\ln\xi\right)\right], \tag{3}$$

where $\xi \propto \exp(b\,t^{-\nu})$ is the correlation length, with $\nu = 1/2$ and b being a non-universal positive constant. Very close to T_c eq. (3) cannot hold for a finite system with linear size $L \ll \xi$. Here ξ has to be replaced by L, and we expect to observe a FSS behavior

$$\chi \propto L^{2-\eta}(\ln L)^{-2r}\left[1 + \mathcal{O}\left(\ln(\ln L)/\ln L\right)\right]. \tag{4}$$

In numerical simulations it proved to be very difficult to verify the KT scaling laws unambiguously. However, if one rejects a power-law ansatz with unnaturally large exponents and large confluent correction terms, then, among the two alternatives, a pure power-law or the exponential KT divergence, the KT predictions are clearly favored. This is the conclusion of most numerical studies of the cosine formulation [7, 13] and, with even stronger evidence, also of the Villain formulation [14] considered here. We shall therefore assume the KT scaling behavior to be qualitatively valid and try to determine the exponents η, ν, and r. But even this goal is still far too ambitious, since a precise determination of all three critical exponents together with the (non-universal) value of β_c would require much more accurate data than one can hope to generate with present day techniques. We hence hold the exponents $\nu = 1/2$ and $\eta = 1/4$ fixed at their theoretically predicted values and enquire if any deviation of the data from the leading scaling behavior can be explained by the multiplicative logarithmic corrections in eqs. (3) and (4).

3 Results

The analysis is based on the high-precision MC simulations of the Villain model (1) described in Ref. [14]. By combining the single-cluster update algorithm with improved estimators for the two-point correlation function, data for the correlation length up to $\xi \approx 140$ could be obtained on a 1200^2 square lattice which satisfies $L > 8\xi$ and should therefore be a very good approximation of the thermodynamic limit. Fits of ξ and χ to the leading KT predictions (omitting the logarithmic correction) with four free parameters (the prefactor, b, ν, and β_c) gave $\nu = 0.48(10)$ and $\beta_c = 0.752(5)$. The latter estimate is in very good agreement with the more precise values of $\beta_c = 0.7524(7)$ and $\beta_c = 0.7515(3)$ reported in Ref. [15] from a study of the dual discrete Gaussian model. Including the theoretically predicted correction $t^{-1/16} \propto (\ln\xi)^{1/8}$ did not improve the quality of the fits.

Further data of the susceptibility at criticality on lattices with up to 512^2 sites showed a clear scaling behavior for $L \geq 100$. $\chi \propto L^{2-\eta}$, with $\eta = 0.2495 \approx 1/4$ at $\beta = 0.74$, and $\eta = 0.2389(6) \neq 1/4$ at $\beta = 0.75 \approx \beta_c$. Since the estimates of β_c from two completely independent simulations agreed so well we concluded in Ref. [14] that $\eta(\beta_c) \neq 1/4$, in disagreement with the KT prediction. To reconcile simulations and theory we speculated

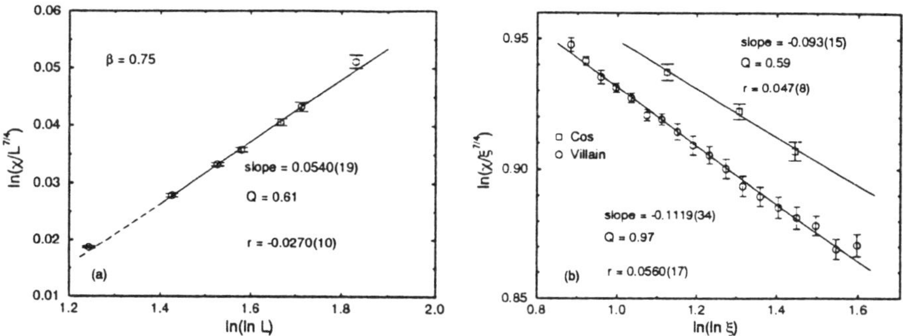

Figure 1: Test of the scaling relations (4) and (3), yielding estimates of the exponent r. The theoretical prediction is $r = -1/16 = -0.0625$.

that the FSS curve for χ might still change for much larger system sizes, but this is of course not very convincing. Mainly based on our negative experience with the $t^{-1/16}$ correction in the $\chi(T)$ fits, we did not try, however, to attribute the observed discrepancy to logarithmic corrections.

In Fig. 1(a) we show the same data, but now fix $\eta = 1/4$ at the theoretical value and assume that (4) *with* the multiplicative logarithmic correction is valid. Since then $\ln(\chi/L^{7/4}) = \text{const.} - 2r\ln(\ln L)$, we expect a straight line when $\ln(\chi/L^{7/4})$ is plotted against $\ln(\ln L)$. This is indeed the case and a linear fit of high statistical quality (goodness-of-fit parameter $Q = 0.61$) yields

$$r = -0.0270 \pm 0.0010, \tag{5}$$

in good agreement with the estimate of $r = -0.02(1)$ from the FSS of Lee-Yang zeros in Ref. [6], but clearly *not* in agreement with the theoretical prediction of $r = -1/16$.

In the analysis of the thermodynamic data near criticality we neglected in Ref. [14] logarithmic corrections in (3) and found in a plot of $\ln(\chi/\xi^{7/4})$ vs $\ln\xi$ a clear negative slope, corresponding to $\eta > 1/4$. We also observed, however, that the data are curved and that for large $\xi \approx 110\ldots140$ the slope decreases. Defining η^{eff} from the local slopes yields an estimate of $\eta^{\text{eff}} \approx 0.267 > 1/4$, while from FSS without logarithmic corrections we concluded that $\eta < 1/4$. In Fig. 1(b) the same data are shown, but similar to Fig. 1(a) we now again fix $\eta = 1/4$ at the theoretical value and assume that (3) *with* the multiplicative logarithmic correction is valid. By plotting $\ln(\chi/\xi^{7/4})$ against $\ln(\ln\xi)$, we see indeed the expected straight line, and a fit over all available data points gives (with $Q = 0.97$)

$$r = 0.0560 \pm 0.0017, \tag{6}$$

now in qualitative agreement with the results in Refs. [8, 9], which are also derived from the approach to criticality in the high-temperature phase, but in striking disagreement with (5), and thus even further apart from the theoretical value of $r = -1/16$. In retrospective this "explains" why we did not observe any improvement when trying fits of $\chi(T)$ *with* the t^r correction fixed to the theoretical prediction $t^{-1/16}$.

Also shown are the three data points of Ref. [14] for the cosine model (with $\xi \approx 21$, 40, and 70) which yield a compatible estimate of $r = 0.047(8)$. Furthermore, using the more extensive data sets of Ref. [7] we find consistent values of $r = 0.050(10)$ and $r = 0.049(10)$, respectively.

Finally it is of course tempting to blame the observed discrepancies between the numerical data and the theoretical expectations on the additive logarithmic corrections in (3)

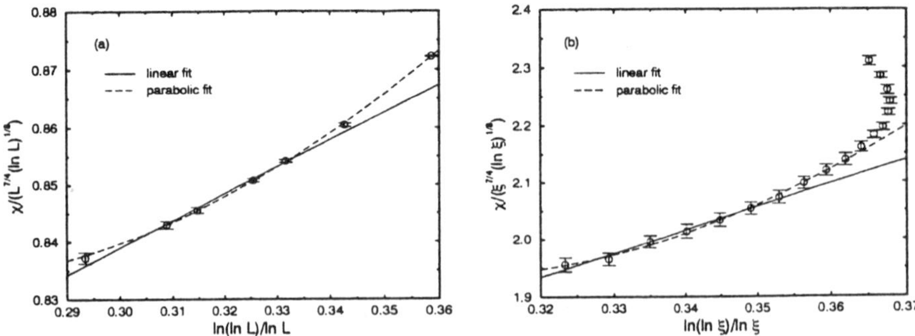

Figure 2: Test for additive logarithmic corrections in (a) the data at criticality and (b) the thermodynamic data. Here the exponents η and r are assumed to take the theoretically predicted values $\eta = 1/4$ and $r = -1/16$.

and (4). In order to test this possibility the data in Fig. 1 are replotted in Fig. 2 in the form $\chi/(L^{2-\eta}(\ln L)^{-2r})$ vs $\ln(\ln L)/\ln L$ and $\chi/(\xi^{2-\eta}(\ln \xi)^{-2r})$ vs $\ln(\ln \xi)/\ln \xi$, respectively, assuming the theoretically predicted values of η and r. The double valuedness in Fig. 2(b) is caused by the fact that $f(\xi) = \ln(\ln \xi)/\ln \xi$ has a maximum $f_{\max} = 1/e \approx 0.3679$ at $\xi_{\max} = e^e \approx 15.15$. We see that both the data for $L > 64$ or $\xi > 40$ can be well fitted with a simple linear function. With a parabolic ansatz the acceptable fit range can even be extended to smaller values of L or ξ. From Fig. 2 it is obvious, however, that we are still too far away from the truly asymptotic region $x \longrightarrow 0$ to take this as a convincing evidence that additive logarithmic corrections can reconcile simulations and theory.

4   Conclusions

When multiplicative logarithmic corrections are taken into account, numerical simulation data of the 2D XY Villain model are quite consistent with the leading KT predictions, assuming the theoretical values of $\nu = 1/2$ and $\eta = 1/4$. Estimates of the logarithmic correction exponent r, however, turn out to be quite inconsistent. Scaling analyses in the FSS region yield a negative ($r \approx -0.03 \ldots -0.02$) and analyses in the high-temperature phase a positive ($r \approx 0.04 \ldots 0.08$) value, both being quite different from the theoretical prediction of $r = -1/16 = -0.0625$. This is obviously related to the fact that analyses neglecting the multiplicative logarithmic correction tend to estimate $\eta < 1/4$ in the FSS region and $\eta > 1/4$ using thermodynamic data. We have no good explanation for this observation other than the common, but unfortunately probably correct statement [16] that the studied system sizes are still much too small to resolve these discrepancies.

Acknowledgments

I would like to thank Ralph Kenna for useful discussions and the DFG for a Heisenberg fellowship.

References

[1] P. Minnhagen, Rev. Mod. Phys. **59** (1987) 1001; H. Kleinert, *Gauge Fields in Condensed Matter* (World Scientific, Singapore, 1989), Vol. I; C. Itzykson and J.-M. Drouffe, *Statistical Field Theory* (University Press, Cambridge, 1989), Vol. I.

[2] V.G. Vaks and A.I. Larkin, Zh. Eksp. Teor. Fiz. **49** (1965) 975 [Sov. Phys. – JETP **22** (1966) 678]; R.G. Bowers and G.S. Joyce, Phys. Rev. Lett. **19** (1967) 630; E. Granato, J.M. Kosterlitz, and J.M. Lee, Phys. Rev. Lett. **66** (1991) 1090; J. Lee, J.M. Kosterlitz, and E. Granato, Phys. Rev. **B43** (1991) 11531; A. Vallat and H. Beck, Phys. Rev. Lett. **68** (1992) 3096; *Physics in Two Dimensions*, Proceedings, Neuchatel, Switzerland, 1991 [Helv. Phys. Acta. **65** (1992) 820-885]; L.J. de Jongh and A.R. Miedema, Advances Physics **23** (1974) 1.

[3] J.M. Kosterlitz and D.J. Thouless, J. Phys. **C6** (1973) 1181; see also V.L. Berezinskii, Zh. Eksp. Teor. Fiz. **61** (1971) 1144 [Sov. Phys.–JETP **34** (1972) 610].

[4] J.M. Kosterlitz, J. Phys. **C7** (1974) 1046.

[5] D.J. Amit, Y.Y. Goldschmidt, and G. Grinstein, J. Phys. **A13** (1980) 585; L.P. Kadanoff and A.B. Zisook, Nucl. Phys. **B180** [FS 2] (1981) 61; C. Itzykson and J.-M. Drouffe, in Ref. [1], p. 218.

[6] R. Kenna and A.C. Irving, Nucl. Phys. **B485** (1997) 485; Phys. Lett. **B351** (1995) 273.

[7] U. Wolff, Nucl. Phys. **B322** (1989) 759; R. Gupta and C.F. Baillie, Phys. Rev. **B45** (1992) 2883.

[8] A. Patrascioiu and E. Seiler, Phys. Rev. **B54** (1996) 7177.

[9] M. Campostrini, A. Pelissetto, P. Rossi, and E. Vicari, Phys. Rev. **B54** (1996) 7301.

[10] J. Villain, J. Phys. (France) **36** (1975) 581.

[11] W. Janke and H. Kleinert, Nucl. Phys. **B270** [FS16] (1986) 135.

[12] W. Janke, Phys. Rev. **B55** (1997) 3580.

[13] J.-K. Kim, preprint hep-lat/9502002 v2 (27 September 1996).

[14] W. Janke and K. Nather, Phys. Rev. **B48** (1993) 7419; Phys. Lett. **A157** (1991) 11.

[15] M. Hasenbusch, M. Marcu, and K. Pinn, Physica **A208** (1994) 124; M. Hasenbusch and K. Pinn, preprint cond-mat/9605019.

[16] J.M. Greif, D.L. Goodstein, and A.F. Silva-Moreira, Phys. Rev. **B25** (1982) 6838; J.L. Cardy, Phys. Rev. **B26** (1982) 6311.

A Chemical Picture of the Dissociation and Thermodynamics of Dense Fluid Hydrogen

A. Bunker, S. Nagel, R. Redmer, and G. Röpke*

Fachbereich Physik, Universität Rostock, Universitätsplatz 3,
D-18051 Rostock, Germany

Recent developments in shock-wave experimental techniques have allowed the 100 GPa range to be probed. The results of new experiments on hydrogen have demonstrated that at 141 GPa fluid hydrogen is in a metallic state. Using analytical calculations as well as Monte Carlo simulations, the pair distribution functions in partly dissociated hydrogen are determined. The equation of state is in good agreement with experiment in this high-pressure region. Estimates for the degree of dissociation are given. Furthermore, we demonstrate the influence of dissociation on the proton-proton pair distribution function.

I. INTRODUCTION

Renewed interest in the study of dense fluid hydrogen has been generated by the first direct evidence of metallization, obtained recently in shock-compression experiments by Weir, Mitchell, and Nellis [1] at 141 GPa. A transition to a fully ionized state with a plasma-like conductivity is expected to occur in nonconducting fluids such as hydrogen at high temperatures $T \geq 10^4$ K and pressures between 10 GPa $\leq P \leq$ 100 GPa but the fundamental question remains as to whether or not this is a first-order phase transition with an instability region and a corresponding critical point. Various theoretical estimates have been given for the location of the critical point of the so far hypothetical plasma phase transition [2–6] which are based upon quantum statistical approaches taking into account many-particle effects such as dynamical screening, self-energy, and polarization forces [7]. However, the neutral components (H, H_2) are usually treated with simple models such as the hard-sphere reference system [8] or perturbation theory [9]. In order to make a comparison with available experimental data for shock-compressed fluid hydrogen and deuterium [1,10–12], these models have to be improved.

Correlations in the neutral, non-ionized state of dense hydrogen can be expressed by the pair distribution functions $g_{ab}(r)$ for the different constituents from which corrections to the ideal pressure are derived. In a first step, we have to take into account pressure dissociation due to the modification of the binding energy of H_2 molecules. The dissociation equilibrium $H_2 \rightleftharpoons H+H$ has to reflect the very subtle changes in the electronic and structural properties

Springer Proceedings in Physics, Volume 83
Computer Simulation Studies in Condensed-Matter Physics X
Eds.: D. P. Landau, K.K. Mon, H. -B. Schüttler
© Springer-Verlag Berlin Heidelberg 1998

at high pressures via correlation contributions which are consistent with the equation of state.

Fluid variational theory (FVT) and the modified hypernetted chain (MHNC) scheme for solving the Ornstein-Zernike equation can both be used to calculate the equation of state for dense fluids [13,14]. Alternatively, Monte Carlo (MC) simulations are a standard tool to study classical systems with short range interaction up to high densities [15]. We apply these three approaches to calculate the equation of state of fluid hydrogen in the experimentally observed pressure region using realistic pair potentials between the components (H and H_2). We compare our results to the experimental data available and with the dissociation model of Holmes, Ross, and Nellis [12,16]. We extract the proton-proton distribution from the partial pair distribution functions in order to make a comparison with recent *ab initio* calculations [17–21].

II. MICROSCOPIC DESCRIPTION OF DENSE FLUID HYDROGEN

We take dense fluid hydrogen to be a mixture of H_2 molecules and H atoms with a dissociation degree $\beta = n_H/(n_H + 2n_{H_2})$. The molecules and atoms interact via effective two-body potentials which approximate the effects of the real many-body interactions. Ross, Ree, and Young [13] use data from single shock experiments up to pressures of 10 GPa to model such an effective, two-body potential between hydrogen molecules. They propose both a highly accurate fifteen-parameter potential and a more approximate three-parameter (exponential-six) potential. We used the latter in our calculations. While no experimental data exists for the effective potential between hydrogen atoms, Ree [22] has suggested a three parameter potential of the same form, but with different parameters, and a resultant core repulsion that is considerably weaker. The parameters for the atom-molecule potential are derived from the Berthelot mixing rule.

The MC simulations have been performed using these potentials with a fixed number of H_2 molecules and H atoms. The degree of dissociation β and the density of particles are thus input variables and fixed for the simulation. In order to make a comparison with our MC simulations, the previously mentioned analytical methods (FVT and MHNC) have been applied to the pure molecular phase of dense fluid hydrogen ($\beta=0$).

III. RESULTS

The calculated pressures for pure molecular hydrogen with the potential we used are compared with the experimental pressures in Table. I. We have first taken as input parameters the densities and temperatures given in the experimental papers [1,11] which, however, should be considered as estimates derived from more simplified models. The present theories yield nearly identical results

TABLE I. Pressures reached in single [11] and multiple shock experiments [1] compared with various theoretical models. The parameter sets (a) are estimates given in the experimental papers neglecting dissociation. The temperature (b) of about 2600 K results from the dissociation model of Holmes, Ross, and Nellis [12]. The densities (c) are those which match the experimental pressures in the present dissociation model.

| EXPERIMENT | | | MOLECULAR HYDROGEN | | | DISSOCIATED MODEL | |
| | | | FVT | MHNC | MC | MC | |
P [GPa]	T [K]	ϱ [g/cm^3]		P [GPa]		P [GPa]	β [%]
9.96 [11]	3020[a]	0.221[a]	9.99	9.58	9.58	9.61	0.13
100 [1]	2275[a]	0.586[a]	91.0	89.4	89.6	93.92	7.4
	2600[b]	0.594[c]				100	10.1
123 [1]	2850[a]	0.628[a]	109.8	108.4	108.5	116.9	14.8
	2600[b]	0.644[c]				123	14.9
141 [1]	3000[a]	0.664[a]	125.5	124.1	124.4	135.1	19.1
	2600[b]	0.678[c]				141	18.2

up to the 140 GPa region. While in very good agreement with the available single shock data up to 10 GPa [11], the theoretical results are systematically too low by about 10% in the pressure range of 140 GPa reached by the multiple shock experiments [1].

In order to study the behavior of fluid hydrogen at high pressures in a more general model, we have to allow for the dissociation of hydrogen molecules into atoms via the chemical equilibrium $H_2 \rightleftharpoons H + H$. The fraction β of dissociated molecules is determined by the correlation contributions to the chemical potential. Taking the input parameters temperature and density as given in the experimental papers, the dissociation degree reaches 19% at the highest pressure, (see Table. I).

Ross [16] has calculated the final temperatures of the shock experiments with an alternative model [12]. An almost constant temperature of about 2600 K results for the region above 100 GPa. The corresponding densities in Table. I are those where our MC simulations for partly dissociated hydrogen match the experimental pressures. The final dissociation degrees change only very little.

From the MC simulations for the mixture of hydrogen molecules and atoms with dissociation degree β, three different pair distribution functions g_{HH}, g_{HH_2}, and $g_{H_2H_2}$ [23,24] are obtained using the respective pair potentials. We used these pair distribution functions to calculate the proton-proton pair distribution function g_{pp} by determining the proton distribution in the H_2 molecule. We compared our results directly to those obtained from *ab initio* path-integral Monte Carlo [17] and quantum molecular dynamics simulations [18–21] and found a reasonable agreement in the characteristic features [23,24]

IV. CONCLUSION

We have calculated the equation of state for dense fluid hydrogen, using effective two-body interaction potentials, and several different theoretical approaches. The single shock experiments up to 10 GPa [11] indicate that hydrogen exists in the pure molecular phase. In the recent multiple shock experiments up to the 140 GPa domain [1], dissociation of hydrogen molecules is found to be important. For instance, at the metallization pressure of 141 GPa as found in the experiment, our model yields a dissociation degree of 18%. Using MC calculations with this dissociation degree taken into account it is shown that the agreement with experiment is markedly improved over the results of MC calculations assuming no dissociation, thus further validating our model. These findings are in agreement with the dissociation model of Holmes, Ross, and Nellis [12] which describes shock-compression experiments up to 83 GPa.

Having obtained a remarkable fraction of monomers in the dense fluid at high pressures due to dissociation, the problem of ionization will be addressed in future work. The problem of the existence of a plasma phase transition at higher temperatures should be treated combining the present improved equation of state for the neutral fluid component with the existing theories for the charged plasma component.

A first comparison of the proton-proton pair distribution function derived from the present chemical picture with *ab initio* computer simulations shows reasonable agreement. Further efforts will be made to systematically study how the results of simulations that make use of a physical picture which starts with ensembles of electrons and protons are related to those obtained with a chemical picture involving bound states such as atoms and molecules as basic elements of the statistical description.

ACKNOWLEDGMENTS

We thank W.D. Kraeft, D. Klug, D. Kremp, A.A. Likalter, J. Meyer-ter-Vehn, M. Ross, M. Schlanges, and C. Toepffer for stimulating discussions. One of us (A.B.) thanks the University of Rostock for the kind hospitality. This work has been supported by the Deutsche Forschungsgemeinschaft within the SFB 198 *Kinetics of partially ionized plasmas*.

* Permanent address: Center for Simulational Physics, University of Georgia, Dept. of Phys. and Astronomy, Athens, GA 30602-2451, USA.

[1] S.T. Weir, A.C. Mitchell, and W.J. Nellis, Phys. Rev. Lett. **76**, 1860 (1996).

[2] W. Ebeling and W. Richert, Phys. Stat. Sol. B **128**, 467 (1985); Phys. Lett. A **108**, 80 (1985); Contr. Plasma Phys. **25**, 1 (1985).

[3] D. Saumon and G. Chabrier, Phys. Rev. Lett. **62**, 2397 (1989); Phys. Rev. A **44**, 5122 (1991); *ibid.* **46**, 2084 (1992).

[4] D. Kremp, W.D. Kraeft, and M. Schlanges, Contrib. Plasma Phys. **33**, 567 (1993); see also M. Schlanges, M. Bonitz, and A. Tschttschjan, *ibid.* **35**, 109 (1995).

[5] H. Reinholz, R. Redmer, and S. Nagel, Phys. Rev. E **52**, 5368 (1995).

[6] W. Ebeling, A. Förster, H. Hess, and M.Yu. Romanovsky, Plasma Phys. & Contr. Fusion **38**, A31 (1996).

[7] W.D. Kraeft, D. Kremp, W. Ebeling, and G. Röpke, *Quantum Statistics of Charged Particle Systems* (Akademie-Verlag, Berlin, 1986).

[8] N.F. Carnahan and K.E. Starling, J. Chem. Phys. **51**, 635 (1969); G.A. Mansoori, N.F. Carnahan, K.E. Starling, and T.W. Leland, *ibid.* **54**, 1523 (1971).

[9] J.D. Weeks, D. Chandler, and H.C. Andersen, J. Chem. Phys. **54**, 5237 (1971).

[10] W.J. Nellis, A.C. Mitchell, M. van Thiel, G.J. Devine, R.J. Trainor, and N. Brown, J. Chem. Phys. **79**, 1480 (1983).

[11] W.J. Nellis, A.C. Mitchell, P.C. McCandless, D.J. Erskine, and S.T. Weir, Phys. Rev. Lett. **68**, 2937 (1992).

[12] N.C. Holmes, M. Ross, and W.J. Nellis, Phys. Rev. B **52**, 15835 (1995).

[13] M. Ross, F.H. Ree, and D.A. Young, J. Chem. Phys. **79**, 1487 (1983).

[14] Y. Rosenfeld and N.W. Ashcroft, Phys. Rev. A **20**, 1208 (1979). For the numerical code, see A. Malijevsky and S. Labik, Mol. Phys. **60**, 663 (1987).

[15] For a review, see K. Binder (Editor), *The Monte Carlo Method in Condensed Matter Physics* (Springer, Berlin, 1992).

[16] M. Ross, Phys. Rev. B **54**, R9589 (1996).

[17] C. Pierleoni, D.M. Ceperley, B. Bernu, and W.R. Magro, Phys. Rev. Lett. **73**, 2145 (1994); W.R. Magro, D.M. Ceperley, C. Pierleoni, and B. Bernu, *ibid.* **76**, 1240 (1996).

[18] D. Hohl, V. Natoli, D.M. Ceperley, and R.M. Martin, Phys. Rev. Lett. **71**, 541 (1993).

[19] D. Klakow, C. Toepffer, and P.-G. Reinhard, Phys. Lett. A **192**, 55 (1994).

[20] J. Kohanoff and J.-P. Hansen, Phys. Rev. Lett. **74**, 626 (1995); Phys. Rev. E **54**, 768 (1996).

[21] L. Collins, I. Kwon, J. Kress, N. Troullier, and D. Lynch, Phys. Rev. E **52**, 6202 (1995).

[22] F.H. Ree, in *Shock Waves in Condensed Matter - 1987*, edited by S.C. Schmidt and N.C. Holmes (Elesevier, New York, 1988), p. 125.

[23] A. Bunker, S. Nagel, R. Redmer, G. Röpke, Contrib. Plasma Phys. (submitted)

[24] A. Bunker, S. Nagel, R. Redmer, G. Röpke, Phys. Rev. B (submitted)

Index of Contributors

Printing: Weihert-Druck GmbH, Darmstadt
Binding: Buchbinderei Schäffer, Grünstadt

Springer Proceedings in Physics

Managing Editor: H. K. V. Lotsch

Springer Proceedings in Physics

Managing Editor: H. K. V. Lotsch